Introductory Algebra

Introductory Algebra

Barbara Poole

North Seattle Community College

Prentice Hall, Englewood Cliffs, New Jersey 07632

Library of Congress Cataloging-in-Publication Data

POOLE, BARBARA, date-
  Introductory algebra.

  Bibliography: p.
  Includes index.
  1. Algebra.   I. Title.
QA154.2.P66   1989   5129   88-28893
ISBN 0-13-500083-1   ISBN 0-13-500091-2

Editorial/production supervision: Virginia R. Huebner
Interior design: Judith A. Matz-Coniglio
Cover design: Judith A. Matz-Coniglio
Manufacturing buyer: Paula Massenaro
Page layout: Meryl Poweski and Karen Noferi
Cover Art: 1975.313 *Alioth*, 1962. Morris Louis, American, 1912–1962.
        Acrylic resin (Magna) or Canvas: $79\frac{1}{2} \times 22\frac{1}{2}$ in. The Tompkins Collection,
        Courtesy, Museum of Fine Arts, Boston.

© 1989 by Prentice-Hall, Inc.
A Division of Simon & Schuster
Englewood Cliffs, New Jersey 07632

Printed in the United States of America
10 9 8 7 6 5 4 3 2 1

ISBN 0-13-500083-1

Prentice-Hall International (UK) Limited, *London*
Prentice-Hall of Australia·Pty. Limited, *Sydney*
Prentice-Hall Canada Inc., *Toronto*
Prentice-Hall Hispanoamericana, S.A., *Mexico*
Prentice-Hall of India Private Limited, *New Delhi*
Prentice-Hall of Japan, Inc., *Tokyo*
Simon & Schuster Asia Pte. Ltd., *Singapore*
Editora Prentice-Hall do Brasil, Ltda., *Rio de Janeiro*

To my family—
Don, Julie, Jim, and Lisa

Contents

5

Polynomials *199*

6

Graphing Straight Lines *255*

7

Graphing and Linear Systems of Equations *315*

8

Rational Expressions *361*

9

Rational Equations and Complex Fractions *401*

10

Radical Expressions *431*

11

Quadratic Equations *469*

Preface

This book was written in order to give my students an algebra book which they could read and understand. The explanations are carefully written in language that is familiar to the general population as well as those students for whom English is a second language. I have used applications of algebra to the real world throughout the book, in order to illustrate to students why this subject is important to them. I believe that I have been successful in providing my students with a book which they understand and find useful, and I hope that my approach will prove equally successful at other colleges and universities throughout the country.

In addition, the following features of the text will make it easier for you to learn algebra:

Four-Color Format: Color has been found to enhance the learning process for many students through highlighting of important procedures within examples, keying particular features of a text to certain colors for easier identification when reviewing material, emphasizing essential rules and definitions, and through simple eye appeal, which adds interest. *Introductory Algebra* takes full pedagogical advantage of its four-color format in all these ways.

Realistic Word Problems: Real-life situations are used to illustrate the many applications of algebra. Geometry, one of the most common applications of mathematics, is used extensively throughout the book. Word problems are interspersed *throughout* the book wherever appropriate. Worked-out solutions to most odd-numbered word problems are contained in the back of the book.

Definitions: The most important definitions are given twice—once in the text, where the concept is introduced, and again in the top margin of the page, as an easy study reference for students who are preparing for an examination.

Historical Notes: Historical notes, labeled "Did You Know?", appear throughout the book, increasing the students' understanding and giving them an appreciation of the *spirit* of mathematics.

Step-by-Step Examples: Each example is worked out in a very clear and *step-by-step* manner. Often, an explanation of each step is given at the side of the example.

Exercises: Examples are followed by margin exercises, and are keyed directly to them, so that students can practice new skills *immediately* after learning them. Answers to margin exercises appear at the end of the section, for immediate feedback. Each section is followed by a problem set, carefully graded for difficulty, and paired so that odd- and even-numbered problems are very similar. Finally, each chapter contains additional exercises which can be used as extra drill or to prepare for a test.

Checkup Problems: These problems provide a review of all prior sections and often help prepare students for the next section. For example, since reciprocal is used to define perpendicular lines, reciprocals are reviewed in the previous section's Checkup Problems.

Writing in Mathematics: Each section contains "Reading and Writing" problems which help students strengthen their verbal skills.

Chapter Tests: Each set of additional exercises is followed by a chapter test, which can be used as practice.

Cumulative Reviews: Cumulative review exercises after chapters 2, 4, 6, 8 and 10 provide additional review of previously covered topics and may help students prepare for examinations.

Final Examination: At the end of the book, a comprehensive Final Examination is provided.

Glossary: A glossary is contained on the inside cover of the book for easy reference.

Objectives: Each section begins with a list of objectives, which outline the concepts to be learned in that section. These may be useful for the instructor preparing a lesson as well as for the student previewing or reviewing material.

Supplementary Materials

For the students, we have provided the following supplementary materials. Ask your instructor whether they are available for your use.

Student Solutions Manual provides fully worked-out solutions for all of the odd-numbered exercises in the book.

Answers to the Even-Numbered Exercises for student use are available only at the request of your instructor.

"How to Study Math", also available at the request of your instructor, provides extra hints about studying mathematics.

Interactive Algebra Tutor for Apple and IBM is keyed to the objectives in the book, and provides tutorial and diagnostic help in the context of additional drill.

Videotapes, free upon adoption of 100 or more copies of the book, cover all major topics in Introductory Algebra.

Audiotapes, free upon adoption of 100 copies or more of the text, provide additional learning assistance.

Acknowledgments

I would like to thank my daughter, Julia, for typing and editing the manuscript. My husband, Donald, and my son, James, and his wife, Lisa, also gave me their encouragement and support.

My colleague at North Seattle Community College, Vicky Ringen, gave me many helpful suggestions. I would also like to thank the college for their support of this project.

The following people deserve special thanks due to their help in preparing the supplements:

Annotated Instructor's Edition:

Agnes Azzolino, Middlesex County College
Richard Clark, Portland Community College
H. Joan Dykes, Edison Community College
Ara Sullenberger, Tarrant County Junior College

Student Solutions Manual:

Helen Burrier, Kirkwood Community College (Intermediate Algebra)
Cathy Pace, Louisiana Tech University (Introductory Algebra)

Videotapes:

Roger Breen and Margaret Greene, Florida Community College at Jacksonville

Audiotapes:

Margaret Greene, Florida Community College at Jacksonville
I would also like to thank the following reviewers:

Intermediate Algebra

James Arnold, University of Wisconsin at Milwaukee
James Blackburn, Tulsa Junior College
Martin Brown, Jefferson Community College
Pat C. Cook, Weatherford College
Richard J. Easton, Indiana State University
Sandra T. Gum, Salem College
Martin W. Johnson, Montgomery County Community College
Charles R. Luttrell, Frederick Community College
Patricia McCann, Franklin University

Philip R. Montgomery, University of Kansas
William R. Neal, Fresno City College
Louis P. Pushkarsky, Trenton Junior College
Emilio Roxin, University of Rhode Island
John Wenger, Harold Washington College
Jerry Wisnieski, Des Moines Area Community College
Ben F. Zirkle, Virginia Western Community College

Introductory Algebra

Larry L. Blevins, University of Northern Colorado
Ronald Bohuslov, College of Alameda
Patricia Deamer, Skyline College
Dorothy Fujimura, California State University at Hayward
Dauhrice Gibson, Gulf Coast Community College
Curtis L. Gooden, Cuyahoga Community College
Dorothy Gotway, University of Missouri at St. Louis
Peter Herron, Suffolk County Community College
Lou Hoelzle, Bucks County Community College
Rudy Maglio, Oakton Community College
Patricia McCann, Franklin University
Lilian Metlitzky, California State Polytechnic University
Vicky Ringen, North Seattle Community College
Carlos R. Rodriguez, San Antonio College
August A. Ruggiero, Essex County College
Susanne M. Shelley, Sacramento City College
Ara Sullenberger, Tarrant Junior College
Jennie Thompson, University of Hawaii
Tommy Thompson, Brookhaven College
Wilma N. Whitaker, Florence-Darlington TEC

My editors at Prentice Hall have been very helpful. They are Priscilla
McGeehon, Christine Peckaitis, editor-in-chief Robert Sickles, and pro-
duction editor, Virginia Huebner. Thanks also to Judith Matz-Coniglio
for her excellent design and layout.

BARBARA POOLE
North Seattle Community College

Introductory
Algebra

Arithmetic

1

1.1 SIMPLIFYING, MULTIPLYING, AND DIVIDING FRACTIONS

Many of the techniques that we use in algebra are based on arithmetic. We begin this chapter by reviewing fractions. Some examples of fractions are $\frac{1}{3}, \frac{3}{5}, \frac{11}{17}, \frac{5}{2},$ and $\frac{15}{9}$. The top number is the **numerator** and the bottom number is the **denominator**. If the numerator is smaller than the denominator, the fraction is a **proper fraction**. If the numerator is greater than or equal to the denominator, the fraction is an **improper fraction**. For example,

$$\frac{1}{3}, \frac{5}{8}, \frac{25}{26} \quad \text{are proper fractions}$$

and

$$\frac{7}{5}, \frac{18}{17}, \frac{105}{98}, \frac{6}{6} \quad \text{are improper fractions}$$

Improper fractions may also be written as whole or mixed numbers. The fraction

$$\frac{5}{2} = 2\frac{1}{2} \quad \text{since} \quad 2\frac{1}{2} = \frac{4}{2} + \frac{1}{2} = \frac{5}{2}$$

1 Factors

Factors of a number are numbers that are multiplied together to give that number. The following are factors of 12.

1 and 12	Since $1 \cdot 12 = 12$
2 and 6	
2 and 2 and 3	Since $2 \cdot 2 \cdot 3 = 12$
3 and 4	

☐ **Exercise 1**   Find the prime factorization.

a. 6 $2 \cdot 3$

We say that one number is divisible by another number if the first number can be divided by the second number and there is no remainder. It is often important to find the prime factors of a number.

> **Prime numbers** are numbers *greater than* 1 that are divisible only by themselves and 1.

The first few prime numbers are 2, 3, 5, 7, and 11. Notice that 4 is not prime since 4 is divisible by 2. Nine is not prime since it is divisible by 3. The prime number factors of 12 are 2, 2, and 3.

EXAMPLE 1

a. Find the prime factorization of 16.

Divide 16 by the first prime number, 2. The result is 8, so 2 and 8 are factors of 16.

b. 8 $2 \cdot 4$

$$16 = 2 \cdot 8$$

Are these factors prime? No, 8 is not prime, so divide it by 2. The result is 4. Then 2 and 4 are factors of 8.

$$16 = 2 \cdot (2 \cdot 4)$$

But 4 is not prime, so divide it by 2. The result is 2, so 2 and 2 are factors of 4.

$$16 = 2 \cdot 2 \cdot (2 \cdot 2)$$

These factors are all prime. The prime factorization of 16 is $2 \cdot 2 \cdot 2 \cdot 2$

c. 27

b. Find the prime factorization of 21.

21 is not divisible by 2, so try dividing by the second prime, 3.

$$21 \div 3 = 7$$

So 3 and 7 are factors of 21. These numbers are both prime, so

$$21 = 3 \cdot 7$$

The prime factorization of 21 is $3 \cdot 7$.   ■

☐ **DO EXERCISE 1.** Check your answers in the Answers to Exercises at the end of Section 1.1.

d. 36

We use prime factors to simplify a fraction. We also use the fact that any number divided by itself is 1. For example, $\frac{3}{3}$ is 1. Notice also that the following is true for fractions.

> $$\frac{a \cdot c}{b \cdot d} = \frac{a}{b} \cdot \frac{c}{d} \qquad \text{For any numbers } a, b, c, \text{ and } d$$

2 Simplifying Fractions

We *simplify a fraction* by factoring the numerator and denominator into primes and then removing the 1.

EXAMPLE 2 Simplify by factoring and removing a 1.

a. $\dfrac{3}{9}$

$$\dfrac{3}{9} = \dfrac{3 \cdot 1}{3 \cdot 3} \qquad \text{Factoring the denominator into primes}$$

$$= \dfrac{3}{3} \cdot \dfrac{1}{3} \qquad \text{Since } \dfrac{a \cdot c}{b \cdot d} = \dfrac{a}{b} \cdot \dfrac{c}{d}$$

$$= \dfrac{1}{3} \qquad \text{Removing the 1}$$

b. $\dfrac{18}{24}$

$$\dfrac{18}{24} = \dfrac{2 \cdot 9}{2 \cdot 12} = \dfrac{2 \cdot 3 \cdot 3}{2 \cdot 2 \cdot 6} = \dfrac{2 \cdot 3 \cdot 3}{2 \cdot 2 \cdot 2 \cdot 3}$$

$$= \dfrac{2 \cdot 3 \cdot 3}{2 \cdot 3 \cdot 2 \cdot 2} \qquad \text{Rearranging factors in the denominator}$$

$$= \dfrac{2 \cdot 3}{2 \cdot 3} \cdot \dfrac{3}{2 \cdot 2} = \dfrac{3}{2 \cdot 2} = \dfrac{3}{4} \qquad \text{Removing the 1} \quad \blacksquare$$

☐ **DO EXERCISE 2.**

3 Multiplying Fractions

When we multiply numbers we find their product. Finding a product is often stated as "multiply."

> To *multiply fractions* multiply the numerators and multiply the denominators.
>
> $$\dfrac{a}{b} \cdot \dfrac{c}{d} = \dfrac{a \cdot c}{b \cdot d}$$

EXAMPLE 3 Multiply $\dfrac{2}{5} \cdot \dfrac{3}{7}$.

$$\dfrac{2}{5} \cdot \dfrac{3}{7} = \dfrac{2 \cdot 3}{5 \cdot 7} = \dfrac{6}{35} \quad \blacksquare$$

☐ **DO EXERCISE 3.**

In a multiplication problem, we may simplify before completing the multiplication.

☐ **Exercise 2** Simplify by factoring and removing a 1.

a. $\dfrac{3}{15}$

b. $\dfrac{12}{36}$

c. $\dfrac{6}{36}$

d. $\dfrac{15}{45}$

☐ **Exercise 3** Multiply.

a. $\dfrac{7}{8} \cdot \dfrac{1}{5}$

b. $\dfrac{5}{9} \cdot \dfrac{8}{3}$

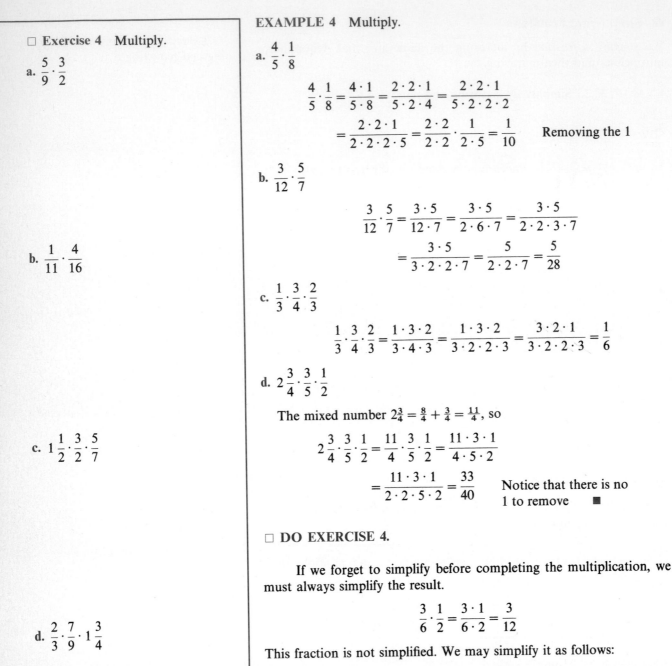

□ **Exercise 4**   Multiply.

a. $\dfrac{5}{9} \cdot \dfrac{3}{2}$

b. $\dfrac{1}{11} \cdot \dfrac{4}{16}$

c. $1\dfrac{1}{2} \cdot \dfrac{3}{2} \cdot \dfrac{5}{7}$

d. $\dfrac{2}{3} \cdot \dfrac{7}{9} \cdot 1\dfrac{3}{4}$

EXAMPLE 4   Multiply.

a. $\dfrac{4}{5} \cdot \dfrac{1}{8}$

$$\frac{4}{5} \cdot \frac{1}{8} = \frac{4 \cdot 1}{5 \cdot 8} = \frac{2 \cdot 2 \cdot 1}{5 \cdot 2 \cdot 4} = \frac{2 \cdot 2 \cdot 1}{5 \cdot 2 \cdot 2 \cdot 2}$$

$$= \frac{2 \cdot 2 \cdot 1}{2 \cdot 2 \cdot 2 \cdot 5} = \frac{2 \cdot 2}{2 \cdot 2} \cdot \frac{1}{2 \cdot 5} = \frac{1}{10} \qquad \text{Removing the 1}$$

b. $\dfrac{3}{12} \cdot \dfrac{5}{7}$

$$\frac{3}{12} \cdot \frac{5}{7} = \frac{3 \cdot 5}{12 \cdot 7} = \frac{3 \cdot 5}{2 \cdot 6 \cdot 7} = \frac{3 \cdot 5}{2 \cdot 2 \cdot 3 \cdot 7}$$

$$= \frac{3 \cdot 5}{3 \cdot 2 \cdot 2 \cdot 7} = \frac{5}{2 \cdot 2 \cdot 7} = \frac{5}{28}$$

c. $\dfrac{1}{3} \cdot \dfrac{3}{4} \cdot \dfrac{2}{3}$

$$\frac{1}{3} \cdot \frac{3}{4} \cdot \frac{2}{3} = \frac{1 \cdot 3 \cdot 2}{3 \cdot 4 \cdot 3} = \frac{1 \cdot 3 \cdot 2}{3 \cdot 2 \cdot 2 \cdot 3} = \frac{3 \cdot 2 \cdot 1}{3 \cdot 2 \cdot 2 \cdot 3} = \frac{1}{6}$$

d. $2\dfrac{3}{4} \cdot \dfrac{3}{5} \cdot \dfrac{1}{2}$

The mixed number $2\frac{3}{4} = \frac{8}{4} + \frac{3}{4} = \frac{11}{4}$, so

$$2\frac{3}{4} \cdot \frac{3}{5} \cdot \frac{1}{2} = \frac{11}{4} \cdot \frac{3}{5} \cdot \frac{1}{2} = \frac{11 \cdot 3 \cdot 1}{4 \cdot 5 \cdot 2}$$

$$= \frac{11 \cdot 3 \cdot 1}{2 \cdot 2 \cdot 5 \cdot 2} = \frac{33}{40} \qquad \begin{array}{l}\text{Notice that there is no} \\ \text{1 to remove} \quad \blacksquare\end{array}$$

□ **DO EXERCISE 4.**

If we forget to simplify before completing the multiplication, we must always simplify the result.

$$\frac{3}{6} \cdot \frac{1}{2} = \frac{3 \cdot 1}{6 \cdot 2} = \frac{3}{12}$$

This fraction is not simplified. We may simplify it as follows:

$$\frac{3}{12} = \frac{3 \cdot 1}{2 \cdot 6} = \frac{3 \cdot 1}{2 \cdot 2 \cdot 3} = \frac{3 \cdot 1}{3 \cdot 2 \cdot 2} = \frac{1}{2 \cdot 2} = \frac{1}{4}$$

4 Dividing Fractions

The division problem $\frac{3}{5} \div \frac{1}{7}$ may be written in another way as

$$\frac{\dfrac{3}{5}}{\dfrac{1}{7}}$$

We know that any number divided by 1 is just that number. We want to make the denominator of the complex fraction above a 1 to simplify the fraction. We may do this by multiplying the numerator and denominator by $\frac{7}{1}$.

$$\frac{\dfrac{3}{5} \cdot \dfrac{7}{1}}{\dfrac{1}{7} \cdot \dfrac{7}{1}} = \frac{\dfrac{21}{5}}{\dfrac{7}{7}} = \frac{\dfrac{21}{5}}{1} = \frac{21}{5}$$

Notice that this is the same operation as inverting the denominator (divisor) and multiplying it times the numerator.

$$\frac{\dfrac{3}{5}}{\dfrac{1}{7}} = \frac{3}{5} \div \frac{1}{7} = \frac{3}{5} \cdot \frac{7}{1} = \frac{21}{5}$$

> To divide two fractions, invert the divisor and multiply.
>
> $$\frac{a}{b} \div \frac{c}{d} = \frac{a}{b} \cdot \frac{d}{c} = \frac{a \cdot d}{b \cdot c}$$

EXAMPLE 5   Divide.

a. $\dfrac{3}{4} \div \dfrac{2}{7}$

$$\frac{3}{4} \div \frac{2}{7} = \frac{3}{4} \cdot \frac{7}{2} = \frac{3 \cdot 7}{4 \cdot 2} = \frac{21}{8}$$

b. $\dfrac{5}{9} \div \dfrac{1}{3}$

$$\frac{5}{9} \div \frac{1}{3} = \frac{5}{9} \cdot \frac{3}{1} = \frac{5 \cdot 3}{9 \cdot 1} = \frac{5 \cdot 3}{3 \cdot 3 \cdot 1} = \frac{3 \cdot 5}{3 \cdot 3 \cdot 1} = \frac{5}{3}$$

c. $4\dfrac{3}{4} \div 1\dfrac{1}{2}$

The mixed numbers

$$4\frac{3}{4} = \frac{16}{4} + \frac{3}{4} = \frac{19}{4} \quad \text{and} \quad 1\frac{1}{2} = \frac{2}{2} + \frac{1}{2} = \frac{3}{2}$$

$$4\frac{3}{4} \div 1\frac{1}{2} = \frac{19}{4} \div \frac{3}{2} = \frac{19}{4} \cdot \frac{2}{3} = \frac{19 \cdot 2}{4 \cdot 3}$$

$$= \frac{2 \cdot 19}{2 \cdot 2 \cdot 3} = \frac{19}{6} \quad \blacksquare$$

☐ **DO EXERCISE 5.**

Recall that a number may be written with a denominator of 1.

☐ **Exercise 5**   Divide.

a. $\dfrac{1}{5} \div \dfrac{2}{3}$

b. $\dfrac{3}{10} \div \dfrac{7}{20}$

c. $\dfrac{2}{9} \div \dfrac{5}{18}$

d. $\dfrac{\dfrac{1}{6}}{\dfrac{5}{36}}$

□ **Exercise 6** Divide.

a. $\dfrac{6}{5} \div 3$

b. $\dfrac{3}{8} \div 10$

EXAMPLE 6 Divide: $\dfrac{3}{5} \div 10$.

$$\frac{3}{5} \div 10 = \frac{3}{5} \div \frac{10}{1} = \frac{3}{5} \cdot \frac{1}{10} = \frac{3}{50} \quad \blacksquare$$

□ **DO EXERCISE 6.**

We may need to use fractions to solve applied problems.

EXAMPLE 7 How many $2\frac{1}{2}$-inch pieces of candy can be cut from an 11-inch strip of candy?

We must divide the 11-inch strip of candy into $2\frac{1}{2}$-inch pieces.

$$2\frac{1}{2} = \frac{4}{2} + \frac{1}{2} = \frac{5}{2}$$

$$11 \div 2\frac{1}{2} = 11 \div \frac{5}{2} = \frac{11}{1} \cdot \frac{2}{5}$$

$$= \frac{11 \cdot 2}{1 \cdot 5} = \frac{22}{5} = 4\frac{2}{5} \qquad \text{Dividing 22 by 5 to get the mixed number}$$

We can cut four pieces of candy from the 11-inch strip. $\quad \blacksquare$

□ **DO EXERCISE 7.**

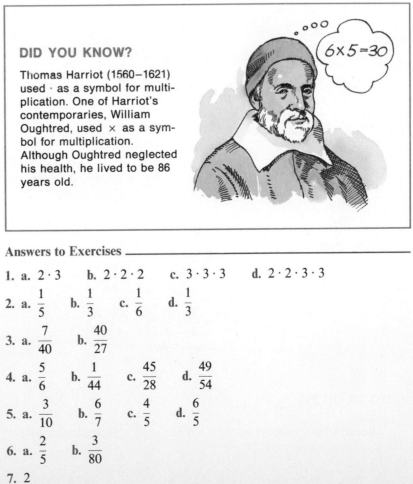

DID YOU KNOW?

Thomas Harriot (1560–1621) used · as a symbol for multiplication. One of Harriot's contemporaries, William Oughtred, used × as a symbol for multiplication. Although Oughtred neglected his health, he lived to be 86 years old.

□ **Exercise 7** How many $3\frac{5}{8}$-ft boards can be cut from a 10-foot board?

Answers to Exercises

1. **a.** $2 \cdot 3$　**b.** $2 \cdot 2 \cdot 2$　**c.** $3 \cdot 3 \cdot 3$　**d.** $2 \cdot 2 \cdot 3 \cdot 3$

2. **a.** $\dfrac{1}{5}$　**b.** $\dfrac{1}{3}$　**c.** $\dfrac{1}{6}$　**d.** $\dfrac{1}{3}$

3. **a.** $\dfrac{7}{40}$　**b.** $\dfrac{40}{27}$

4. **a.** $\dfrac{5}{6}$　**b.** $\dfrac{1}{44}$　**c.** $\dfrac{45}{28}$　**d.** $\dfrac{49}{54}$

5. **a.** $\dfrac{3}{10}$　**b.** $\dfrac{6}{7}$　**c.** $\dfrac{4}{5}$　**d.** $\dfrac{6}{5}$

6. **a.** $\dfrac{2}{5}$　**b.** $\dfrac{3}{80}$

7. 2

PROBLEM SET 1.1

In all the problem sets, do the odd-numbered exercises and check your answers in the back of the book. Do the even-numbered problems only if you had difficulty with the odd-numbered exercises.

A. *Find the prime factorization.*

1. 28   **2.** 16   **3.** 15   **4.** 25

5. 32   **6.** 48   **7.** 22   **8.** 35

B. *Simplify by factoring and removing a 1.*

9. $\dfrac{4}{12}$ **10.** $\dfrac{7}{21}$ **11.** $\dfrac{9}{6}$ **12.** $\dfrac{15}{25}$

13. $\dfrac{27}{9}$ **14.** $\dfrac{28}{7}$ **15.** $\dfrac{33}{22}$ **16.** $\dfrac{63}{45}$

C. *Multiply and simplify if possible, by factoring and removing a 1.*

17. $\dfrac{3}{4} \cdot \dfrac{1}{7}$ **18.** $\dfrac{5}{8} \cdot \dfrac{1}{6}$ **19.** $\dfrac{5}{8} \cdot \dfrac{3}{10}$ **20.** $\dfrac{7}{9} \cdot \dfrac{3}{10}$

21. $\dfrac{12}{5} \cdot \dfrac{3}{6}$ **22.** $\dfrac{8}{7} \cdot \dfrac{14}{3}$ **23.** $\dfrac{5}{10} \cdot \dfrac{6}{8}$ **24.** $\dfrac{6}{7} \cdot \dfrac{14}{3}$

25. $\dfrac{5}{9} \cdot \dfrac{9}{25}$ **26.** $\dfrac{3}{7} \cdot \dfrac{7}{9}$ **27.** $\dfrac{7}{8} \cdot \dfrac{24}{7}$ **28.** $\dfrac{11}{21} \cdot \dfrac{42}{11}$

29. $\dfrac{3}{4} \cdot \dfrac{1}{7} \cdot \dfrac{5}{2}$ **30.** $\dfrac{1}{2} \cdot \dfrac{3}{5} \cdot \dfrac{7}{8}$ **31.** $\dfrac{11}{8} \cdot \dfrac{4}{5} \cdot \dfrac{5}{3}$ **32.** $\dfrac{1}{2} \cdot \dfrac{5}{3} \cdot \dfrac{21}{10}$

33. $1\dfrac{1}{2} \cdot \dfrac{2}{3} \cdot \dfrac{5}{3}$ **34.** $\dfrac{9}{8} \cdot 2\dfrac{2}{3} \cdot \dfrac{1}{5}$

D. *Divide and simplify, if possible, by factoring and removing a 1.*

35. $\dfrac{3}{5} \div \dfrac{1}{6}$ **36.** $\dfrac{5}{9} \div \dfrac{1}{3}$ **37.** $\dfrac{7}{8} \div \dfrac{4}{5}$ **38.** $\dfrac{3}{11} \div \dfrac{8}{12}$

39. $\dfrac{5}{9} \div \dfrac{4}{3}$ **40.** $\dfrac{7}{5} \div \dfrac{9}{8}$ **41.** $\dfrac{2}{5} \div 10$ **42.** $\dfrac{9}{8} \div 8$

43. $\dfrac{20}{3} \div \dfrac{5}{6}$ **44.** $\dfrac{25}{7} \div \dfrac{14}{5}$ **45.** $\dfrac{7}{9} \div \dfrac{1}{3}$ **46.** $\dfrac{16}{25} \div \dfrac{1}{5}$

47. $12 \div \dfrac{5}{8}$ **48.** $1 \div \dfrac{1}{2}$ **49.** $\dfrac{6}{7} \div \dfrac{3}{14}$ **50.** $\dfrac{7}{11} \div \dfrac{21}{22}$

51. $\dfrac{8}{3} \div 5$ **52.** $\dfrac{11}{5} \div 5$ **53.** $\dfrac{75}{3} \div \dfrac{25}{6}$ **54.** $\dfrac{18}{5} \div \dfrac{9}{20}$

55. $3 \div \dfrac{2}{3}$ **56.** $6 \div \dfrac{3}{2}$ **57.** $\dfrac{21}{7} \div \dfrac{3}{7}$ **58.** $\dfrac{54}{5} \div \dfrac{9}{10}$

59. A recipe calls for $\frac{3}{4}$ cup of sugar. How many cups of sugar are needed to triple the recipe?

60. If there are 12 sheets of plywood in a stack, find the height of the stack if the plywood is $\frac{5}{8}$ inch thick.

61. How many dresses containing $2\frac{3}{4}$ yards of material can be made from $5\frac{1}{2}$ yards of fabric?

62. How many boards of length $3\frac{3}{4}$ feet can be cut from a 12-foot board?

63. If the numerator is smaller than the denominator, a fraction is a _____ fraction.

64. If the numerator is larger than or equal to the denominator, a fraction is an _____ fraction.

65. Prime numbers are numbers _____ than 1 that are divisible only by themselves and 1.

66. To simplify a fraction, _____ the numerator and denominator into primes and remove the 1.

67. To multiply fractions, multiply the _____ and multiply the _____.

68. To divide two fractions, invert the _____ and multiply.

Equivalent fractions   fractions with the same value.

1.2   ADDING AND SUBTRACTING FRACTIONS

When we add or subtract fractions, we find the result of combining fractions of the same type. Fractions with the same denominators are of the same type. Hence when we add three-eighths of a pie and one-eighth of a pie, we are adding eighths of a pie.

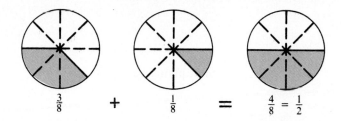

$$\frac{3}{8} \quad + \quad \frac{1}{8} \quad = \quad \frac{4}{8} = \frac{1}{2}$$

1 **Adding Fractions with the Same Denominators**

> *To add fractions with the same denominators,* add only the numerators and keep the same denominator.

EXAMPLE 1   Add. Simplify, if possible, by removing a 1.

a. $\dfrac{5}{12} + \dfrac{1}{12}$

$$\frac{5}{12} + \frac{1}{12} = \frac{6}{12} = \frac{1}{2}$$

b. $\dfrac{3}{4} + \dfrac{6}{4}$

$$\frac{3}{4} + \frac{6}{4} = \frac{9}{4} \quad \blacksquare$$

You may have learned to change $\frac{9}{4}$ to $2\frac{1}{4}$. In algebra the improper fraction where the numerator is greater than or equal to the denominator is more useful.

☐ DO EXERCISE 1.

2 **Multiplying by 1**

To add fractions with different denominators, we must convert to fractions with the same denominator. To do this, we must find a way to make the denominators the same.

*The numerator and the denominator of a fraction may be multiplied by the same number to give an **equivalent fraction**.* This is called multiplying by 1.

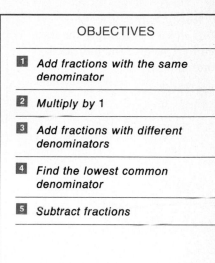

OBJECTIVES

1 *Add fractions with the same denominator*

2 *Multiply by 1*

3 *Add fractions with different denominators*

4 *Find the lowest common denominator*

5 *Subtract fractions*

☐ **Exercise 1**   Add and simplify, if possible, by removing a 1.

a. $\dfrac{1}{10} + \dfrac{3}{10}$

b. $\dfrac{5}{6} + \dfrac{3}{6}$

□ **Exercise 2** Multiply the fraction by 1 to give the indicated denominator.

a. $\dfrac{1}{3} \cdot \dfrac{?}{?} = \dfrac{?}{9}$

b. $2 \cdot \dfrac{?}{?} = \dfrac{?}{10}$

c. $\dfrac{3}{8} \cdot \dfrac{?}{?} = \dfrac{?}{24}$

d. $\dfrac{4}{9} \cdot \dfrac{?}{?} = \dfrac{?}{18}$

□ **Exercise 3** Add and simplify, if possible, by removing a 1.

a. $\dfrac{1}{2} + \dfrac{3}{10}$

b. $\dfrac{1}{2} + \dfrac{5}{6}$

EXAMPLE 2 Multiply the fraction by 1 to give a denominator of 8.

a. $\dfrac{3}{4}$

$$\dfrac{3}{4} \cdot \dfrac{2}{2} = \dfrac{6}{8} \qquad \text{Multiplying by } \tfrac{2}{2} \text{ to give a denominator of 8}$$

Notice that $\tfrac{3}{4}$ and $\tfrac{6}{8}$ are equivalent fractions.

b. $\dfrac{1}{2}$

$$\dfrac{1}{2} \cdot \dfrac{4}{4} = \dfrac{4}{8} \qquad \text{Multiplying by } \tfrac{4}{4} \text{, which is 1}$$

c. 5

$$5 = \dfrac{5}{1} = \dfrac{5}{1} \cdot \dfrac{8}{8} = \dfrac{40}{8} \qquad ■$$

□ **DO EXERCISE 2.**

3 Adding Fractions with Different Denominators

Now we may *add fractions with different denominators*. We multiply each fraction by 1 to give them the same denominators and then add the numerators.

EXAMPLE 3 Add and simplify, if possible, by removing a 1.

a. $\dfrac{1}{3} + \dfrac{1}{4}$

$$\dfrac{1}{3} + \dfrac{1}{4} = \dfrac{1}{3} \cdot \dfrac{4}{4} + \dfrac{1}{4} \cdot \dfrac{3}{3}$$
$$= \dfrac{4}{12} + \dfrac{3}{12} = \dfrac{7}{12}$$

Notice that this fraction is already simplified.

b. $\dfrac{5}{6} + \dfrac{2}{3}$

$$\dfrac{5}{6} + \dfrac{2}{3} = \dfrac{5}{6} \cdot \dfrac{3}{3} + \dfrac{2}{3} \cdot \dfrac{6}{6}$$
$$= \dfrac{15}{18} + \dfrac{12}{18} = \dfrac{27}{18}$$
$$= \dfrac{3 \cdot 3 \cdot 3}{2 \cdot 3 \cdot 3} = \dfrac{3 \cdot 3 \cdot 3}{3 \cdot 3 \cdot 2} = \dfrac{3}{2} \qquad \text{Simplifying} \qquad ■$$

□ **DO EXERCISE 3.**

Lowest Common Denominator smallest number that all denominators will divide into without a remainder.

4 Lowest Common Denominator

When we add fractions with different denominators we may make our work easier by using the **lowest common denominator** (LCD).

> Use the following steps to find the LCD.
>
> 1. Write the prime factors of the number.
> 2. For the LCD use each factor the greatest number of times that it occurs in any factorization.

EXAMPLE 4 Find the LCD.

a. $\frac{1}{8}$ and $\frac{5}{12}$

In the following chart, if a factor of the second number is the same as a factor of the first number, place this factor beneath the factor of the first number.

Prime Factorization

8	2	2	2	
12	2	2		3
LCD	2	2	2	3

Listing the numbers in each column once

$$LCD = 2 \cdot 2 \cdot 2 \cdot 3 = 24$$

b. $\frac{7}{15}$ and $\frac{4}{27}$

Prime Factorization

15	3	5		
27	3		3	3
LCD	3	5	3	3

$$LCD = 3 \cdot 5 \cdot 3 \cdot 3 = 135$$

c. $\frac{1}{3}$ and $\frac{1}{8}$

Prime Factorization

3	3			
8		2	2	2
LCD	3	2	2	2

$$LCD = 3 \cdot 2 \cdot 2 \cdot 2 = 24$$

The LCD is 24 since 3 and 8 have no common factors. ∎

□ DO EXERCISE 4.

□ **Exercise 4** Find the LCD.

a. $\frac{1}{5}$ and $\frac{1}{3}$

b. $\frac{3}{8}$ and $\frac{7}{24}$

c. $\frac{5}{16}$ and $\frac{1}{36}$

d. $\frac{7}{12}$ and $\frac{5}{42}$

a. $\dfrac{1}{5} + \dfrac{1}{3}$

b. $\dfrac{1}{4} + \dfrac{3}{8}$

c. $\dfrac{9}{7} + \dfrac{2}{3}$

d. $\dfrac{1}{12} + \dfrac{3}{8}$

We may now do Example 3b as shown in Example 5a.

EXAMPLE 5   Add.

a. $\dfrac{5}{6} + \dfrac{2}{3}$

Find the LCD.

Prime Factorization

	2	3		
6	2	3		
3		3		
LCD	2	3		

The LCD is 2 · 3 or 6.

$$\dfrac{5}{6} + \dfrac{2}{3} = \dfrac{5}{6} + \dfrac{2}{3} \cdot \dfrac{2}{2} \qquad \text{Multiplying by 1}$$

$$= \dfrac{5}{6} + \dfrac{4}{6} = \dfrac{9}{6} = \dfrac{3}{2}$$

b. $\dfrac{4}{9} + \dfrac{7}{30}$

Prime Factorization

9	3	3		
30	3		5	2
LCD	3	3	5	2

$$\text{LCD} = 3 \cdot 3 \cdot 5 \cdot 2 = 90$$

$$\dfrac{4}{9} + \dfrac{7}{30} = \dfrac{4}{9} \cdot \dfrac{10}{10} + \dfrac{7}{30} \cdot \dfrac{3}{3} \qquad \begin{array}{l}\text{Multiplying by 1's to give}\\ \text{the fractions the LCD}\end{array}$$

$$= \dfrac{40}{90} + \dfrac{21}{90} = \dfrac{61}{90} \qquad \blacksquare$$

□ **DO EXERCISE 5.**

5 Subtracting Fractions

Subtraction of fractions is done in the same way as addition of fractions. If the fractions have the same denominator, subtract the numerators and keep the same denominator. If they have different denominators, multiply by 1 to get the LCD and then subtract the numerators and keep the same denominator. Always simplify the result.

EXAMPLE 6 Subtract and simplify if possible, by removing a 1.

a. $\dfrac{6}{7} - \dfrac{1}{2}$

$$\dfrac{6}{7} - \dfrac{1}{2} = \dfrac{6}{7} \cdot \dfrac{2}{2} - \dfrac{1}{2} \cdot \dfrac{7}{7} \qquad \text{Multiplying by 1's to get the LCD of 14}$$

$$= \dfrac{12}{14} - \dfrac{7}{14} = \dfrac{5}{14}$$

b. $\dfrac{15}{12} - \dfrac{5}{6}$

$$\dfrac{15}{12} - \dfrac{5}{6} = \dfrac{15}{12} - \dfrac{5}{6} \cdot \dfrac{2}{2} \qquad \text{The LCD is 12}$$

$$= \dfrac{15}{12} - \dfrac{10}{12} = \dfrac{5}{12}$$

c. $7 - \dfrac{3}{5}$

$$7 - \dfrac{3}{5} = \dfrac{7}{1} - \dfrac{3}{5} = \dfrac{7}{1} \cdot \dfrac{5}{5} - \dfrac{3}{5}$$

$$= \dfrac{35}{5} - \dfrac{3}{5} = \dfrac{32}{5}$$

d. $2\dfrac{1}{4} - 1\dfrac{1}{5}$

$$2\dfrac{1}{4} - 1\dfrac{1}{5} = \dfrac{9}{4} - \dfrac{6}{5} = \dfrac{9}{4} \cdot \dfrac{5}{5} - \dfrac{6}{5} \cdot \dfrac{4}{4} \qquad \text{The LCD is 20}$$

$$= \dfrac{45}{20} - \dfrac{24}{20} = \dfrac{21}{20}$$

e. $\dfrac{18}{5} - 3$

$$\dfrac{18}{5} - 3 = \dfrac{18}{5} - \dfrac{3}{1} = \dfrac{18}{5} - \dfrac{3}{1} \cdot \dfrac{5}{5}$$

$$= \dfrac{18}{5} - \dfrac{15}{5} = \dfrac{3}{5}$$

□ **Exercise 6** Subtract and simplify, if possible, by removing a 1.

a. $\dfrac{1}{3} - \dfrac{1}{4}$

b. $\dfrac{7}{12} - \dfrac{1}{24}$

c. $\dfrac{25}{8} - 2$

d. $9 - \dfrac{2}{7}$

e. $3\dfrac{3}{8} - 2\dfrac{1}{12}$

f. $\dfrac{5}{18} - \dfrac{1}{45}$

f. $\dfrac{3}{20} - \dfrac{1}{36}$

Find the LCD.

Prime Factorization

	2	2	5		
20	2	2	5		
36	2	2		3	3
LCD	2	2	5	3	3

The LCD is $2 \cdot 2 \cdot 5 \cdot 3 \cdot 3$ or 180.

$$\frac{3}{20} - \frac{1}{36} = \frac{3}{20} \cdot \frac{9}{9} - \frac{1}{36} \cdot \frac{5}{5}$$

$$= \frac{27}{180} - \frac{5}{180} = \frac{22}{180}$$

$$= \frac{11}{90} \qquad \text{Subtracting and simplifying} \qquad ■$$

□ **DO EXERCISE 6.**

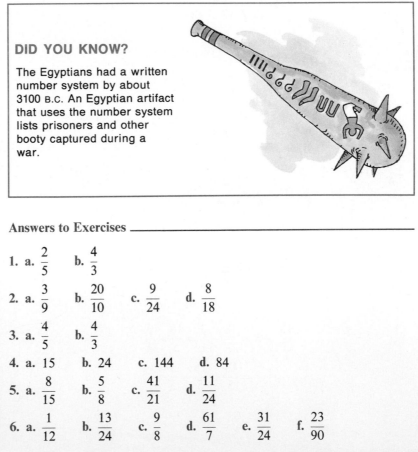

DID YOU KNOW?

The Egyptians had a written number system by about 3100 B.C. An Egyptian artifact that uses the number system lists prisoners and other booty captured during a war.

Answers to Exercises _____

1. a. $\dfrac{2}{5}$ b. $\dfrac{4}{3}$

2. a. $\dfrac{3}{9}$ b. $\dfrac{20}{10}$ c. $\dfrac{9}{24}$ d. $\dfrac{8}{18}$

3. a. $\dfrac{4}{5}$ b. $\dfrac{4}{3}$

4. a. 15 b. 24 c. 144 d. 84

5. a. $\dfrac{8}{15}$ b. $\dfrac{5}{8}$ c. $\dfrac{41}{21}$ d. $\dfrac{11}{24}$

6. a. $\dfrac{1}{12}$ b. $\dfrac{13}{24}$ c. $\dfrac{9}{8}$ d. $\dfrac{61}{7}$ e. $\dfrac{31}{24}$ f. $\dfrac{23}{90}$

PROBLEM SET 1.2

A. *Add and simplify, if possible, by removing a 1.*

1. $\dfrac{5}{6} + \dfrac{3}{6}$

2. $\dfrac{4}{8} + \dfrac{2}{8}$

3. $\dfrac{11}{5} + \dfrac{4}{5}$

4. $\dfrac{9}{20} + \dfrac{11}{20}$

B. *Multiply by 1 to give the fraction the denominator indicated.*

5. $3 \cdot \dfrac{?}{?} = \dfrac{?}{10}$

6. $\dfrac{5}{9} \cdot \dfrac{?}{?} = \dfrac{?}{45}$

7. $\dfrac{9}{8} \cdot \dfrac{?}{?} = \dfrac{?}{72}$

8. $\dfrac{11}{6} \cdot \dfrac{?}{?} = \dfrac{?}{54}$

C. *Find the LCD. Do not add.*

9. $\dfrac{1}{7}$ and $\dfrac{3}{14}$

10. $\dfrac{1}{8}$ and $\dfrac{5}{16}$

11. $\dfrac{11}{5}$ and $\dfrac{8}{3}$

12. $\dfrac{9}{8}$ and $\dfrac{7}{9}$

13. $\dfrac{5}{24}$ and $\dfrac{7}{36}$

14. $\dfrac{1}{18}$ and $\dfrac{5}{48}$

15. $\dfrac{7}{20}$ and $\dfrac{9}{32}$

16. $\dfrac{1}{10}$ and $\dfrac{4}{75}$

D. *Add and simplify, if possible, by removing a 1.*

17. $\dfrac{1}{4} + \dfrac{1}{6}$

18. $\dfrac{1}{5} + \dfrac{3}{10}$

19. $\dfrac{5}{7} + \dfrac{4}{3}$

20. $\dfrac{7}{8} + \dfrac{6}{5}$

21. $\dfrac{5}{12} + \dfrac{9}{24}$

22. $\dfrac{3}{10} + \dfrac{2}{20}$

23. $\dfrac{9}{8} + \dfrac{5}{6}$

24. $\dfrac{15}{7} + \dfrac{5}{3}$

25. $\dfrac{1}{3} + \dfrac{5}{21}$

26. $\dfrac{3}{8} + \dfrac{2}{5}$

27. $\dfrac{5}{3} + \dfrac{1}{6}$

28. $\dfrac{8}{9} + \dfrac{1}{27}$

E. *Subtract and simplify, if possible, by removing a 1.*

29. $\dfrac{2}{3} - \dfrac{4}{9}$

30. $\dfrac{7}{11} - \dfrac{1}{4}$

31. $\dfrac{7}{6} - \dfrac{1}{3}$

32. $\dfrac{12}{5} - \dfrac{1}{4}$

33. $\dfrac{3}{10} - \dfrac{1}{6}$

34. $\dfrac{4}{5} - \dfrac{2}{3}$

35. $\dfrac{5}{12} - \dfrac{1}{6}$

36. $\dfrac{8}{24} - \dfrac{1}{12}$

37. $5 - \dfrac{3}{4}$

38. $8 - \dfrac{11}{3}$

39. $\dfrac{15}{8} - 1$

40. $\dfrac{27}{10} - 2$

41. $9\dfrac{4}{5} - \dfrac{2}{3}$

42. $4\dfrac{3}{4} - \dfrac{5}{7}$

43. $\dfrac{7}{12} - \dfrac{1}{32}$

44. $\dfrac{9}{15} - \dfrac{1}{20}$

45. John walked for $\frac{3}{4}$ of a mile and jogged for $\frac{3}{7}$ of a mile. How far did he go?

46. Susan must buy $\frac{1}{8}$ yard of ribbon plus $\frac{1}{4}$ yard of ribbon. How much must she buy?

47. Jennifer bought $\frac{1}{4}$ yard less fabric than the $3\frac{5}{8}$ yards that the pattern recommended. How much fabric did she buy?

48. Greg caught two fish. One weighed $7\frac{3}{4}$ pounds and the other weighed $5\frac{1}{3}$ pounds. What is the difference in the weights of the two fish?

49. To add fractions with the same denominators, add only the _____ and keep the same denominator.

50. The numerator and denominator may be multiplied by _____ to give an equivalent fraction.

51. To add fractions with different denominators first multiply each fraction by one to give it the _____ _____ _____.

52. Subtraction of fractions is similar to addition of fractions except that we _____ the numerators.

1.3 DECIMAL AND EXPONENTIAL NOTATION

Fractions may be converted to decimals and some decimals may be changed to fractions. In this section we review these processes. We also discuss exponential notation.

OBJECTIVES

1 Write decimals in words

2 Convert from decimals to fractions

3 Convert from fractions to decimals

4 Write exponential notation for numbers such as $3 \cdot 3 \cdot 3$

5 Write expanded notation for numbers such as 2^5

6 Give the meaning of expressions with exponents of 0 and 1

1 Place Value

The place values for our number system are shown below.

Millions	Hundred-Thousands	Ten-Thousands	Thousands	Hundreds	Tens	Ones		Tenths	Hundredths	Thousandths	Ten-Thousandths	Hundred-Thousandths	Millionths
3,	5	6	8,	9	2	1.		5	3	7	2	6	8

We use this chart to help us read and write decimals.

EXAMPLE 1 Write in words.

a. 0.05

The last number to the right of the decimal point is in the hundredths place. The decimal is written *five hundredths*.

b. 0.0425

The last number to the right of the decimal place is in the ten-thousandths place. The number is written *four hundred twenty-five ten-thousandths*.

c. 7.309

The decimal point is translated to an "and" when a number other than zero is written to its left. The number is written *seven and three hundred nine thousandths*. ■

☐ **DO EXERCISE 1.**

2 Decimals to Fractions

To change *decimals to fractions*, we recall the place values used in our number system.

EXAMPLE 2 Change to a fraction and simplify, if possible, by removing a 1.

a. 0.18

$$0.18 = 18 \text{ hundredths} = \frac{18}{100} = \frac{2 \cdot 3 \cdot 3}{2 \cdot 2 \cdot 5 \cdot 5}$$

$$= \frac{9}{50} \quad \text{Simplifying}$$

☐ **Exercise 1** Write in words.

a. 0.02

b. 0.0314

c. 6.053

d. 71.8

□ **Exercise 2** Change to a fraction and simplify, if possible, by removing a 1.

a. 0.38

b. 0.025

c. 0.3179

d. 16.8

□ **Exercise 3** Change to decimals and round to the nearest thousandth.

a. $\dfrac{5}{8}$

b. $\dfrac{3}{7}$

c. $\dfrac{2}{9}$

d. $\dfrac{11}{3}$

b. 0.179

$$0.179 = 179 \text{ thousandths} = \frac{179}{1000}$$

c. 3.2

$$3.2 = 3 \text{ and } 2 \text{ tenths} = 3\frac{2}{10} = 3\frac{1}{5} \quad ■$$

□ **DO EXERCISE 2.**

3 **Fractions to Decimals**

The fraction $\dfrac{a}{b}$ may be written as $a \div b$.

EXAMPLE 3 Change to decimal notation.

a. $\dfrac{1}{8}$

$$
\begin{array}{r}
0.125 \\
8\,)\,\overline{1.000} \\
\underline{8} \\
20 \\
\underline{16} \\
40 \\
\underline{40} \\
\end{array}
$$

b. $\dfrac{5}{6}$

$$
\begin{array}{r}
0.8333 \\
6\,)\,\overline{5.0000} \\
\underline{4\,8} \\
20 \\
\underline{18} \\
20 \\
\underline{18} \\
20 \\
\underline{18} \\
2 \\
\end{array}
$$
 Notice that we may continue to divide ■

In this book we carry the division out to four decimal places and round the answer to three places. Recall that to round the answer, if the number in the fourth place is 5 or larger, we add one to the digit in the third place and eliminate all digits to the right of the third place. If the number in the fourth place is less than 5, we eliminate all digits to the right of the third place. This is called rounding to the nearest thousandth. We will use the symbol $\approx$, which means approximately equal to, to show that we have rounded the answer. Hence $\frac{5}{6} \approx 0.833$.

□ **DO EXERCISE 3.**

Number squared   a number to the second power.
Number cubed   a number to the third power.

4   **Exponents**

We want to find a shorter notation for numbers such as

$$3 \cdot 3 \cdot 3 \cdot 3$$

Recall that *factors* of a number are numbers that are multiplied together to give that number. Thus if $a \cdot b \cdot c = d$, then a, b, and c are factors of d. The number

$$3 \cdot 3 \cdot 3 \cdot 3$$

has four factors that are all the same, so we write it as 3^4. Three is called the *base* and 4 is the *exponent*. The number 3^4 is written in exponential notation. It is called an exponential number.

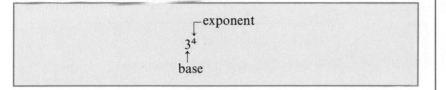

Notice that the exponent shows how many times the base is written as a factor. The number is read "three to the fourth power." When the exponent is a 2 or a 3, the numbers are given special names.

$$3^2 = \textbf{three squared} \text{ or three to the second power}$$

$$3^3 = \textbf{three cubed} \text{ or three to the third power}$$

$$3^4 = \text{three to the fourth power}$$

EXAMPLE 4   Write in exponential notation.

a. $4 \cdot 4 \cdot 4 \cdot 4 \cdot 4 \cdot 4 = 4^6$

b. $10 \cdot 10 \cdot 10 \cdot 10 = 10^4$

c. If we use x to mean a number,

$$x \cdot x \cdot x = x^3 \quad \blacksquare$$

☐ **DO EXERCISE 4.**

5   **Expanded Notation**

Notice that we can reverse the procedure.

EXAMPLE 5   Expand.

a. $8^3 = 8 \cdot 8 \cdot 8$

b. $10^2 = 10 \cdot 10$

c. $x^4 = x \cdot x \cdot x \cdot x \quad \blacksquare$

☐ **DO EXERCISE 5.**

☐ **Exercise 4**   Write in exponential notation.

a. $5 \cdot 5 \cdot 5$

b. $7 \cdot 7 \cdot 7 \cdot 7 \cdot 7$

c. $x \cdot x$

d. $x \cdot x \cdot x \cdot x \cdot x$

☐ **Exercise 5**   Expand.

a. 2^4

b. 9^2

c. x^3

d. x^6

□ **Exercise 6** What is the meaning of each of the following?

a. 2^1 $= 2$

b. 7^0 $= 1$

c. 8^1 $= 8$

d. 4^0 $= 1$

---

6 ■ **Exponents of 1 and Zero**

One and zero are also used as exponents. Consider the following.

$$3 \cdot 3 \cdot 3 = 3^3$$

$$3 \cdot 3 = 3^2$$

We are dividing the left side of the first expression by 3 to get the left side of the second expression. Hence, continuing the pattern, the next expressions are

$$3 = 3^1$$

$$1 = 3^0$$

This means that $3 = 3^1$ and $1 = 3^0$. We have the following definitions:

> $a^1 = a$ for any number a
>
> $a^0 = 1$ for any number a except zero.

The expression 0^0 is not defined.

EXAMPLE 6 What is the meaning of each of the following?

a. 5^1

$$5^1 = 5$$

b. 9^0

$$9^0 = 1$$

c. 6^0

$$6^0 = 1 \qquad ■$$

□ **DO EXERCISE 6.**

Answers to Exercises

1. a. Two hundredths   **b.** Three hundred fourteen ten-thousandths
c. Six and fifty-three thousandths   **d.** Seventy-one and eight tenths
2. a. $\dfrac{19}{50}$ **b.** $\dfrac{1}{40}$ **c.** $\dfrac{3179}{10,000}$ **d.** $16\dfrac{4}{5}$
3. a. 0.625   **b.** 0.429   **c.** 0.222   **d.** 3.667
4. a. 5^3 **b.** 7^5 **c.** x^2 **d.** x^5
5. a. $2 \cdot 2 \cdot 2 \cdot 2$ **b.** $9 \cdot 9$ **c.** $x \cdot x \cdot x$ **d.** $x \cdot x \cdot x \cdot x \cdot x \cdot x$
6. a. 2   **b.** 1   **c.** 8   **d.** 1

PROBLEM SET 1.3

A. *Change to a fraction and simplify, if possible, by removing a 1.*

1. 0.68

2. 0.26

3. 0.035

4. 0.088

5. 0.317

6. 0.519

7. 0.03

8. 0.047

9. 15.08

10. 3.005

11. 121.2

12. 17.75

B. *Change to decimal notation. Round to the nearest thousandth.*

13. $\dfrac{1}{2}$

14. $\dfrac{3}{4}$

15. $\dfrac{7}{8}$

16. $\dfrac{4}{9}$

17. $\dfrac{2}{5}$

18. $\dfrac{3}{10}$

19. $\dfrac{4}{11}$

20. $\dfrac{6}{7}$

21. $\dfrac{32}{5}$

22. $\dfrac{17}{2}$

23. $\dfrac{54}{7}$

24. $\dfrac{57}{9}$

C. *Write in exponential notation.*

25. $3 \cdot 3$

26. $8 \cdot 8 \cdot 8 \cdot 8$

27. $2 \cdot 2 \cdot 2 \cdot 2 \cdot 2$

28. $7 \cdot 7 \cdot 7$

29. $x \cdot x \cdot x \cdot x$

30. $x \cdot x$

D. *Expand.*

31. 5^4

32. 3^3

33. 7^2

34. 9^5

35. x^3

36. x^5

E. *What is the meaning of each of the following?*

37. 7^0

38. 10^0

39. 6^1

40. 8^1

41. $x^0, x \neq 0$

42. $y^0, y \neq 0$

43. b^1

44. z^1

45. The symbol $\approx$ means "_____ equal to."

46. In the expression 2^5, 2 is called the _____.

47. The number 4^3 may be read "four _____" or "four to the third power."

48. The expression 0^0 is not _____.

Percent   per one hundred.

1.4   PERCENT

Percent, denoted by %, means "per hundred." We use this fact to change from percents to decimals or fractions and from decimals or fractions to percents.

1   Changing from Percents to Decimals

We want to write 28% in decimal notation, so we write

$$28\% = \frac{28}{100} \qquad \text{Since percent means per hundred}$$

This may also be written $\frac{28}{100} = 28 \cdot \frac{1}{100}$ or $28 \cdot (0.01)$. So

$$x\% \qquad \text{means} \qquad x \cdot \frac{1}{100} \quad \text{or} \quad x \cdot (0.01)$$

Therefore, $28\% = 28 \cdot (0.01) = 0.28$.

EXAMPLE 1   Change to a decimal.

a. $74\% = 74 \cdot (0.01) = 0.74$

b. $3\% = 3 \cdot (0.01) = 0.03$

c. $\frac{3}{4}\% = 0.75\% = (0.75) \cdot (0.01) = 0.0075$

> Notice that when we change from percents to decimals, we move the decimal point two places to the left.

☐ **DO EXERCISE 1.**

OBJECTIVES

1 *Change from percent notation to decimal notation*

2 *Change from decimal notation to percent notation*

3 *Convert from a percent to a fraction*

4 *Convert from a fraction to a percent*

☐ **Exercise 1**   Change to a decimal.

a. 35%

b. 7%

c. 41.5%

d. 100%

e. $\frac{1}{2}\%$

f. $\frac{1}{3}\%$

2 Changing from Decimals to Percents

When we change from a percent to a decimal, we use the fact that percent means "per hundred." We use this same fact to change from a decimal to a percent. We need to change the decimal to a fraction with 100 as the denominator.

EXAMPLE 2   Change to percent.

a. 0.72

$$0.72 = \frac{72}{100} = 72\%$$

b. 0.06

$$0.06 = \frac{6}{100} = 6\%$$

Notice that we are multiplying the numerator and denominator of the decimal by 100. When we multiply a decimal by 100, we move the decimal point two places to the right.

c. 0.875

$$0.875 = 0.875 \times \frac{100}{100} = \frac{87.5}{100} = 87.5\%$$

or

$$0.875 = 87.5\%$$

> To change from decimals to percents, move the decimal point two places to the right.

■

□ **DO EXERCISE 2.**

3 Changing from Percents to Fractions

To change from a *percent* to a *fraction*, use the fact that $x\%$ means $x \cdot \frac{1}{100}$.

EXAMPLE 3   Change to a fraction and simplify, if possible, by removing a 1.

a. $74\% = 74 \cdot \frac{1}{100} = \frac{74}{100} = \frac{37}{50}$

b. $3.5\% = 3.5 \cdot \frac{1}{100} = \frac{3.5}{100} = \frac{3.5}{100} \cdot \frac{10}{10}$ Multiplying by 1 to remove the decimal point from the numerator

$$= \frac{35}{1000} = \frac{7}{200} \quad ■$$

□ **DO EXERCISE 3.**

4 Changing from Fractions to Percents

To change from a *fraction* to a *percent*, convert the fraction to a decimal and then change the decimal to a percent.

EXAMPLE 4 Change to percent.

a. $\dfrac{1}{8} = 0.125 = 12.5\%$ Moving the decimal point two places to the right

b. $\dfrac{3}{7} \approx 0.4285 \approx 0.429 \approx 42.9\%$ ■

□ DO EXERCISE 4.

The following conversions from fractions to decimals and percents should be memorized.

Fraction	Decimal	Percent
$\dfrac{1}{10}$	0.1	10%
$\dfrac{1}{5}$	0.2	20%
$\dfrac{3}{10}$	0.3	30%
$\dfrac{2}{5}$	0.4	40%
$\dfrac{1}{2}$	0.5	50%
$\dfrac{3}{5}$	0.6	60%
$\dfrac{7}{10}$	0.7	70%
$\dfrac{4}{5}$	0.8	80%
$\dfrac{9}{10}$	0.9	90%
1	1.0	100%

Additional Important Fractions

$\dfrac{1}{4}$	0.25	25%
$\dfrac{1}{3}$	0.333	33.3%
$\dfrac{2}{3}$	0.667	66.7%
$\dfrac{3}{4}$	0.75	75%

□ **Exercise 4**   Change to percent.

a. $\dfrac{3}{4}$

b. $\dfrac{1}{3}$

c. $\dfrac{5}{7}$

d. $\dfrac{7}{8}$

Answers to Exercises

1. **a.** 0.35   **b.** 0.07   **c.** 0.415   **d.** 1.00   **e.** 0.005   **f.** 0.003

2. **a.** 38%   **b.** 1%   **c.** 100%   **d.** 75.4%

3. **a.** $\dfrac{13}{25}$ **b.** $\dfrac{37}{100}$ **c.** $\dfrac{87}{1000}$ **d.** $\dfrac{73}{125}$

4. **a.** 75%   **b.** 33.3%   **c.** 71.4%   **d.** 87.5%

PROBLEM SET 1.4

A. *Change to a decimal.*

1. 60%

2. 25%

3. 7%

4. 9%

5. 32.3%

6. 7.8%

7. 100%

8. 20.6%

9. $\frac{1}{4}$%

10. $\frac{2}{5}$%

11. $\frac{2}{3}$%

12. $\frac{2}{9}$%

B. *Change to percent.*

13. 0.26

14. 0.96

15. 0.06

16. 0.01

17. 1.00

18. 0.387

19. 0.045

20. 0.009

21. 3.4

22. 7.8

23. 2.45

24. 3.29

C. *Change to a fraction and simplify, if possible, by removing a 1.*

25. 28%

26. 75%

27. 39%

28. 61%

29. 38.5%

30. 42.1%

31. 2.4%

32. 7.9%

33. 0.42%

34. 0.68%

35. 8%

36. 6%

D. *Change to percent.*

37. $\dfrac{1}{2}$

38. $\dfrac{4}{5}$

39. $\dfrac{5}{9}$

40. $\dfrac{5}{8}$

41. $\dfrac{7}{10}$

42. $\dfrac{3}{10}$

43. $\dfrac{1}{3}$

44. $\dfrac{2}{3}$

45. $\dfrac{11}{5}$

46. $\dfrac{12}{3}$

47. $\dfrac{8}{7}$

48. $\dfrac{5}{3}$

49. Percent means per _____.

50. To change from a percent to a decimal, move the decimal point two places to the _____.

51. A percent may be converted to a fraction by using the fact that $x\%$ mean x times _____.

52. To change from a decimal to a percent, move the decimal point two places to the _____.

Natural numbers the counting numbers: 1, 2, 3, 4, . . .

Whole numbers the natural numbers and zero: 0, 1, 2, 3, 4, . . .

Integers . . . $-4, -3, -2, -1, 0, 1, 2, 3, 4, . . .$

Rational numbers numbers of the form a/b where a and b are integers and b cannot be zero.

Irrational numbers numbers that cannot be written as the ratio of two integers.

1.5 PROPERTIES OF THE REAL NUMBERS

OBJECTIVES

1 *Recognize the commutative laws of addition and multiplication*

2 *Do calculations in problems with parentheses*

3 *Recognize the associative laws of addition and multiplication*

4 *Recognize the distributive law*

The numbers used for counting are called the **natural numbers**. They are

$$1, \quad 2, \quad 3, \quad 4, \quad 5, \quad 6, \quad 7, \quad 8, \quad 9, \quad 10, \quad 11, \quad 12, \quad . . .$$

where "..." means that we continue on indefinitely. When we include zero with this set of numbers, we get the whole numbers. The **whole numbers** are

$$0, \quad 1, \quad 2, \quad 3, \quad 4, \quad 5, \quad 6, \quad 7, \quad 8, \quad 9, \quad 10, \quad 11, \quad . . .$$

The **integers** are

$$. . ., \quad -4, \quad -3, \quad -2, \quad -1, \quad 0, \quad 1, \quad 2, \quad 3, \quad 4, \quad . . .$$

Some integers are shown on the real number line:

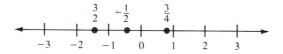

Rational numbers are numbers of the form a/b, where a and b are integers and b cannot be zero. We will explain later why b cannot be zero. Examples of rational numbers are $\frac{3}{5}, -\frac{8}{3}, 0, 6, 0.3$, and -2.7. Notice that 0 and 6 are rational numbers since they can be written $\frac{0}{1}$ and $\frac{6}{1}$. The numbers 0.3 and -2.7 are rational numbers since they can be written as $\frac{3}{10}$ and $-2\frac{7}{10}$ or $-\frac{27}{10}$. Rational numbers may also be shown on the real number line:

For every point on the real number line there is a unique number and for every number there is a point on the line. The numbers that are associated with points on the real number line are called **real numbers**. The real numbers are composed of the rational numbers and the irrational numbers. **Irrational numbers** are numbers that cannot be written in the form a/b, where a and b are integers and b is not zero.

Notice that the natural numbers are a subset of the whole numbers (if a number is a natural number, it is also a whole number). Similarly, the whole numbers are a subset of the integers, the integers are a subset of the rational numbers, and the rational numbers and irrational numbers are subsets of the real numbers. The following chart shows the relationship of these sets of numbers.

Real numbers are composed of the rational numbers and irrational numbers.
Commutative Law of Addition $a + b = b + a$ where a and b are real numbers.
(Changing the order of the numbers does not change the sum.)

☐ **Exercise 1** Add.

a. $8 + 4$

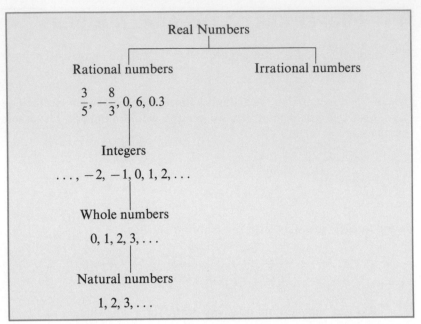

The real numbers have some special properties which we discuss next.

■ The Commutative Laws of Addition and Multiplication

The commutative law of addition tells us that changing the order of two numbers does not change the sum.

Commutative Law of Addition

$$a + b = b + a$$

where a and b are real numbers.

---

b. $4 + 8$

EXAMPLE 1 Show that $5 + 6 = 6 + 5$.

$5 + 6 = 11$
$6 + 5 = 11$ Notice that the result is the same   ■

☐ **DO EXERCISE 1.**

Commutative Law of Multiplication $a \cdot b = b \cdot a$ where a and b are real numbers. (Numbers may be multipied in any order.)

Parentheses The pair of symbols ( ) that tell us to do the operation inside them first.

> **Commutative Law of Multiplication**
>
> $$a \cdot b = b \cdot a$$
>
> where a and b are real numbers.

EXAMPLE 2 Show that $7 \cdot 5 = 5 \cdot 7$.

$$7 \cdot 5 = 35$$
$$5 \cdot 7 = 35 \quad \blacksquare$$

□ **DO EXERCISE 2.**

Notice that in both the commutative law for addition and the commutative law for multiplication, the order of the numbers a and b changes as we move from one side of the equal sign to the other.

$$a + b = b + a$$
$$a \cdot b = b \cdot a \qquad (a \text{ and } b \text{ are in reverse order})$$

An easy way to recognize the commutative law is to notice that the numbers are in reverse order.

2 **Parentheses**

> **Parentheses** tell us to do the operation inside them first.

EXAMPLE 3 Calculate.

a. $4 + (5 + 6) = 4 + 11 = 15$

b. $(7 \cdot 8) \cdot 2 = 56 \cdot 2 = 112 \quad \blacksquare$

□ **DO EXERCISE 3.**

□ **Exercise 2** Multiply.

a. $8 \cdot 3$

b. $3 \cdot 8$

□ **Exercise 3** Calculate.

a. $3 + (7 + 8)$

b. $4 \cdot (6 \cdot 2)$

Associative law of addition $(a + b) + c = a + (b + c)$ where a, b, and c are real numbers. (This law allows us to move the parentheses and regroup the numbers in an addition problem.)

Associative law of multiplication $(a \cdot b) \cdot c = a \cdot (b \cdot c)$ where a, b, and c are real numbers. (This law allows us to move the parentheses and regroup the numbers in a multiplication.)

▣ The Associative Laws of Addition and Multiplication

The commutative laws of addition and multiplication allow us to add or multiply in any order. The associative laws allow us to move the parentheses to regroup the numbers.

Associative Law of Addition
$$(a + b) + c = a + (b + c)$$
where a, b, and c are real numbers.

EXAMPLE 4   Show that $(3 + 5) + 7 = 3 + (5 + 7)$

$(3 + 5) + 7 = 8 + 7 = 15$

$3 + (5 + 7) = 3 + 12 = 15$ Notice that the result is the same   ■

□ **DO EXERCISE 4.**

Associative Law of Multiplication
$$(a \cdot b) \cdot c = a \cdot (b \cdot c)$$
where a, b, and c are real numbers.

Notice that for the associative laws, the order of the numbers does not change.

EXAMPLE 5   Show that $(7 \cdot 9) \cdot 2 = 7 \cdot (9 \cdot 2)$.

$$(7 \cdot 9) \cdot 2 = 63 \cdot 2 = 126$$
$$7 \cdot (9 \cdot 2) = 7 \cdot 18 = 126 \quad ■$$

□ **DO EXERCISE 5.**

In an expression such as $a \cdot b + c \cdot d$, we often omit the "·" and write $ab + cd$. *We also assume that the multiplications are to be done before the addition.*

□ **Exercise 4**   Calculate.

a. $(7 + 9) + 3$

b. $7 + (9 + 3)$

□ **Exercise 5**   Calculate.

a. $(6 \cdot 2) \cdot 2$

b. $6 \cdot (2 \cdot 2)$

Distributive Law of Multiplication over Addition $a \cdot (b + c) = a \cdot b + a \cdot c$ or $(b + c) \cdot a = b \cdot a + c \cdot a$ where a, b, and c are real numbers.

■4 The Distributive Law

Our next property allows us to do a problem in two ways.

> **Distributive Law of Multiplication over Addition**
>
> $$a \cdot (b + c) = a \cdot b + a \cdot c \quad \text{or} \quad (b + c) \cdot a = b \cdot a + c \cdot a$$
>
> where a, b, and c are real numbers.

EXAMPLE 6 Show that $3 \cdot (4 + 5) = 3 \cdot 4 + 3 \cdot 5$

$3 \cdot (4 + 5) = 3 \cdot 9 = 27$ Do the operation inside the parentheses first

$3 \cdot 4 + 3 \cdot 5 = 12 + 15 = 27$ Multiply before adding   ■

□ **DO EXERCISE 6.**

Notice that the distributive law involves two operations, $+$ and $\cdot$, in the same problem. The commutative and associative laws involve only one operation in each problem. We must be able to distinguish among the distributive, associative, and commutative laws. Recall that the order of the numbers changes for the commutative laws. For the associative laws the order of the numbers does not change.

EXAMPLE 7 Which laws are illustrated below?

a. $3 \cdot (5 \cdot 2) = (3 \cdot 5) \cdot 2$ Associative law of multiplication

The order does not change and one operation is involved.

b. $4 \cdot (6 + 7) = 4 \cdot 6 + 4 \cdot 7$ Distributive law

Two operations are used.

c. $6 + 5 = 5 + 6$ Commutative law of addition

One operation is involved and the order changes.

d. $(2 + 4) + 6 = 6 + (2 + 4)$ Commutative law of addition

One operation is used and the order changes.   ■

□ **DO EXERCISE 7.**

□ **Exercise 6** Calculate.

a. $2 \cdot (7 + 9)$

b. $2 \cdot 7 + 2 \cdot 9$

c. $4 \cdot (6 + 3)$

d. $4 \cdot 6 + 4 \cdot 3$

□ **Exercise 7** Which laws are illustrated below?

a. $(3 \cdot 4) \cdot 8 = 3 \cdot (4 \cdot 8)$

b. $7 \cdot 6 = 6 \cdot 7$

c. $(8 + 7) + 6 = 8 + (7 + 6)$

d. $4 \cdot (3 + 2) = 4 \cdot 3 + 4 \cdot 2$

e. $45 \cdot 64 = 64 \cdot 45$

f. $\frac{1}{2} \cdot \left(5 + \frac{3}{8}\right) = \frac{1}{2} \cdot 5 + \frac{1}{2} \cdot \frac{3}{8}$

g. $37 + 101 = 101 + 37$

h. $3 + (5 + 12) = (5 + 12) + 3$

i. $6 \cdot 20 + 6 \cdot 14 = 6 \cdot (20 + 14)$

j. $\frac{5}{9} + 4 = 4 + \frac{5}{9}$

k. $(6 \cdot 3) \cdot 1 = 1 \cdot (6 \cdot 3)$

The following chart summarizes the commutative, associative, and distributive laws.

Commutative law of addition	$a + b = b + a$
Commutative law of multiplication	$a \cdot b = b \cdot a$
Associative law of addition	$(a + b) + c = a + (b + c)$
Associative law of multiplication	$(a \cdot b) \cdot c = a \cdot (b \cdot c)$
Distributive laws of multiplication over addition	$a(b + c) = ab + ac$
	$(b + c) \cdot a = ba + ca$

Answers to Exercises _____

1. a. 12   b. 12
2. a. 24   b. 24
3. a. 18   b. 48
4. a. 19   b. 19
5. a. 24   b. 24
6. a. 32   b. 32   c. 36   d. 36
7. a. Associative law of multiplication   b. Commutative law of multiplication   c. Associative law of addition   d. Distributive law   e. Commutative law of multiplication   f. Distributive law   g. Commutative law of addition   h. Commutative law of addition   i. Distributive law   j. Commutative law of addition   k. Commutative law of multiplication

PROBLEM SET 1.5

A. *Calculate.*

1. (a) $5 + 7$ **2.** (a) $6 + 4$ **3.** (a) $9 \cdot 8$ **4.** (a) $4 \cdot 10$

    (b) $7 + 5$ (b) $4 + 6$ (b) $8 \cdot 9$ (b) $10 \cdot 4$

B. *Combine by doing the operation in the parentheses first.*

5. $6 + (4 + 8)$ **6.** $(9 + 7) + 6$ **7.** $4 \cdot (3 \cdot 5)$ **8.** $(2 \cdot 7) \cdot 3$

C. *Calculate.*

9. (a) $4 + (7 + 3)$ **10.** (a) $8 + (2 + 10)$ **11.** (a) $(20 \cdot 3) \cdot 4$ **12.** (a) $(1 \cdot 6) \cdot 8$

    (b) $(4 + 7) + 3$ (b) $(8 + 2) + 10$ (b) $20 \cdot (3 \cdot 4)$ (b) $1 \cdot (6 \cdot 8)$

D. *Calculate.*

13. (a) $5 \cdot (2 + 4)$ **14.** (a) $7 \cdot (3 + 1)$ **15.** (a) $10 \cdot (2 + 4)$ **16.** (a) $9 \cdot (1 + 6)$

    (b) $5 \cdot 2 + 5 \cdot 4$ (b) $7 \cdot 3 + 7 \cdot 1$ (b) $10 \cdot 2 + 10 \cdot 4$ (b) $9 \cdot 1 + 9 \cdot 6$

E. *Which laws are illustrated below?*

17. $6 + (3 + 2) = (6 + 3) + 2$

18. $7 \cdot (3 \cdot 5) = (7 \cdot 3) \cdot 5$

19. $4 \cdot (3 + 2) = 4 \cdot 3 + 4 \cdot 2$

20. $8 \cdot (6 + 7) = 8 \cdot 6 + 8 \cdot 7$

21. $77 + 109 = 109 + 77$

22. $1 + 0 = 0 + 1$

23. $(54 + 3) + 2 = 2 + (54 + 3)$

24. $17 + (9 + 6) = (17 + 9) + 6$

25. $21 \cdot 3 + 21 \cdot 8 = 21 \cdot (3 + 8)$

26. $5 \cdot (6 + 14) = 5 \cdot 6 + 5 \cdot 14$

27. $(8 \cdot 9) \cdot 4 = 8 \cdot (9 \cdot 4)$

28. $27 \cdot (8 \cdot 3) = (27 \cdot 8) \cdot 3$

29. $(3 \cdot 7) \cdot 6 = 6 \cdot (3 \cdot 7)$

30. $5 \cdot (9 \cdot 11) = (9 \cdot 11) \cdot 5$

31. The counting numbers are the _____ numbers.

32. If zero is included with the set of natural numbers, this is the set of _____ numbers.

33. Numbers of the form a/b, where a and b are integers and $b \neq 0$, are called _____ numbers.

34. The _____ law of addition says that changing the order of two numbers does not change the sum.

35. To do calculation such as $3 \cdot 2 + 4 \cdot 5$, do multiplications before _____.

36. The distributive law involves two _____, + and .

Variable a letter chosen to represent any number.
Formula relationship between variables.
Perimeter of a geometric figure the distance around the figure.

1.6 VARIABLES AND GEOMETRIC FORMULAS

1 Variables

When we write a letter such as b or x, the letter may stand for any number. We call these letters **variables**. To evaluate an expression such as bx means to substitute numbers for variables and calculate the answer. Recall that bx means $b \cdot x$.

EXAMPLE 1 Evaluate.

a. bx when $b = 3$ and $x = 7$

$$bx = b \cdot x = 3 \cdot 7 = 21$$

b. ayz when $a = 2$, $y = 4$, and $z = 3$

$$ayz = a \cdot y \cdot z = 2 \cdot 4 \cdot 3 = 24 \quad \blacksquare$$

☐ **DO EXERCISE 1.**

Recall that in an expression such as $ab + cd$, multiplications are to be done before additions.

EXAMPLE 2 Evaluate $2x + 3y$ when $x = 4$ and $y = 6$.

$$2x + 3y = 2 \cdot 4 + 3 \cdot 6 = 8 + 18 = 26 \quad \blacksquare$$

☐ **DO EXERCISE 2.**

Many word problems can be solved if we know a relationship between certain variables. This relationship is often called a **formula**. Some formulas for perimeter are given in this section.

2 Perimeter

The **perimeter** of a geometric figure is the distance around it. The opposite sides of a rectangle are equal in length. Therefore:

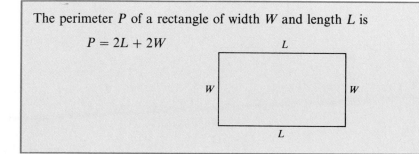

The perimeter P of a rectangle of width W and length L is

$$P = 2L + 2W$$

☐ **Exercise 1** Evaluate.

a. abc when $a = 2$, $b = 6$, and $c = 4$

b. LWH when $L = 4$, $W = 2$, and $H = 5$

☐ **Exercise 2** Evaluate.

a. $3a + 4b$ when $a = 5$ and $b = 6$

b. $5x + 6z$ when $x = 1$ and $z = \dfrac{1}{2}$

□ **Exercise 3** Find the perimeter of the following rectangles.

a.

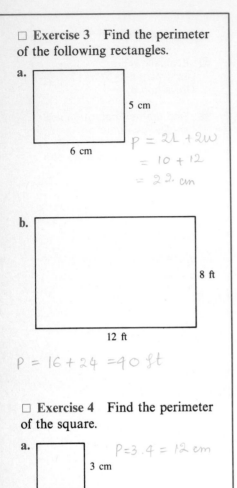

5 cm

6 cm

$P = 2L + 2W$
$= 10 + 12$
$= 22\ cm$

b.

8 ft

12 ft

$P = 16 + 24 = 40\ ft$

□ **Exercise 4** Find the perimeter of the square.

a.

$P = 3 \cdot 4 = 12\ cm$

3 cm

3 cm

b.

8 in.

8 in.

$P = 4 \cdot 8 = 32\ in$

We will use the following abbreviations.

centimeters: cm   inches: in.

meters: m   feet: ft

EXAMPLE 3 Find the perimeter of the rectangle.

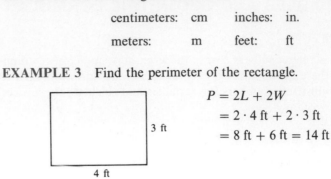

3 ft

4 ft

$P = 2L + 2W$
$= 2 \cdot 4\ \text{ft} + 2 \cdot 3\ \text{ft}$
$= 8\ \text{ft} + 6\ \text{ft} = 14\ \text{ft}$ ∎

□ **DO EXERCISE 3.**

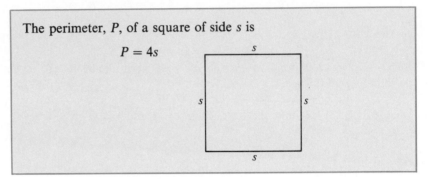

The perimeter, P, of a square of side s is

$$P = 4s$$

s

s   s

s

EXAMPLE 4 Find the perimeter of the square.

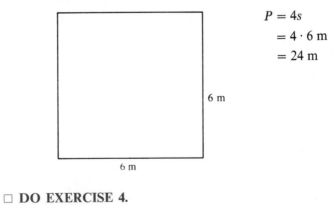

6 m

6 m

$P = 4s$
$= 4 \cdot 6\ \text{m}$
$= 24\ \text{m}$ ∎

□ **DO EXERCISE 4.**

Circumference   the perimeter of a circle.

Diameter   The distance from one point on circle to another point on a circle through the center of the circle.

Radius   one-half the diameter.

Pi   the ratio of the circumference of any circle to its diameter. (Pi is an irrational number, denoted by the symbol π.)

Closed figures with three or more sides are called *polygons*. The perimeter of any polygon is the distance around it.

EXAMPLE 5   Find the perimeter.

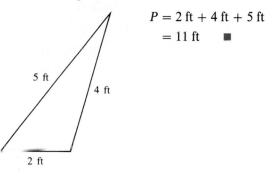

$$P = 2 \text{ ft} + 4 \text{ ft} + 5 \text{ ft}$$
$$= 11 \text{ ft} \quad \blacksquare$$

☐ **DO EXERCISE 5.**

3 Circumference

The perimeter of a circle is called the **circumference**. Consider the following circle, with center O. The **radius** is labeled r and the **diameter** is denoted by d. *The radius is one-half the diameter.*

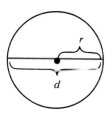

To calculate the circumference of a circle, we need a special number, **pi**, which is an irrational number. This number pi, denoted π, is the ratio of the circumference, C, of any circle to its diameter d. We write $C/d = \pi$.

> The value of π is approximately 3.14. The circumference, C, of a circle of radius r is
>
> $$C = 2\pi r$$

☐ **Exercise 5**   Find the perimeter.

a.

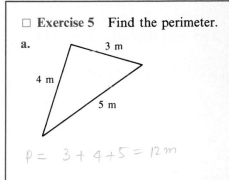

$P = 3 + 4 + 5 = 12 \, m$

b.

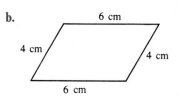

$P = 6 + 4 + 6 + 4$
$= 20 \, cm$

□ **Exercise 6** Find the circumference of the circles. Use 3.14 for π.

a. With radius of 8 cm

$C = 2 \times 3.14 \times 8$
$= 50.24\,cm$

b. With diameter of 14 m

$C = 2 \times 3.14 \times 7$
$= 43.96\,m$

c. With radius of 9 ft

$C = 2 \times 3.14 \times 9$
$= 56.52\,ft$

d. With diameter of 17 in.

$C = 53.38\,in$

EXAMPLE 6

a. Find the circumference of the circle with radius 4 cm.

$$C = 2\pi r$$
$$\approx 2 \cdot 3.14 \cdot 4\ \text{cm}$$
$$\approx 25.12\ \text{cm}$$

b. Find the circumference of the circle with diameter 10 m. To find the circumference, we need to know the radius. The radius is one-half the diameter.

$$r = \frac{1}{2} \cdot 10\ \text{m} = 5\ \text{m}$$

The circumference is

$$C = 2\pi r \approx 2 \cdot 3.14 \cdot 5\ \text{m} \approx 31.4\ \text{m}$$

Notice that since $C = 2\pi r = \pi(2r)$ and the diameter of a circle is twice the radius, we may use the formula

$$C = \pi d$$

to work the problem above.

$$C = \pi d \approx 3.14 \cdot 10\ \text{m} \approx 31.4\ \text{m} \quad \blacksquare$$

□ **DO EXERCISE 6.**

DID YOU KNOW?

The value used for pi varied in accuracy. One Chinese approximation for pi, found in 480 A.D., was 355/113. The Bible uses 3 as a value for pi. Archimedes used the value 3.14 in 240 B.C.

Answers to Exercises

1. a. 48   **b.** 40
2. a. 39   **b.** 8
3. a. 22 cm   **b.** 40 ft
4. a. 12 cm   **b.** 32 in.
5. a. 12 m   **b.** 20 cm
6. a. 50.24 cm   **b.** 43.96 m   **c.** 56.52 ft   **d.** 53.38 in.

PROBLEM SET 1.6

A. *Evaluate when* $x = 3$ *and* $y = 7$.

1. xy **2.** $3x + 4y$ **3.** $6x + 2y$ **4.** $5x$

B. *Evaluate when* $a = 3$, $b = 6$, *and* $c = 5$.

5. $5b$ **6.** abc **7.** $3b + 5c$ **8.** $4a + b + 2c$

C. *Find the perimeter of the following rectangles and squares.*

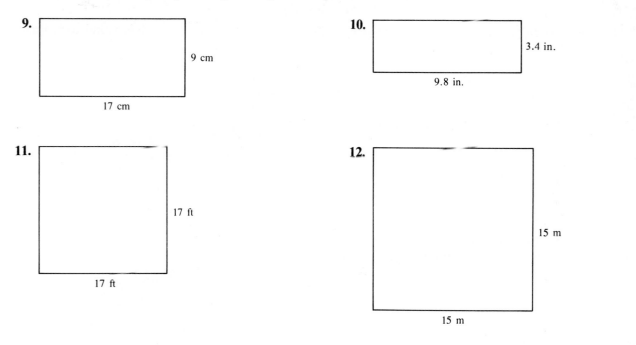

9. 9 cm, 17 cm

10. 3.4 in., 9.8 in.

11. 17 ft, 17 ft

12. 15 m, 15 m

D. *Find the perimeter of the following figures.*

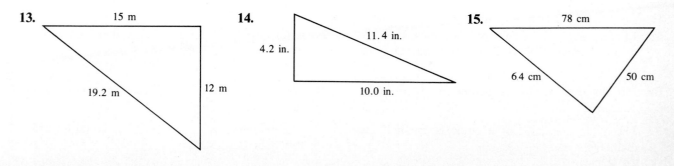

13. 15 m, 19.2 m, 12 m

14. 4.2 in., 11.4 in., 10.0 in.

15. 78 cm, 64 cm, 50 cm

E. *Find the circumference of the following circles. Use 3.14 for* π.

16.

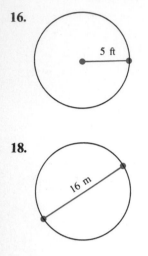

5 ft

17.

12 in.

18.

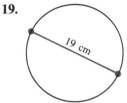

16 m

19.

19 cm

20. A rectangle has length 8 cm and width 6 cm. Find the perimeter.

$$P = 2L + 2w$$
$$= 16 + 12 = 28 \, cm$$

21. A triangle has sides of length 6, 8, and 10 ft. Find the perimeter.

$$P = 6 + 8 + 10$$
$$= 24 \, ft$$

22. The length of the side of a square is 11 m. Find the perimeter.

$$P = 4S$$
$$= 4 \cdot 11 = 44 \, m$$

23. John wishes to build a rectangular fence with a length of 25 ft. The width of the fence is to be 21 ft. How much fence does he need?

$$P = 2 \cdot 25 + 2 \cdot 21 =$$
$$= 50 + 42 = 92 \, ft$$

24. A triangular lot has sides of length 90, 100, and 137 ft. What is the perimeter of the lot?

$$P = 327 \, ft$$

25. Find the distance around a circle of radius 2 in.

$$C = 2\pi r$$
$$= 2 \times 3.14 \times 2 = 12.56 \, in.$$

26. Find the circumference of a circle with radius 10 m.

27. A letter that stands for any number is a

_____ .

28. To _____ an expression means to substitute numbers for variables and calculate the answer.

29. A formula is a _____ between certain quantities.

30. The distance around a geometric figure is its

_____ .

31. The perimeter of a circle is given a specific name, the _____ .

32. The _____ of a circle is one-half the diameter.

33. The number π is _____ equal to 3.14. the _____ .

Area of a geometric figure   the number of unit squares contained in a figure.

1.7 MORE GEOMETRIC FORMULAS

<table>
<tr><td colspan="2">OBJECTIVES</td></tr>
</table>

1 **Areas of Rectangles and Squares**

The **area** of a geometric figure is the number of unit squares contained in it. Some examples of unit squares are shown.

1 *Find the area of a rectangle or square*

2 *Find the area of a parallelogram or triangle*

3 *Find the area of a circle*

4 *Find the lateral area of a right circular cylinder*

5 *Find the volume of a rectangular solid or a right circular cylinder*

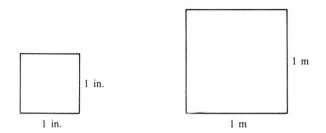

To find the area of a rectangle, we could count the number of unit squares.

☐ **Exercise 1**   Find the areas of the following rectangles.

a.
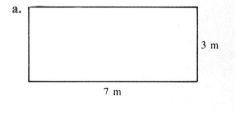

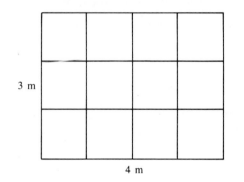

The area of the rectangle is 12 square meters, since there are 12 square units contained in it. It is easier to use the formula for the area of a rectangle:

The area A of a rectangle of length L and width W is

$$A = LW$$

EXAMPLE 1   Find the area of the following rectangle.

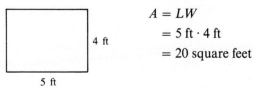

$A = LW$
$= 5\text{ ft} \cdot 4\text{ ft}$
$= 20$ square feet

b.
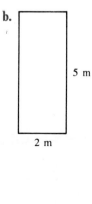

When we multiply 5 ft by 4 ft, we get 20 square feet. "Square feet" is often abbreviated ft^2.   ■

☐ **DO EXERCISE 1.**

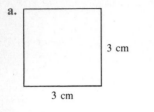
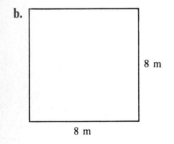
If the rectangle is a square, the length and the width are the same. If we let the length and the width both equal s, then $A = LW = s \cdot s = s^2$.

> The area, A, of a square of side s is
> $$A = s^2$$

EXAMPLE 2   Find the area of the square.

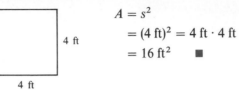

4 ft

4 ft

$A = s^2$

$= (4 \text{ ft})^2 = 4 \text{ ft} \cdot 4 \text{ ft}$

$= 16 \text{ ft}^2$ ■

□ **DO EXERCISE 2.**

2 **Areas of Parallelograms and Triangles**

A four-sided figure with two pairs of parallel sides is called a *parallelogram*. We can convert a parallelogram to a rectangle by cutting off one triangular end and attaching it to the other side as shown.

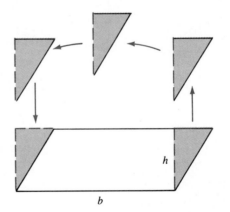

h

b

Therefore, the area of the parallelogram is the same as the area of the rectangle.

> The area A of a parallelogram of base b and height h is
> $$A = bh$$

EXAMPLE 3   Find the area of the parallelogram.

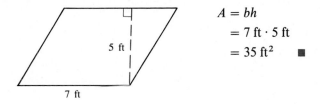

$A = bh$
$= 7\text{ ft} \cdot 5\text{ ft}$
$= 35\text{ ft}^2$ ■

□ **DO EXERCISE 3.**

The area of a triangle is one-half the area of a parallelogram.

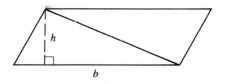

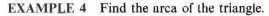

The area A of a triangle of base b and height h is

$$A = \frac{1}{2}bh$$

EXAMPLE 4   Find the area of the triangle.

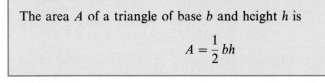

$A = \frac{1}{2}bh$

$= \frac{1}{2} \cdot 6\text{ cm} \cdot 12\text{ cm}$

$= 36\text{ cm}^2$ ■

□ **DO EXERCISE 4.**

3   Area of a Circle

We must also be able to find the area of a circle.

The area A of a circle of radius r is
$$A = \pi r^2$$

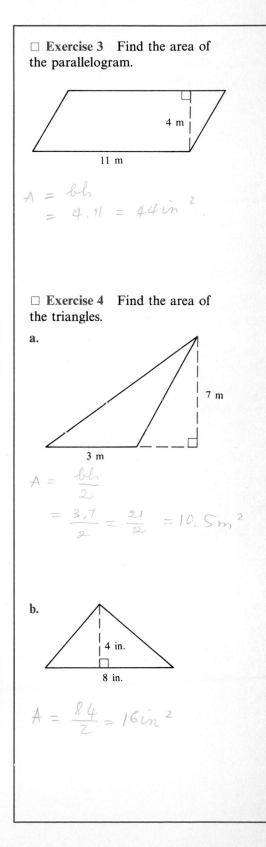

□ **Exercise 3**   Find the area of the parallelogram.

4 m

11 m

$A = bh$
$= 4 \cdot 11 = 44\text{ in}^2$

□ **Exercise 4**   Find the area of the triangles.

a.

7 m

3 m

$A = \frac{bh}{2}$

$= \frac{3 \cdot 7}{2} = \frac{21}{2} = 10.5\text{ m}^2$

b.

4 in.

8 in.

$A = \frac{8 \cdot 4}{2} = 16\text{ in}^2$

□ **Exercise 5** Find the area of the following circles.

a.

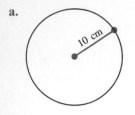

b.

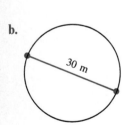

EXAMPLE 5 Find the area of the following circles.

a. $A = \pi r^2$

$\approx 3.14 \cdot 7 \text{ cm} \cdot 7 \text{ cm}$

$\approx 153.86 \text{ cm}^2$

b. $A = \pi r^2$

$\approx 3.14 \cdot 3 \text{ ft} \cdot 3 \text{ ft}$

$\approx 28.26 \text{ ft}^2$

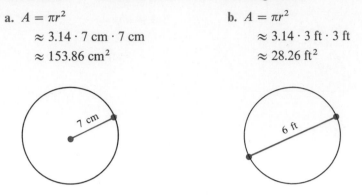

Notice that in the example, r was raised to the power, or expanded, before we multiplied by π. We always *raise numbers to powers before multiplying.* This is part of the order of operations agreement, which we will study in Section 2.4. ■

□ **DO EXERCISE 5.**

4 **Lateral Area of a Right Circular Cylinder**

It is helpful in more advanced mathematics to know the lateral area of a right circular cylinder. The lateral area may be found by opening and smoothing out the cylinder so that one side of the rectangle is the circumference of the circular bottom and the other side is the height of the cylinder.

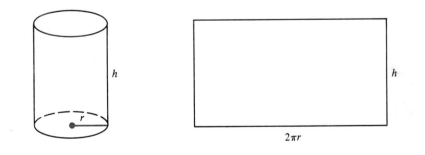

The lateral area LA of a right circular cylinder is

$$LA = 2\pi rh$$

where r is the radius of the circular bottom and h is the height of the cylinder.

Volume of a solid   the number of unit cubes contained in a solid.

EXAMPLE 6   Find the lateral area of the following right circular cylinder.

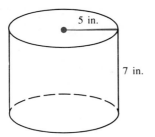

$$LA = 2\pi rh$$
$$\approx 2 \cdot 3.14 \cdot 5 \text{ in.} \cdot 7 \text{ in.}$$
$$\approx 219.8 \text{ in.}^2 \quad \blacksquare$$

☐ DO EXERCISE 6.

5   Volumes of Rectangular Solids and Right Circular Cylinders

The **volume** of a solid is the number of unit cubes contained in it. Some examples of unit cubes are shown here.

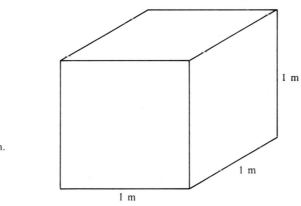

The following figure has width 2 m, length 3 m, and height 2 m.

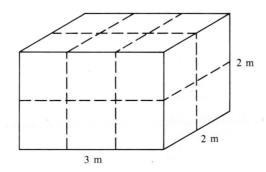

We could count the number of unit cubes contained in it to find the volume. However, there are $2 \cdot 3$ cubes in the bottom layer and the same number of cubes in the top layer. Hence the total number of cubes is

☐ **Exercise 6**   Find the lateral area of the following figures.

a.

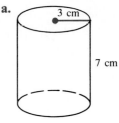

b.

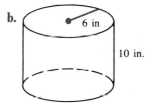

□ **Exercise 7** Find the volume of the following.

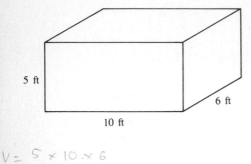

5 ft

6 ft

10 ft

$V = 5 \times 10 \times 6$

$= 300 \, ft^3$

□ **Exercise 8** Find the volume of the right circular cylinders.

a.

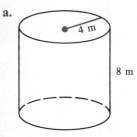

4 m

8 m

$V = 3.14 \times 16 \times 8$

$= 401.92 \, m^3$

b.

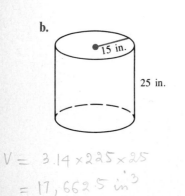

15 in.

25 in.

$V = 3.14 \times 225 \times 25$

$= 17,662.5 \, in^3$

twice that in the bottom layer, or $2 \cdot 2 \cdot 3 = 12$. The volume is 12 cubic meters. We may use the following formula:

The volume V of a rectangular solid of width W, length L, and height H is

$$V = LWH$$

EXAMPLE 7 Find the volume of the solid shown.

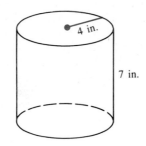

2 cm

3 cm

5 cm

$V = LWH$

$= 5 \, cm \cdot 3 \, cm \cdot 2 \, cm$

$= 30 \, cm^3$ ■

□ **DO EXERCISE 7.**

The volume of a right circular cylinder may be found by multiplying the area of the circular base times the height of the cylinder.

The volume V of a right circular cylinder of radius r and height h is

$$V = \pi r^2 h$$

EXAMPLE 8 Find the volume of the right circular cylinder.

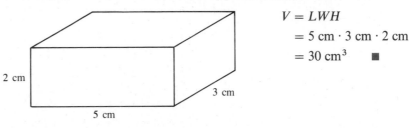

4 in.

7 in.

$V = \pi r^2 h$

$\approx 3.14 \cdot (4 \, in.)^2 \cdot 7 \, in.$

$\approx 351.68 \, in.^3$ ■

□ **DO EXERCISE 8.**

Answers to Exercises

1. a. $21 \, m^2$ b. $10 \, m^2$
2. a. $9 \, cm^2$ b. $64 \, m^2$
3. $44 \, m^2$
4. a. $10.5 \, m^2$ b. $16 \, in.^2$
5. a. $314 \, cm^2$ b. $706.5 \, m^2$
6. a. $131.88 \, cm^2$ b. $376.8 \, in.^2$
7. $300 \, ft^3$
8. a. $401.92 \, m^3$ b. $17,662.5 \, in.^3$

PROBLEM SET 1.7

A. *Find the area.*

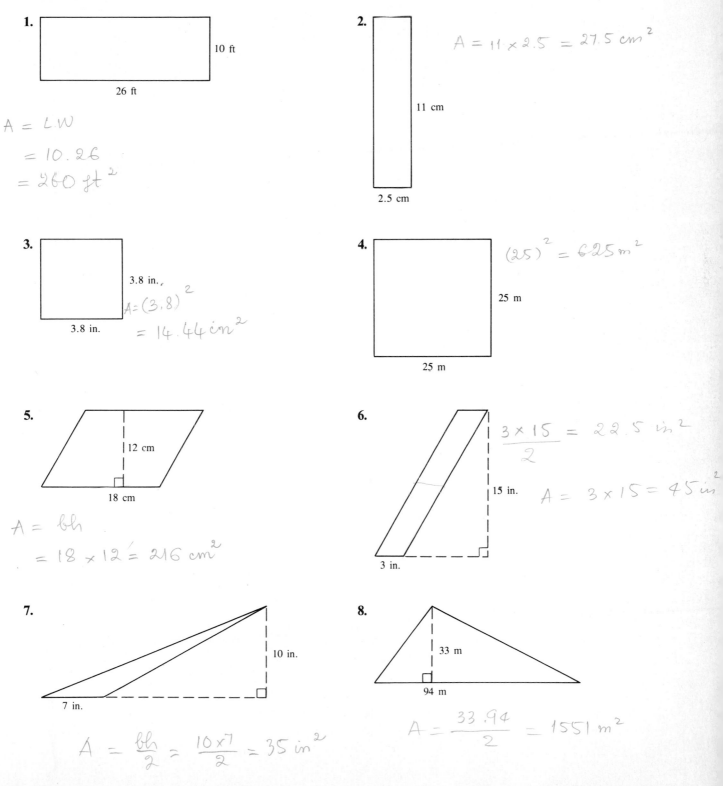

1.

10 ft

26 ft

$A = L.W$

$= 10 . 26$

$= 260 \, ft^2$

2.

11 cm

2.5 cm

$A = 11 \times 2.5 = 27.5 \, cm^2$

3.

3.8 in.

3.8 in.

$A = (3.8)^2$

$= 14.44 \, in^2$

4.

25 m

25 m

$(25)^2 = 625 \, m^2$

5.

12 cm

18 cm

$A = bh$

$= 18 \times 12 = 216 \, cm^2$

6.

15 in.

3 in.

$\dfrac{3 \times 15}{2} = 22.5 \, in^2$

$A = 3 \times 15 = 45 \, in^2$

7.

10 in.

7 in.

$A = \dfrac{bh}{2} = \dfrac{10 \times 7}{2} = 35 \, in^2$

8.

33 m

94 m

$A = \dfrac{33 . 94}{2} = 1551 \, m^2$

B. *Find the area of the circles.*

9.

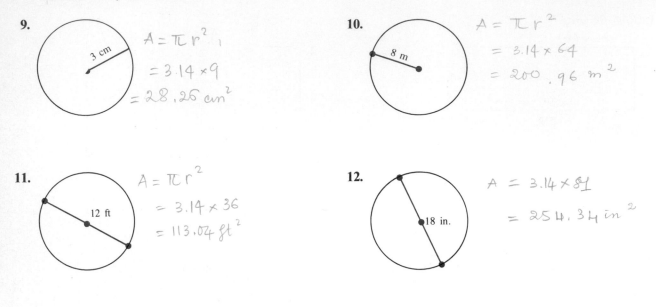

3 cm

$A = \pi r^2$

$= 3.14 \times 9$

$= 28.26$ cm^2

10.

8 m

$A = \pi r^2$

$= 3.14 \times 64$

$= 200.96$ m^2

11.

12 ft

$A = \pi r^2$

$= 3.14 \times 36$

$= 113.04$ ft^2

12.

18 in.

$A = 3.14 \times 81$

$= 254.34$ in^2

C. *Find the lateral area of the right circular cylinders.*

13.

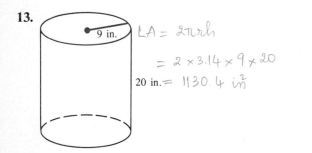

9 in.

20 in.

$LA = 2\pi r h$

$= 2 \times 3.14 \times 9 \times 20$

$= 1130.4$ in^2

14.

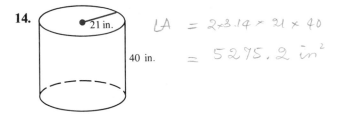

21 in.

40 in.

$LA = 2 \times 3.14 \times 21 \times 40$

$= 5275.2$ in^2

15.

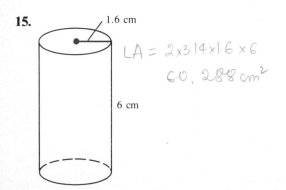

1.6 cm

6 cm

$LA = 2 \times 3.14 \times 1.6 \times 6$

$= 60.288$ cm^2

16.

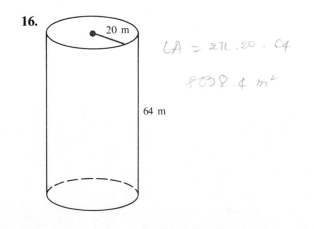

20 m

64 m

$LA = 2\pi \cdot 20 \cdot 64$

$= 8038.4$ m^2

D. *Find the volume of the solids.*

17.
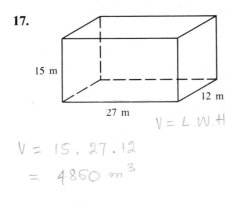
$V = L.W.H$

$V = 15, 27.12$
$= 4850 \ m^3$

18.
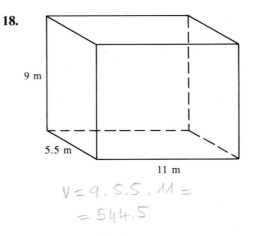

$V = 9.5.5.11 =$
$= 544.5$

E. *Find the volume of the right circular cylinders.*

19.
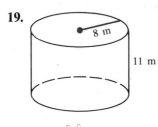
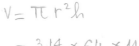

$V = \pi r^2 h$

$= 3.14 \times 64 \times 11$
$= 2210.56 \ m^2$

20.
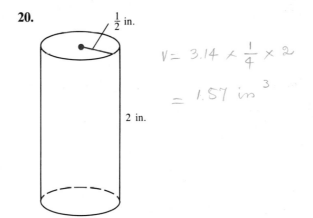

$V = 3.14 \times \frac{1}{4} \times 2$

$= 1.57 \ in^3$

21.
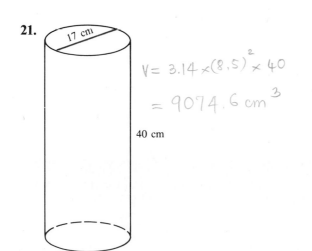

$V = 3.14 \times (8.5)^2 \times 40$

$= 9074.6 \ cm^3$

22.
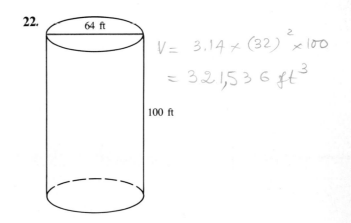

$V = 3.14 \times (32)^2 \times 100$

$= 321,536 \ ft^3$

23. Find the area of a rectangle of length 28 m and width 12 m.

24. How much carpet should be puchased for a room of length 6 yards and width $3\frac{2}{3}$ yards?

25. Find the area of a triangle with base 17 cm and height 6 cm.

26. Janet wishes to paint a triangular sign that has a base of 7 ft and a height of 8 ft. What is the area of the sign?

27. Find the area of a circle of radius 12 in.

28. What is the area of a 15-in. pizza?

29. Find the volume of a box that is 15 cm long, 12 cm wide, and 8 cm high.

30. Find the volume of a metal can with radius 2 in. and height 7 in.

31. The _____ of a geometric figure is the number of unit squares that it contains.

32. A four-sided figure with two pairs of parallel sides is a _____ .

33. The area of a _____ is one-half the area of a parallelogram.

34. The lateral area of a right circular cylinder does not include the _____ .

35. The volume of a solid is the number of _____ cubes contained in it.

36. The volume of a right circular cylinder may be found by multiplying the area of the circular base times the _____ of the cylinder.

CHAPTER 1   ADDITIONAL EXERCISES (OPTIONAL)

Section 1.1

Find the prime factorization.

1. 45   **2.** 27   **3.** 64   **4.** 36

Multiply and simplify.

5. $\dfrac{11}{7} \cdot \dfrac{49}{22}$

6. $\dfrac{10}{17} \cdot \dfrac{51}{4}$

7. $2\dfrac{1}{4} \cdot 3\dfrac{1}{5}$

8. $6\dfrac{2}{3} \cdot 4\dfrac{1}{10}$

9. $\dfrac{3}{5} \cdot 5\dfrac{1}{3} \cdot \dfrac{10}{21}$

10. $3\dfrac{1}{6} \cdot \dfrac{12}{7} \cdot \dfrac{4}{3}$

Divide and simplify.

11. $\dfrac{3}{4} \div 9$

12. $7 \div \dfrac{5}{14}$

13. A recipe calls for $2\frac{1}{4}$ cups of flour. How many cups of flour are necessary to double the recipe?

14. How many lengths of pipe $2\frac{1}{4}$ ft long can be cut from a 12-ft pipe?

Section 1.2

Add and simplify, if possible.

15. $\dfrac{2}{3} + \dfrac{5}{8}$

16. $\dfrac{4}{5} + \dfrac{5}{7}$

17. $\dfrac{5}{9} + \dfrac{7}{6}$

18. $\dfrac{5}{8} + \dfrac{7}{12}$

19. $\dfrac{3}{5} + \dfrac{9}{4} + \dfrac{3}{10}$

20. $\dfrac{2}{3} + \dfrac{5}{12} + \dfrac{5}{8}$

Subtract and simplify, if possible.

21. $\dfrac{11}{4} - \dfrac{3}{7}$

22. $\dfrac{8}{5} - \dfrac{2}{3}$

23. $4 - \dfrac{5}{2}$

24. $7 - \dfrac{8}{3}$

25. $2\dfrac{1}{6} - \dfrac{5}{12}$

26. $5\dfrac{1}{8} - \dfrac{3}{4}$

Section 1.3

Write in words.

27. 0.37

28. 0.876

Change to decimal notation. Round to the nearest thousandth.

29. $\dfrac{17}{7}$

30. $\dfrac{29}{3}$

Change to a fraction and simplify.

31. 0.068

32. 0.095

Write in exponential notation.

33. $2 \cdot 2 \cdot 3 \cdot 3 \cdot 3$

34. $x \cdot x \cdot y \cdot y \cdot y \cdot y$

Expand.

35. 3^4

36. 6^8

What is the meaning of the following?

37. 5^0

38. 7^1

Section 1.4

Change to a decimal.

39. 54.8%

40. 3.2%

41. $\frac{1}{2}$%

42. $\frac{3}{4}$%

Change to a fraction and simplify, if possible.

43. 25.8%

44. 17.2%

Change to percent.

45. 0.7

46. 0.4

47. 5

48. 9

Section 1.5

Which laws are illustrated below?

49. $56 + 30 = 30 + 56$

50. $7 \cdot 9 = 9 \cdot 7$

51. $4 \cdot (5 \cdot 6) = (4 \cdot 5) \cdot 6$

52. $3 + (8 + 9) = (3 + 8) + 9$

53. $7 + (2 + 4) = (2 + 4) + 7$

54. $6 \cdot (5 \cdot 3) = (5 \cdot 3) \cdot 6$

55. Is there a commutative law for division? That is, does $a \div b = b \div a$ for all real numbers a and b? Give an example to illustrate your answer.

56. Is there an associative law for division? Does $a \div (b \div c) = (a \div b) \div c$ for all real numbers a, b, and c? Give an example to illustrate your answer.

Section 1.6

Evaluate when $a = 2$, $b = 4$, and $c = 5$.

57. ab

58. $2a + 3c$

59. $5b + 4c$

60. $3abc$

61. Find the perimeter of the rectangle with length 18 ft and width 15 ft.

62. Find the perimeter of the square with side 28 cm.

Find the circumference of the circle with the given diameter. Use 3.14 for π.

63. Diameter 25 cm

64. Diameter 37 m

Section 1.7

65. Find the area of a rectangle with width 6.8 in. and length 15 in.

66. Find the area of a square with side 23.4 m.

67. Find the area of a triangle with base 12 in. and height 35.4 in.

68. Find the area of a parallelogram with base 7.5 cm and height 42 cm.

For the following problems, use 3.14 for π.
Find the area of the circles with following diameters.

69. Diameter: 16 in.

70. Diameter: 12 cm

Find the lateral area of the right circular cylinder with given height and radius.

71. Height: 5 in.; radius: 3.2 in.

72. Height: 6.3 cm; radius: 11 cm

73. Find the volume of a rectangular solid that has a width of 2 cm, a height of 5.4 cm, and a length of 3.1 cm.

74. Find the volume of a right circular cylinder with a radius of 7 ft and a height of 3 ft.

CHAPTER 1   PRACTICE TEST

1. Multiply: $\dfrac{3}{8} \cdot \dfrac{16}{9}$

2. Divide: $\dfrac{3}{4} \div \dfrac{12}{15}$

3. Add: $\dfrac{5}{12} + \dfrac{1}{6}$

4. Subtract: $\dfrac{5}{7} - \dfrac{1}{3}$

5. Change to a decimal: $\dfrac{7}{9}$.

6. Change to a fraction: 4.05.

7. Expand: 8^3.

8. Change to a decimal: 39.2%.

9. Change to percent: $\frac{5}{8}$.

10. Evaluate: $3 \cdot (4 + 5)$.

11. What law is illustrated by the following?

$$3 \cdot (4 + 9) = 3 \cdot 4 + 3 \cdot 9$$

12. Evaluate when $a = 3$, $b = 8$, and $c = \frac{1}{2}$: $2a + bc$.

13. Find the perimeter of the triangle.

14. Find the circumference of the circle.

1. _____

2. _____

3. _____

4. _____

5. _____

6. _____

7. _____

8. _____

9. _____

10. _____

11. _____

12. _____

13. _____

14. _____

Find the area of each figure.

15. _____

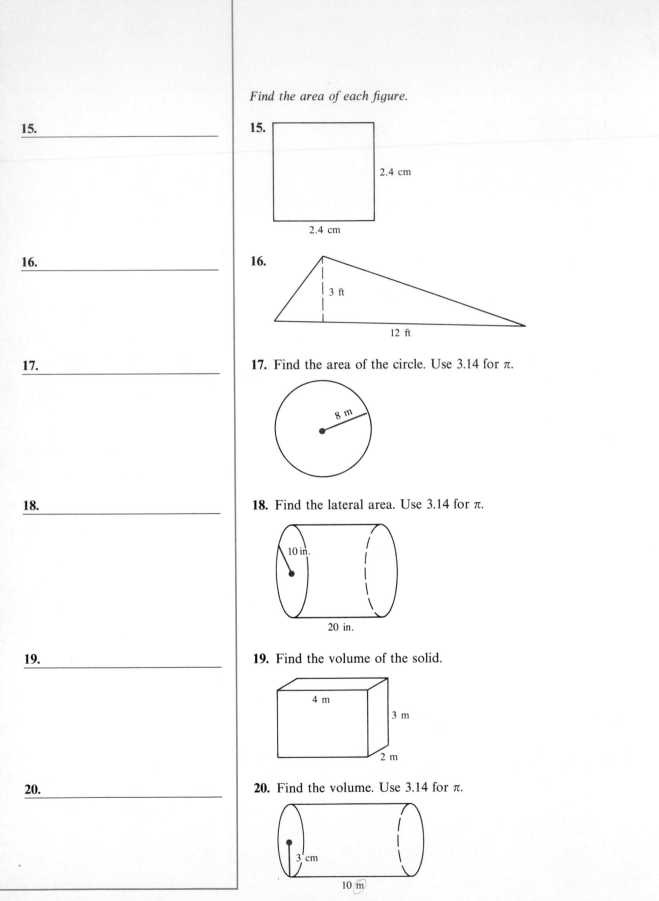

15.

2.4 cm

2.4 cm

16. _____

16.

3 ft

12 ft

17. _____

17. Find the area of the circle. Use 3.14 for π.

8 m

18. _____

18. Find the lateral area. Use 3.14 for π.

10 in.

20 in.

19. _____

19. Find the volume of the solid.

4 m

3 m

2 m

20. _____

20. Find the volume. Use 3.14 for π.

3 cm

10 m

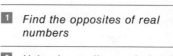

2 Real Numbers

2.1 THE REAL NUMBERS AND ABSOLUTE VALUE

⑤ The Real Number Line and Opposites

The real number line is shown below. The opposites of the positive real numbers are called the negative real numbers. Similarly, the opposites of the negative real numbers are the positive real numbers. Zero is neither positive nor negative.

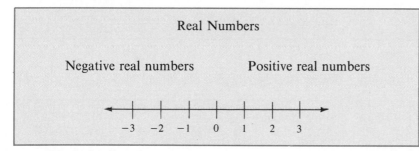

EXAMPLE 1

a. Find the opposite of 3.
The opposite of 3 is -3.

b. Find the opposite of -2.
The opposite of -2 is 2.

c. Find the opposite of $\frac{3}{5}$.
The opposite of $\frac{3}{5}$ is $-\frac{3}{5}$.

d. Find the opposite of -2.4.
The opposite of -2.4 is 2.4 ∎

☐ **DO EXERCISE 1.**

Sometimes we indicate that we want the opposite of a number by placing a negative sign in front of the number in parentheses. To *simplify* an expression often means to write the expression with the least number of symbols.

☐ **Exercise 1** Find the opposite.

a. 2

b. -4

c. -8.7

d. $-\left(\dfrac{5}{8}\right)$

□ **Exercise 2** Simplify, by writing with the least number of symbols.

a. $-(-7)$

b. $-(8)$

c. $-(-15)$

d. $-\left(\dfrac{5}{8}\right)$

e. $-(-2.3)$

f. $-(-9.1)$

EXAMPLE 2 Simplify, by writing with the least number of symbols.

a. $-(-6)$

$$-(-6) = 6$$

b. $-(3)$

$$-(3) = -3$$

c. $-\left(\dfrac{6}{5}\right)$

$$-\left(\dfrac{6}{5}\right) = -\dfrac{6}{5}$$

d. $-(-5.1)$

$$-(-5.1) = 5.1 \quad \blacksquare$$

□ **DO EXERCISE 2.**

2 **Inequalities**

When two numbers are not equal, we have an inequality. If a number is to the right of another number on the number line, it is the greater of the two numbers.

Notice that 3 is greater than 1. This is denoted by $3 > 1$.

> For any numbers a and b,
> $$a > b$$
> means that a is to the right of b on the number line.

This is read "a is greater than b."

If a number is to the left of another number on the number line, it is the smaller of the two numbers. Notice that -2 is less than -1. We write $-2 < -1$.

> For any numbers a and b
> $$a < b$$
> means that a is to the left of b on the number line.

This is read "a is less than b."

Expressions such as $a < b$ and $b > a$ are called *inequalities*.

Absolute Value a real number's distance from zero on the number line. Absolute value is symbolized by | |.

EXAMPLE 3 Replace the ? with < or >.

a. 5 ? 3

$$5 > 3$$

b. 3 ? 5

$$3 < 5$$

c. −3 ? −5

$$-3 > -5$$

d. −5 ? −3

$$-5 < -3$$

Notice that the inequality symbol points to the smaller number. Therefore, $a > b$ has the same meaning as $b < a$. ■

□ **DO EXERCISE 3.**

Inequalities may also be written for fractions and decimals. Consider the following number line.

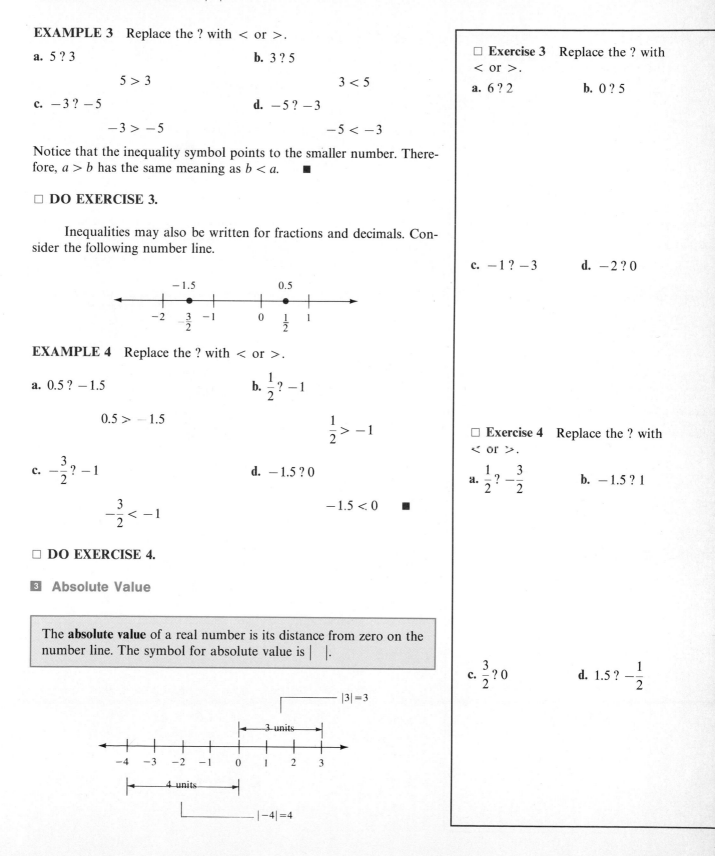

EXAMPLE 4 Replace the ? with < or >.

a. 0.5 ? −1.5

$$0.5 > -1.5$$

b. $\frac{1}{2}$? −1

$$\frac{1}{2} > -1$$

c. $-\frac{3}{2}$? −1

$$-\frac{3}{2} < -1$$

d. −1.5 ? 0

$$-1.5 < 0$$ ■

□ **DO EXERCISE 4.**

3 **Absolute Value**

The **absolute value** of a real number is its distance from zero on the number line. The symbol for absolute value is | |.

□ **Exercise 3** Replace the ? with < or >.

a. 6 ? 2

b. 0 ? 5

c. −1 ? −3

d. −2 ? 0

□ **Exercise 4** Replace the ? with < or >.

a. $\frac{1}{2}$? $-\frac{3}{2}$

b. −1.5 ? 1

c. $\frac{3}{2}$? 0

d. 1.5 ? $-\frac{1}{2}$

□ **Exercise 5**   Find the following.

a. $|7| = 7$ **b.** $|-6| = 6$

c. $|-10| = 10$ **d.** $|5.9| = 5.9$

e. $\left|-\dfrac{1}{2}\right| \ \frac{1}{2}$ **f.** $-|-3| \ -3$

g. $-|8.3| \ -8.3$ **h.** $-\left|-\dfrac{7}{4}\right| \ -7.4$

EXAMPLE 5   Find the following.

a. $|3| = 3$

b. $|-4| = 4$ Since -4 is 4 units from zero

c. $|0| = 0$

d. $|-8| = 8$ Since -8 is 8 units from zero

e. $\left|\dfrac{3}{8}\right| = \dfrac{3}{8}$

f. $|-7.5| = 7.5$ ■

□ **DO EXERCISE 5.**

DID YOU KNOW?

The English mathematician Thomas Harriot introduced the symbols < and > for "less than" and "greater than." In 1585, Harriot was sent by Sir Walter Raleigh to survey the new land of America.

Answers to Exercises _____

1. a. -2 **b.** 4   **c.** 8.7   **d.** $-\dfrac{8}{5}$

2. a. 7   **b.** -8 **c.** 15   **d.** $-\dfrac{5}{8}$ **e.** 2.3   **f.** 9.1

3. a. $6 > 2$ **b.** $0 < 5$ **c.** $-1 > -3$ **d.** $-2 < 0$

4. a. $\dfrac{1}{2} > -\dfrac{3}{2}$ **b.** $-1.5 < 1$ **c.** $\dfrac{3}{2} > 0$ **d.** $1.5 > -\dfrac{1}{2}$

5. a. 7   **b.** 6   **c.** 10   **d.** 5.9   **e.** $\dfrac{1}{2}$ **f.** -3 **g.** -8.3

h. $-\dfrac{7}{4}$

PROBLEM SET 2.1

A. *Find the opposite.*

1. 5

2. -4

3. -10

4. 9

5. $-\dfrac{3}{5}$

6. $-\dfrac{8}{7}$

7. 2.4

8. 9.5

9. $\dfrac{11}{7}$

10. $\dfrac{15}{4}$

11. -10.4

12. -21.6

B. *Simplify, by writing with the least number of symbols.*

13. $-(3)$

14. $-(7)$

15. $-(-4)$

16. $-(-6)$

17. $-(2)$

18. $-(8)$

19. $-(-99)$

20. $-(-10)$

21. $-\left(\dfrac{2}{5}\right)$

22. $-\left(\dfrac{1}{7}\right)$

23. $-\left(-\dfrac{12}{7}\right)$

24. $-\left(-\dfrac{34}{5}\right)$

25. $-(-7.8)$

26. $-(-3.2)$

27. $-(71.4)$

28. $-(85.3)$

C. *Replace the ? with $<$ or $>$.*

29. 9 ? 5

30. 7 ? 2

31. -4 ? -3

32. -7 ? -6

33. 0 ? -3

34. 0 ? -5

35. -7 ? 3

36. -6 ? 5

37. 1.5 ? 0

38. $\dfrac{1}{2}$? 0

39. -1.5 ? 1

40. $-\dfrac{1}{2}$? 2

D. *Find the following.*

41. $|5|$
42. $|6|$
43. $|-3|$
44. $|-4|$

45. $|0|$
46. $|-99|$
47. $|15|$
48. $|10|$

49. $\left|-\dfrac{1}{3}\right|$
50. $\left|\dfrac{5}{6}\right|$
51. $|3.4|$
52. $|5.8|$

53. $|-10.1|$
54. $|-19.7|$
55. $\left|\dfrac{15}{8}\right|$
56. $\left|\dfrac{14}{3}\right|$

57. $-|-8.6|$
58. $-|-7.3|$
59. $-\left|\dfrac{11}{5}\right|$
60. $-\left|\dfrac{3}{7}\right|$

61. The opposites of the positive real numbers are the _____ real numbers.

62. Zero is neither _____ nor negative.

63. To _____ an expression often means to write with the least number of operation signs and grouping symbols.

64. If a number is to the left of another number on the number line, it is the _____ of the two numbers.

65. The _____ of a number is its distance from zero on the number line.

Checkup

The following problems provide a review of some of Section 1.3.
Change to decimal notation. Round to the nearest thousandth.

66. $\dfrac{3}{8}$
67. $\dfrac{7}{5}$
68. $-\dfrac{5}{12}$
69. $\dfrac{9}{7}$

Change to a fraction and simplify.

70. 0.54
71. 0.36
72. 0.039
73. 0.084

2.2 ADDITION OF INTEGERS

OBJECTIVE

■ Addition of integers may be shown on the number line. The following are rules for adding two integers a and b on the number line. We begin at a.

■ *Add integers*

1. If b is positive, we move to the right.
2. If b is negative, we move to the left.
3. If b is zero, we remain at a.

The sum is the integer where we end.

□ **Exercise 1** Add, using a number line.

a. $6 + (-8)$

EXAMPLE 1 Add.

a. $5 + (-7)$

Begin at 5. Move 7 units to the left.

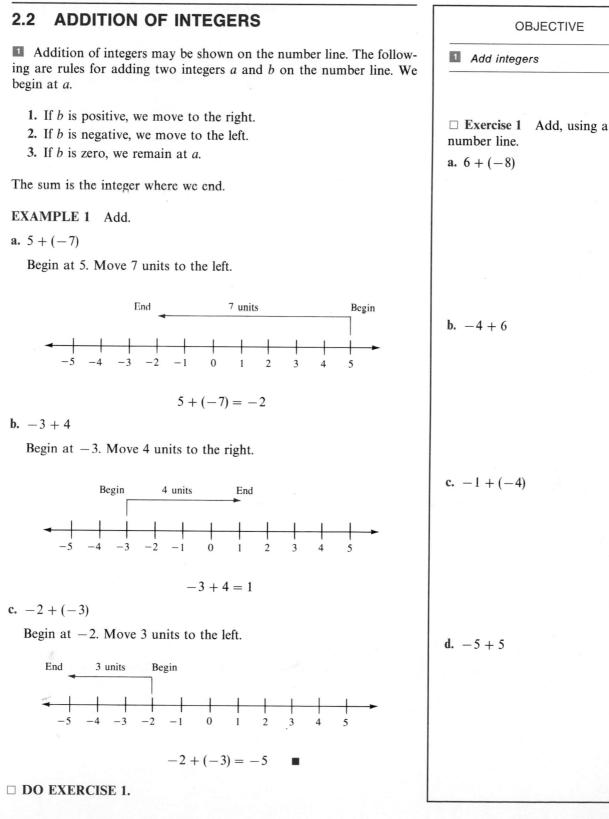

$$5 + (-7) = -2$$

b. $-4 + 6$

b. $-3 + 4$

Begin at -3. Move 4 units to the right.

$$-3 + 4 = 1$$

c. $-1 + (-4)$

c. $-2 + (-3)$

Begin at -2. Move 3 units to the left.

$$-2 + (-3) = -5 \quad ■$$

d. $-5 + 5$

□ **DO EXERCISE 1.**

□ **Exercise 2** Add.

a. $4 + 3$ **b.** $(-3) + (-2)$

The number line helps us to see how integers are added, but we must learn to add them quickly. Integers may be added using the following rules.

> To add two integers with the same sign, add their absolute values and give the answer the common sign.

A number without a sign is positive. Hence $4 = +4$ and $5 = +5$.

EXAMPLE 2 Add.

a. $4 + 5 = (+4) + (+5) = +9 = 9$

c. $(-7) + (-2)$ **d.** $(-1) + (-6)$

The Hindus invented signed numbers to symbolize money:

$$\$4 + \$5 = \$9$$

b. $(-7) + (-3) = -10$ Adding $|-7|$ and $|-3|$ and giving the answer the common sign

A debt of \$7 and a debt of \$3 is a debt of \$10.

□ **Exercise 3** Add.

a. $(9) + (-3)$ **b.** $8 + (-2)$

c. $(-2) + (-4) = -6$ ■

□ **DO EXERCISE 2.**

> To add two integers with different signs, subtract the smaller absolute value from the larger absolute value. Give the answer the sign of the number with the larger absolute value.

EXAMPLE 3 Add.

c. $(-4) + (+2)$ **d.** $-7 + 5$

a. $(+8) + (-5) = +(8 - 5) = +3$

$\qquad |+8| > |-5|$ so the answer is positive

If we have \$8 and we spend \$5, we have \$3 left.

b. $(-7) + (+2) = -(7 - 2) = -5$

$\qquad |-7| > |+2|$ so the answer is negative

If we owe \$7 and we pay \$2 on the debt, we now owe \$5.

e. $8 + (-3)$ **f.** $-5 + 2$

c. $-6 + 2 = (-6) + (+2) = -4$

d. $-5 + 9 = 4$

Notice that to *add integers*, we must *sometimes subtract*. ■

□ **DO EXERCISE 3.**

Notice that when we add an integer to its opposite, we get zero. The opposite of zero is zero.

$$a + (-a) = 0 \qquad \text{for any integer } a$$

EXAMPLE 4 Add.

a. $6 + (-6) = 0$

b. $-12 + 12 = 0$ ∎

□ **DO EXERCISE 4.**

To add several integers, we may add the positive integers and add the negative integers, then add the results.

EXAMPLE 5 Add.

a. $(-3) + 4 + (-6) + 1 = (4 + 1) + [(-3) + (-6)] = 5 + (-9) = -4$
The grouping symbols $[\ ]$ are called *brackets*.

b. $-8 + 6 + (-3) + 4 = (6 + 4) + [(-8 + (-3)] = 10 + (-11) = -1$ ∎

□ **DO EXERCISE 5.**

□ **Exercise 4** Add.

a. $3 + (-3)$ **b.** $7 + (-7)$

c. $-20 + 20$ **d.** $-48 + 48$

□ **Exercise 5** Add.

a. $(-2) + 7 + (-4) + 8$

b. $3 + (-10) + 20 + (-5)$

Answers to Exercises _____

1. a. -2 **b.** 2   **c.** -5 **d.** 0

2. a. 7   **b.** -5 **c.** -9 **d.** -7

3. a. 6   **b.** 6   **c.** -2 **d.** -2 **e.** 5   **f.** -3

4. a. 0   **b.** 0   **c.** 0   **d.** 0

5. a. 9   **b.** 8

PROBLEM SET 2.2

Add. Do not use a number line except to check.

1. $(+2) + (+4)$ **2.** $(+5) + (+7)$ **3.** $2 + 4$ **4.** $5 + 7$

5. $(-3) + (-4)$ **6.** $(-5) + (-1)$ **7.** $-7 + (-8)$ **8.** $-6 + (-4)$

9. $(-8) + (-3)$ **10.** $(-5) + (-9)$ **11.** $-3 + (-5)$ **12.** $-9 + (-4)$

13. $(-9) + 6$ **14.** $(-7) + 2$ **15.** $-9 + 6$ **16.** $-7 + 2$

17. $5 + (-4)$ **18.** $2 + (-6)$ **19.** $6 + (-4)$ **20.** $3 + (-8)$

21. $8 + (-7)$ **22.** $1 + (-4)$ **23.** $3 + (-9)$ **24.** $9 + (-5)$

25. $6 + (-6)$ **26.** $4 + (-4)$ **27.** $-8 + 8$ **28.** $-3 + 3$

29. $5 + 0$ **30.** $8 + 0$ **31.** $0 + 4$ **32.** $0 + 7$

33. $-2 + 0$ **34.** $-7 + 0$ **35.** $0 + (-10)$ **36.** $0 + (5)$

37. $(-10) + 4$ **38.** $(-12) + 6$ **39.** $3 + (-11)$ **40.** $8 + (-9)$

41. $(-9) + (-2)$ **42.** $(-7) + (-3)$ **43.** $4 + 8$ **44.** $9 + 4$

45. $-6 + (-7)$ **46.** $-3 + (-2)$ **47.** $-4 + 8$ **48.** $-6 + 7$

49. $15 + (-8)$ **50.** $8 + (-6)$ **51.** $-9 + (-6)$ **52.** $-7 + (-7)$

53. $-6 + 5$ **54.** $-10 + 2$ **55.** $-20 + 20$ **56.** $-8 + 8$

57. $(-4) + 7 + (-1) + 3$ **58.** $6 + (-3) + 2 + (-9)$ **59.** $-6 + 3 + 5 + (-1)$ **60.** $7 + (-3) + 5 + (-4)$

61. To add two integers with the same sign, add their absolute values and give the answer the _____ sign.

62. To add two integers with different signs, _____ the smaller absolute value from the larger absolute value and give the answer the sign of the number with the larger absolute value.

63. An integer added to its opposite is _____ .

Checkup

The following problems provide a review of some of Section 1.4.
Change to a decimal.

64. 32.5% **65.** 240% **66.** 7% **67.** 9%

Change to percent.

68. 0.479 **69.** 0.3 **70.** 2.34 **71.** 0.56

Change to a fraction.

72. 31% **73.** 8%

Change to percent.

76. $\dfrac{2}{3}$ **77.** $\dfrac{1}{2}$ **78.** $\dfrac{7}{5}$ **79.** $\dfrac{9}{7}$

2.3 SUBTRACTION OF INTEGERS

1

> To subtract one integer from another, add the opposite of the second integer to the first integer.
>
> $$a - b = a + (-b)$$
>
> where a and b are integers

Notice that we change the operation to addition and change the sign of the second number, then follow the rules for addition of signed numbers.

EXAMPLE 1   Subtract.

a. $(5) - (+8) = 5 + (-8) = -3$

If we have $5 and we spend $8, we owe $3.

b. $(-3) - (+6) = (-3) + (-6) = -9$

If we spend $3 and we spend $6, we have spent $9.

c. $(-4) - (-5) = -4 + (+5) = 1$

If we have a debt of $4 and someone takes away a debt of $5, we have gained $1. ■

□ DO EXERCISE 1.

Sometimes we do not write the parentheses in the original problem.

EXAMPLE 2   Subtract.

a. $6 - 8 = (+6) - (+8) = (+6) + (-8) = -2$

b. $-7 - 9 = (-7) - (+9) = (-7) + (-9) = -16$

c. $-4 - (-7) = (-4) - (-7) = (-4) + (+7) = 3$

In Example c the parentheses are needed in the original problem to separate the two negative signs. ■

□ DO EXERCISE 2.

OBJECTIVE

1 *Subtract integers*

□ **Exercise 1**   Subtract.

a. $(3) - (+4)$ **b.** $(7) - (+2)$

c. $(-8) - (+2)$ **d.** $(-6) - (+5)$

e. $(-3) - (-8)$ **f.** $(9) - (-1)$

g. $(8) - (-8)$ **h.** $(-8) - (-8)$

i. $(0) - (3)$ **j.** $(0) - (-3)$

□ **Exercise 2**   Subtract.

a. $5 - 3$ **b.** $9 - 10$

c. $-5 - 6$ **d.** $-7 - 8$

e. $3 - (-3)$ **f.** $-9 - (-7)$

g. $0 - 4$ **h.** $-4 - 0$

□ **Exercise 3** Subtract.

a. $8 - 9$ **b.** $5 - 8$

c. $-7 - 10$ **d.** $-1 - 16$

e. $3 - (-4)$ **f.** $6 - (-5)$

□ **Exercise 4** Perform the operations indicated.

a. $-2 + 8 - 3$

b. $9 - (-7) - 4$

c. $-15 - (-3) + 4$

d. $-8 + 6 - (-12)$

We can make our work easier by mentally inserting a plus sign between the two numbers except when two signs occur together. In that case, we use our rule for opposites.

EXAMPLE 3 Subtract.

a. $7 - 9 = 7 + (-9) = -2$

b. $-3 - 15 = -3 + (-15) = -18$

c. $4 - (-25) = 4 + 25 = 29$ Recall that $-(-25) = +25$ ■

□ **DO EXERCISE 3.**

EXAMPLE 4 Perform the operations indicated.

a. $6 - 3 + 5 = 6 + (-3) + 5$ Subtraction rule

$\qquad\qquad\quad = 6 + 5 + (-3)$ Rearranging the integers

$\qquad\qquad\quad = 11 + (-3) = 8$

b. $7 - (-4) - 10 = 7 + 4 + (-10)$

$\qquad\qquad\qquad\quad = 11 + (-10) = 1$ ■

□ **DO EXERCISE 4.**

Answers to Exercises

1. a. -1 **b.** 5   **c.** -10 **d.** -11 **e.** 5   **f.** 10   **g.** 16
h. 0   **i.** -3 **j.** 3
2. a. 2   **b.** -1 **c.** -11 **d.** -15 **e.** 6   **f.** -2
g. -4 **h.** -4
3. a. -1 **b.** -3 **c.** -17 **d.** -17 **e.** 7   **f.** 11
4. a. 3   **b.** 12   **c.** -8 **d.** 10

PROBLEM SET 2.3

Subtract.

1. $7 - (+9)$ **2.** $(3) - (+4)$ **3.** $7 - (+5)$ **4.** $9 - (+3)$

5. $(-7) - (+4)$ **6.** $(-8) - (+3)$ **7.** $(-2) - (+5)$ **8.** $(-1) - (+6)$

9. $(-4) - (-1)$ **10.** $(-5) - (-2)$ **11.** $(-3) - (-7)$ **12.** $(-6) - (-10)$

13. $7 - 5$ **14.** $15 - 10$ **15.** $8 - 9$ **16.** $7 - 8$

17. $-3 - 3$ **18.** $-7 - 7$ **19.** $-9 - 8$ **20.** $-8 - 7$

21. $4 - (-2)$ **22.** $6 - (-1)$ **23.** $-7 - (-8)$ **24.** $-2 - (-5)$

25. $6 - (-8)$ **26.** $7 - (-6)$ **27.** $-2 - (-1)$ **28.** $-5 - (-3)$

29. $6 - 9$ **30.** $5 - 7$ **31.** $-2 - 10$ **32.** $-1 - 9$

33. $0 - 5$ **34.** $0 - 8$ **35.** $-9 - 0$ **36.** $-5 - 0$

37. $-6 - (-6)$ **38.** $-4 - (-2)$ **39.** $1 - (-3)$ **40.** $6 - (-5)$

41. $8 - 9 + 11$ **42.** $2 - 4 + 14$ **43.** $-8 - (-1)$ **44.** $-7 - (-7)$

45. $0 - (-4)$ **46.** $0 - (-7)$ **47.** $0 - (-2)$ **48.** $0 - (-6)$

49. $10 - 21 - 7$ **50.** $-3 - 17 + 4$ **51.** $(-40) - (-60) - 10$ **52.** $80 - (-90) - 120$

53. $-12 - 4 - (-8)$ **54.** $-9 - 7 - (-15)$

55. Sunya had $58 in her checking account. She wrote a check for $74. What signed number represents the amount her checking account is overdrawn?

56. Dennis bought a shirt for $24 and charged it on his credit card. Then he used the card to pay $18 for dinner. What signed number represents how much he owes on his credit card?

57. Victoria owed Mary $14. Then Mary purchased a ring from Victoria for $16, allowing Victoria to pay her debt. What signed number represents the amount of money Mary gave Victoria?

58. To subtract one signed number from another, add the _____ of the second integer to the first integer.

Checkup

The following problems provide a review of some of Section 1.5.
Which laws are illustrated below?

59. $4 \cdot (3 + 2) = 4 \cdot 3 + 4 \cdot 2$

60. $(7 + 5) \cdot 4 = 7 \cdot 4 + 5 \cdot 4$

61. $(8 \cdot 9) \cdot 7 = 8 \cdot (9 \cdot 7)$

62. $(5 + 1) + 6 = 5 + (1 + 6)$

63. $(2 + 3) + 5 = 5 + (2 + 3)$

64. $(9 \cdot 7) \cdot 4 = 4 \cdot (9 \cdot 7)$

65. $7 \cdot 8 = 8 \cdot 7$

66. $3 \cdot (8 + 1) = 3 \cdot 8 + 3 \cdot 1$

2.4 MULTIPLICATION AND DIVISION OF INTEGERS

OBJECTIVES

1 *Multiply or divide integers with different signs*

2 *Multiply or divide integers with the same signs*

3 *Use order of operations to evaluate expressions*

1 Different Signs

If we multiply two integers with different signs, what is the sign of the answer? Look for a pattern.

$$(3) \cdot (4) = 12$$
$$(3) \cdot (3) = 9$$
$$(3) \cdot (2) = 6$$
$$(3) \cdot (1) = 3$$
$$(3) \cdot (0) = 0$$
$$(3) \cdot (-1) = ?$$
$$(3) \cdot (-2) = ?$$

The answer decreases by 3 with each multiplication

It appears that a positive integer times a negative integer is a negative integer. This is true. It is also true that if we divide two integers with different signs, the answer is negative.

> To multiply or divide two integers with different signs, multiply or divide their absolute values. Give the answer a negative sign.

EXAMPLE 1 Multiply or divide.

a. $8 \cdot (-2) = -16$ **b.** $(-3) \cdot 5 = -15$

c. $(-12) \div 3 = -4$ **d.** $(-8) \div 2 = -4$ ■

□ **DO EXERCISE 1.**

2 Like Signs

If we multiply or divide two integers that both have positive signs, the operations are carried out as in arithmetic and the answer is positive. What sign do we give the answer if we multiply two negative integers? Consider the following.

$$(-3) \cdot (4) = -12$$
$$(-3) \cdot (3) = -9$$
$$(-3) \cdot (2) = -6$$
$$(-3) \cdot (1) = -3$$
$$(-3) \cdot (0) = 0$$
$$(-3) \cdot (-1) = ?$$
$$(-3) \cdot (-2) = ?$$

The answer increases by 3 with each multiplication

□ **Exercise 1** Multiply or divide.

a. $3 \cdot (-2)$

b. $(-5) \cdot 2$

c. $(-16) \div 4$

d. $(-3) \div 1$

e. $7 \cdot (-1)$

f. $(-3) \cdot 4$

□ **Exercise 2** Multiply or divide.

a. $(8)(3)$ **b.** $(5) \cdot (2)$

c. $(-8)(-2)$ **d.** $-5(-4)$

e. $16 \div 4$ **f.** $32 \div 8$

g. $(-18) \div (-3)$ **h.** $(-6) \div (-2)$

□ **Exercise 3** Multiply.

a. $7(0)$ **b.** $0(-4)$

□ **Exercise 4** Divide.

a. $4 \div 0$ **b.** $0 \div 3$

c. $0 \div (-5)$ **d.** $1 \div 0$

It appears that a negative integer times a negative integer is a positive integer. This is the case. It is also true that a negative integer divided by a negative integer is a positive integer.

> To multiply or divide two integers with the same sign, multiply or divide their absolute values. Give the answer a positive sign.

EXAMPLE 2 Multiply or divide.

a. $8 \cdot 4 = +32 = 32$ **b.** $(-3) \cdot (-6) = 18$

c. $(-15) \div (-5) = 3$ **d.** $5 \cdot 4 = 20$

Multiplication in algebra may be indicated by two parentheses without a sign between them. It may also be indicated by a number outside parentheses and the other number inside the parentheses.

e. $(-3)(-5) = 15$ **f.** $-7(-3) = 21$ ■

□ **DO EXERCISE 2.**

> Any number times zero is zero.

EXAMPLE 3 Multiply.

a. $0(10) = 0$ **b.** $(-3)0 = 0$ ■

□ **DO EXERCISE 3.**

> Zero divided by any number other than zero is zero. Division by zero is not allowed.
>
> $$\frac{0}{a} = 0, \qquad a \neq 0$$
>
> $$\frac{a}{0} \text{ is not allowed}$$

In Chapter 3 we explain why division by zero is not allowed.

EXAMPLE 4 Divide.

a. $12 \div 0$ Not allowed

b. $0 \div 5 = 0$

If we have zero cookies to divide among five boys, each boy gets zero cookies! ■

□ **DO EXERCISE 4.**

Do not confuse multiplication with addition or subtraction.

EXAMPLE 5 Calculate.

a. $-7 - 8 = -15$ **b.** $(-7)(-8) = 56$

c. $-8 + 4 = -4$ **d.** $(-8)4 = -32$

e. $9 - 4 = 5$ **f.** $9(-4) = -36$ ∎

☐ **DO EXERCISE 5.**

3 The Order of Operations

When we work examples that contain more than one operation we must know which operation to do first or we will get different answers for the same example. The following agreement has been established.

Order of Operations

1. Do the operations *within* parentheses.
2. Raise all numbers to powers, working from left to right.
3. Perform multiplication and division in the order in which they occur, working from left to right.
4. Add or subtract in order from left to right.

EXAMPLE 6 Calculate.

a.
$$9 - 2(3 + 4) = 9 - 2(7) \qquad \text{Adding within the parentheses}$$
$$= 9 - 14 \qquad \text{Multiplying before subtracting}$$
$$= -5$$

b. $7 + \dfrac{8}{2} = 7 + 4 = 11$

c. $8 \div 2(3) = 4(3) = 12$ Remember to multiply or divide in order from left to right

d. $(4)(-2) - 3(6) = -8 - 18 = -26$

e. $\dfrac{7 - 3 \cdot 5}{4} = \dfrac{7 - 15}{4} = -\dfrac{8}{4} = -2$ The fraction bar acts as parentheses in the numerator

f.
$$\dfrac{2(6 + 4) - 3}{3(2) + 1} = \dfrac{2(10) - 3}{3(2) + 1} \qquad \text{Adding within the parentheses}$$
$$= \dfrac{20 - 3}{6 + 1} \qquad \text{Multiplying before subtracting or adding}$$
$$= \dfrac{17}{7}$$

Notice that we do the operations in the numerator and then, if necessary, the operations in the denominator, and then we divide.

☐ **Exercise 5** Calculate.

a. $-6 - 5$

b. $(-6)(-5)$

c. $8 - 2$

d. $8(-2)$

e. $-7 + 9$

f. $(-7)9$

□ **Exercise 6** Calculate.

a. $8 - 3(2)$

b. $5 - (7 + 4)$

c. $12 - \dfrac{9}{3}$

d. $6 \div 3(4)$

e. $\dfrac{14 + 5 \cdot 2}{8}$

f. $(-3)(-7) + (4)(-2)$

g. $\dfrac{4(3 - 1) + 2(2)}{2(5 - 2)}$

h. $\dfrac{5(2 + 4) - 3(3)}{6(2) - 5}$

i. $4 - 3^2 + 2(7)$

j. $4^2 - 5(3) + 2(5 - 7)$

g. $2^2 - 3(2 + 4) - 3^2 = 2^2 - 3(6) - 3^2$ Adding within the parentheses

$= 4 - 3(6) - 9$ Raising to powers

$= 4 - 18 - 9$

$= -23$ ∎

□ **DO EXERCISE 6.**

DID YOU KNOW?

Brahmagupta, the great Hindu mathematician, in about the year 628 knew and used the general rules for multiplication and division by both positive and negative numbers.

$(-3)(-2) = 6$

Answers to Exercises

1. a. -6 **b.** -10 **c.** -4 **d.** -3 **e.** -7 **f.** -12
2. a. 24 **b.** 10 **c.** 16 **d.** 20 **e.** 4 **f.** 4 **g.** 6
h. 3
3. a. 0 **b.** 0
4. a. Not allowed **b.** 0 **c.** 0 **d.** Not allowed
5. a. -11 **b.** 30 **c.** 6 **d.** -16 **e.** 2 **f.** -63
6. a. 2 **b.** -6 **c.** 9 **d.** 8 **e.** 3 **f.** 13 **g.** 2 **h.** 3
i. 9 **j.** -3

PROBLEM SET 2.4

A. *Multiply.*

1. 7(8)

2. 6(7)

3. $(-4)(-2)$

4. $(-3)(-6)$

5. $(-7)3$

6. $(-4)8$

7. $9(-7)$

8. $7(-6)$

9. $(-8)(-4)$

10. $(-7)(-2)$

11. $(-10)(-2)$

12. $(-11)(-3)$

13. 5(8)

14. 5(9)

15. $(-6)9$

16. $(-7)8$

17. $(-3)(-10)$

18. $(-2)(-11)$

19. $(-10)(-40)$

20. $(-7)(-30)$

21. 7(0)

22. 0(8)

23. $(-4)(0)$

24. $(-6)(0)$

B. *Divide.*

25. $28 \div 7$

26. $54 \div 6$

27. $(-9) \div (-3)$

28. $(-12) \div (-2)$

29. $(-14) \div (-7)$

30. $(-35) \div (-5)$

31. $(-15) \div 5$

32. $(-30) \div 6$

33. $60 \div (-10)$

34. $50 \div (-2)$

35. $0 \div 7$

36. $0 \div 10$

37. $\dfrac{-25}{-5}$

38. $\dfrac{-70}{-7}$

39. $\dfrac{-80}{8}$

40. $\dfrac{-90}{3}$

41. $\dfrac{0}{8}$

42. $\dfrac{0}{6}$

43. $\dfrac{-7}{0}$

44. $\dfrac{-9}{0}$

C. *Calculate.*

45. $-2 - 5$

46. $-3 - 8$

47. $(-2)(-5)$

48. $(-3)(-8)$

49. $-6 + 9$

50. $-7 + 9$

51. $(-6)9$

52. $(-7)9$

53. $8 - 6$

54. $7 - 6$

55. $8(-6)$

56. $7(-6)$

D. *Calculate.*

57. $8 + 5(4)$

58. $-7 - 3(-4)$

59. $-8 - (6 + 2)$

60. $9 + 3 - (-1)$

61. $14 - \dfrac{8}{4}$

62. $-3 + \dfrac{14}{7}$

63. $(-8) \div 4(2)$

64. $9 \div 3(5)$

65. $\dfrac{18 - 7 \cdot 2}{2}$

66. $\dfrac{3(-9) + 12}{-5}$

67. $-7(-2) + 5(-3)$

68. $3(-7) - 4(-2)$

69. $\dfrac{-3 \cdot 4 - 2 \cdot 5}{-2}$

70. $\dfrac{7(-3) + 5(3)}{-3}$

71. $\dfrac{2(16 - 10) - 3}{4(2) - 12}$

72. $\dfrac{3(-16 + 12) + 4}{7 - 3(5)}$

73. $\dfrac{4 - (5 - 3)}{8(2) - 3(4)}$

74. $\dfrac{2 - (7 - 10)}{9(2) - 3(6 - 4)}$

75. $6^2 - 3(8 - 4) + 3^2$

76. $5(8 - 2) - 4^2 - 2^2$

77. To multiply two integers with different signs, multiply their absolute values and give the answer a _____ sign.

78. Any number times zero is _____.

79. To divide two integers with the same sign, divide their absolute values and give the number a _____ sign.

80. Division by _____ is not allowed.

81. According to the order of operations agreement, _____ and _____ are done before addition or subtraction.

Checkup

The following problem provide a review of some of Sections 1.1 and 1.2. These problems will help you with the next section.

Do the operation indicated. Simplify, if possible.

82. $\dfrac{3}{4} \cdot \dfrac{8}{5}$

83. $\dfrac{5}{7} \cdot \dfrac{14}{3}$

84. $\dfrac{5}{8} \div \dfrac{3}{10}$

85. $\dfrac{15}{7} \div \dfrac{7}{3}$

86. $\dfrac{1}{3} + \dfrac{5}{9}$

87. $\dfrac{3}{4} + \dfrac{2}{7}$

88. $\dfrac{4}{3} - \dfrac{3}{7}$

89. $\dfrac{3}{8} - \dfrac{1}{24}$

2.5 MORE ON RATIONAL NUMBERS

1 Addition

We may add signed rational numbers in the same way that we add integers.

EXAMPLE 1 Add.

a. $-\dfrac{3}{5} + \dfrac{1}{5} = -\dfrac{2}{5}$

b. $-\dfrac{1}{3} + \left(-\dfrac{3}{4}\right) = -\dfrac{4}{12} + \left(-\dfrac{9}{12}\right) = -\dfrac{13}{12}$ The lowest common denominator is 12

c. $-2.7 + (-1.2) = -3.9$ ∎

☐ DO EXERCISE 1.

2 Subtraction

Subtraction of signed rational numbers is also the same as for integers.

EXAMPLE 2 Subtract.

a. $2.3 - 4 = 2.3 + (-4.0) = -1.7$

b. $-\dfrac{1}{8} - \left(-\dfrac{3}{8}\right) = -\dfrac{1}{8} + \dfrac{3}{8} = \dfrac{2}{8} = \dfrac{1}{4}$

c. $-\dfrac{4}{5} - \dfrac{2}{5} = -\dfrac{4}{5} + \left(-\dfrac{2}{5}\right) = -\dfrac{6}{5}$ ∎

☐ DO EXERCISE 2.

OBJECTIVES

1 Add signed rational numbers

2 Subtract signed rational numbers

3 Multiply and divide signed rational numbers

4 Find the reciprocal of a rational number (except zero)

5 Find three equivalent fractions to a given fraction by changing two signs

☐ **Exercise 1**   Add.

a. $-\dfrac{3}{4} + \dfrac{1}{4}$

b. $-1.6 + (-0.4)$

c. $-\dfrac{3}{8} + \dfrac{1}{12}$

d. $-10.4 + 3.5$

☐ **Exercise 2**   Subtract.

a. $3 - 5.6$ **b.** $6.2 - (-3.8)$

c. $-\dfrac{1}{7} - \left(-\dfrac{3}{7}\right)$ **d.** $-\dfrac{1}{9} - \dfrac{1}{3}$

□ **Exercise 3** Multiply or divide.

a. $-\dfrac{2}{3}\left(\dfrac{1}{7}\right)$ **b.** $-\dfrac{1}{5}\left(-\dfrac{1}{3}\right)$

c. $(-0.5)(2)$ **d.** $-\dfrac{3}{4} \div \left(-\dfrac{1}{3}\right)$

e. $-\dfrac{7}{8} \div \dfrac{8}{3}$ **f.** $3 \div (-0.2)$

□ **Exercise 4** Multiply.

a. $\dfrac{1}{6}(12)(7)$ **b.** $4(0.3)(8)$

❸ Multiplication and Division

The rules for multiplication and division of signed rational numbers are the same as for integers.

EXAMPLE 3 Multiply or divide.

a. $-\dfrac{3}{8}\left(\dfrac{1}{5}\right) = -\dfrac{3}{40}$

b. $-\dfrac{1}{3}\left(-\dfrac{2}{3}\right) = \dfrac{2}{9}$

c. $-0.2(-3) = 0.6$

d. $-\dfrac{5}{6} \div \dfrac{5}{7} = -\dfrac{5}{6} \cdot \dfrac{7}{5} = -\dfrac{5}{5} \cdot \dfrac{7}{6} = -1 \cdot \dfrac{7}{6} = -\dfrac{7}{6}$

e. $-\dfrac{1}{3} \div \left(-\dfrac{3}{2}\right) = -\dfrac{1}{3}\left(-\dfrac{2}{3}\right) = \dfrac{2}{9}$

f. $(-0.8) \div 2 = (-0.8)\left(\dfrac{1}{2}\right) = -0.4$ ■

□ **DO EXERCISE 3.**

If we wish to multiply more than two numbers, we may multiply them two at a time.

EXAMPLE 4 Multiply.

a. $\dfrac{1}{3}(-3)(5) = -\dfrac{3}{3}(5) = -1(5) = -5$

b. $-0.5(2)(-6) = -1(-6) = 6$ ■

□ **DO EXERCISE 4.**

All signed real numbers may be added, subtracted, multiplied, and divided using the rules for integers.

Reciprocal   An inverted real number. The product of a number and its reciprocal equals one. Zero does not have a reciprocal.

4 Reciprocals

> To find the **reciprocal** of a real number, invert the number. Zero does not have a reciprocal.

EXAMPLE 5

Number	*Reciprocal*
$\dfrac{3}{8}$	$\dfrac{8}{3}$
$\dfrac{1}{-3}$	-3
5	$\dfrac{1}{5}$
0	No reciprocal ■

□ **DO EXERCISE 5.**

□ **Exercise 5**   Find the reciprocal.

a. $\dfrac{5}{9}$

b. $\dfrac{-3}{4}$

c. -7

d. $\dfrac{6}{5}$

□ **Exercise 6** Find three equivalent fractions by changing signs.

a. $\dfrac{5}{8}$

b. $\dfrac{-3}{4}$

c. $\dfrac{3}{2}$

d. $\dfrac{-5}{6}$

5 **Equivalent Fractions by Changing Signs**

Every fraction has three signs. There is a sign on the fraction and one on the numerator and one on the denominator. Any two of the signs may be changed at the same time. The result is called an *equivalent fraction.*

EXAMPLE 6 Find three equivalent fractions by changing signs.

a. $+\dfrac{+1}{+2}$

$$+\frac{+1}{+2} = +\frac{-1}{-2} = -\frac{+1}{-2} = -\frac{-1}{2}$$

Notice that $+(+1/+2) \neq -(+1/+2)$ because only one sign has been changed.

b. $+\dfrac{-1}{+3}$

$$+\frac{-1}{+3} = -\frac{+1}{+3} = -\frac{-1}{-3} = +\frac{+1}{-3}$$ ■

□ **DO EXERCISE 6.**

Answers to Exercises

1. a. $-\dfrac{1}{2}$ **b.** -2.0 **c.** $-\dfrac{7}{24}$ **d.** -6.9

2. a. -2.6 **b.** 10.0 **c.** $\dfrac{2}{7}$ **d.** $-\dfrac{4}{9}$

3. a. $-\dfrac{2}{21}$ **b.** $\dfrac{1}{15}$ **c.** -1.0 **d.** $\dfrac{9}{4}$ **e.** $-\dfrac{21}{64}$ **f.** -15.0

4. a. 14 **b.** 9.6

5. a. $\dfrac{9}{5}$ **b.** $\dfrac{4}{-3}$ **c.** $\dfrac{1}{-7}$ **d.** $\dfrac{5}{6}$

6. a. $+\dfrac{-5}{-8}, -\dfrac{+5}{-8}, -\dfrac{-5}{+8}$ **b.** $+\dfrac{+3}{-4}, -\dfrac{-3}{-4}, +\dfrac{-3}{+4}$

c. $+\dfrac{-3}{-2}, -\dfrac{-3}{+2}, -\dfrac{+3}{-2}$ **d.** $-\dfrac{+5}{+6}, -\dfrac{-5}{-6}, +\dfrac{-5}{+6}$

PROBLEM SET 2.5

A. *Add.*

1. $-\dfrac{1}{5} + \dfrac{3}{5}$

2. $-\dfrac{1}{4} + \dfrac{3}{4}$

3. $-\dfrac{1}{8} + \left(-\dfrac{3}{8}\right)$

4. $-\dfrac{2}{9} + \left(-\dfrac{4}{9}\right)$

5. $-\dfrac{3}{5} + \dfrac{1}{15}$

6. $-\dfrac{1}{12} + \dfrac{7}{6}$

7. $-\dfrac{2}{5} + \left(-\dfrac{5}{6}\right)$

8. $-\dfrac{2}{9} + \left(-\dfrac{2}{3}\right)$

9. $-2.7 + 6.4$

10. $-4.8 + 7.5$

11. $-1.2 + (-9.1)$

12. $-2.1 + (-7.4)$

13. $-9.8 + 3.1$

14. $-10.4 + 2.5$

15. $-\dfrac{1}{10} + \left(-\dfrac{1}{50}\right)$

16. $-\dfrac{3}{5} + \left(-\dfrac{1}{25}\right)$

B. *Subtract.*

17. $5.4 - 6.3$

18. $7.8 - 9.1$

19. $7 - 9.1$

20. $6 - 7.3$

21. $3.2 - (-1.1)$

22. $5.8 - (-2.4)$

23. $-2.1 - 3.6$

24. $-8.1 - 1.5$

25. $\dfrac{1}{8} - \dfrac{7}{8}$

26. $\dfrac{1}{3} - \dfrac{5}{3}$

27. $-\dfrac{4}{5} - \left(-\dfrac{3}{5}\right)$

28. $-\dfrac{1}{6} - \left(-\dfrac{5}{6}\right)$

29. $-\dfrac{7}{8} - \dfrac{1}{3}$

30. $-\dfrac{1}{5} - \dfrac{5}{3}$

31. $\dfrac{1}{9} - \left(-\dfrac{3}{2}\right)$

32. $\dfrac{2}{7} - \left(-\dfrac{2}{3}\right)$

C. *Multiply.*

33. $2(-2.4)$

34. $6(-2.1)$

35. $-0.2(-0.3)$

36. $(-0.5)(-0.4)$

37. $(-4)(3.8)$

38. $(-7)(2.5)$

39. $(-4.1)(-1.1)$

40. $(-2.2)(-1.2)$

41. $\dfrac{5}{8}\left(-\dfrac{4}{3}\right)$

42. $\dfrac{3}{7}\left(-\dfrac{14}{5}\right)$

43. $-\dfrac{1}{7}\left(-\dfrac{3}{7}\right)$

44. $-\dfrac{1}{8}\left(-\dfrac{7}{9}\right)$

45. $\dfrac{3}{7}\left(-\dfrac{21}{3}\right)$

46. $\dfrac{5}{9}\left(-\dfrac{18}{5}\right)$

47. $-\dfrac{8}{5}\left(-\dfrac{10}{4}\right)$

48. $-\dfrac{7}{3}\left(-\dfrac{9}{14}\right)$

49. $\dfrac{1}{3}(9)(-5)$

50. $\dfrac{1}{4}(-12)(6)$

51. $(-2.5)(-3)(2)$

52. $4.1(-2)(-3)$

D. *Divide.*

53. $\dfrac{1}{7} \div \left(-\dfrac{3}{5}\right)$

54. $\dfrac{3}{4} \div \left(-\dfrac{1}{9}\right)$

55. $-\dfrac{2}{5} \div \left(-\dfrac{3}{5}\right)$

56. $-\dfrac{3}{8} \div \left(-\dfrac{7}{8}\right)$

57. $-\dfrac{7}{5} \div \left(\dfrac{7}{10}\right)$

58. $-\dfrac{9}{8} \div \left(\dfrac{9}{16}\right)$

59. $-\dfrac{8}{5} \div \left(-\dfrac{16}{25}\right)$

60. $-\dfrac{11}{3} \div \left(-\dfrac{22}{9}\right)$

61. $3 \div (-0.6)$

62. $7 \div (-0.7)$

63. $(-6.3) \div (-3)$

64. $(-5.4) \div (-3)$

E. *Find the reciprocal.*

65. $\dfrac{3}{4}$

66. $\dfrac{1}{7}$

67. $\dfrac{-11}{5}$

68. $\dfrac{-12}{7}$

69. 3

70. 8

71. -10

72. -52

F. *Find three other equivalent fractions by changing signs.*

73. $\dfrac{1}{8}$ $+\dfrac{+1}{+8} \; ; \; \dfrac{-1}{-8} \; ; -\dfrac{-1}{8}$ **74.** $\dfrac{1}{9}$

75. $\dfrac{-2}{7}$

76. $\dfrac{-5}{9}$

$\dfrac{-1}{-8}$

77. $\dfrac{-11}{5}$

78. $\dfrac{17}{-3}$

79. The rules for adding and subtracting signed rational numbers are the same as those for adding and subtracting _____.

80. To find the reciprocal of a number, _____ the number (except zero, which does not have a reciprocal).

81. The rules for multiplication and division of signed rational numbers are the same as those for multiplication and division of _____.

82. Every fraction has _____ signs.

Checkup

The following problems provide a review of some of Sections 2.2, 2.3, and 2.4, and will help you with the next section.

83. $-3 + 4$ **84.** $-2 - 6$ **85.** $-9(8)$ **86.** $(-7)(-7)$ **87.** $-9 - 17$ **88.** $5(-3)$

Term in one variable a number or the product of a number and a variable raised to a power.

Numerical coefficient of a variable in a term the numerical factor in a term.

2.6 COMBINING TERMS AND REMOVING GROUPING SYMBOLS

1 Terms

> A **term** in one variable is a number or the product of a number and a variable raised to a power.

In a mathematical expression, terms are separated by plus or minus signs.

EXAMPLE 1

Expression	Number of Terms	Terms
$x + 2$	2	$x, 2$
x	1	x
$4x^2 + 3x - 2$	3	$4x^2, 3x, -2$
$\dfrac{x}{4} - 5$	2	$\dfrac{x}{4}, -5$

Remember that $x/4$ is $\frac{1}{4}x$. ■

□ **DO EXERCISE 1.**

Coefficients

> The **numerical coefficient** of a variable in a term is the number by which the variable is multiplied. If there is no number beside the variable, the numerical coefficient is plus 1 or negative 1, depending on whether the sign in front of the variable is plus or minus.

EXAMPLE 2 Find the numerical coefficient of each term.

a. $3x^4 + x^2 - 2x + 8$

The terms are $3x^4$, x^2, $-2x$, and 8.
The numerical coefficient of the first term is 3, of the second term is 1, of the third term is -2, and of the last term is 8. (We may think of the term 8 as containing a variable since $8 = 8x^0$.)

b. $2x^3 - x + 7$

The terms of the expression are $2x^3$, $-x$, and 7.
The numerical coefficients are 2, -1, and 7. Notice that the minus sign in front of the x makes its numerical coefficient -1.

The numerical coefficient is often simply called the coefficient. ■

□ **DO EXERCISE 2.**

□ **Exercise 1** Find the terms in the expression.

a. $x + 3$　　　　**b.** y

c. $x - 5$　　　　**d.** $x^2 + 3x - 4$

e. $\dfrac{x}{2} - 3$　　　**f.** $y + 3$

□ **Exercise 2** Find the coefficient of each term.

a. $x^6 - 4x^3 + 5x - 3$

b. $-y^5 + 3y^2 + y + 2$

Like terms   terms that have the same variable or variables raised to exactly the same powers.

◰ Combining Like Terms

> **Like terms** are terms that have the same variable or variables and the variable(s) of each term must have the same exponents.

□ Exercise 3 Identify the like terms.

a. $6a + 4b + 3a$

b. $12xy - 3xy + 4y$

c. $9 + 3x - 6$

d. $4y^2 - 3y^2 + y$

EXAMPLE 3   Identify the like terms.

a. $3x + 4y + 6x$

$3x$ and $6x$ are like terms.

b. $2a - 3ab + 9ab$

$-3ab$ and $9ab$ are like terms. *Notice that the minus sign is included in the term $-3ab$*

c. $6x^2 + x - 3x^2$

$6x^2$ and $-3x^2$ are like terms.

d. $x + 1 - 4$

1 and -4 are like terms. *Notice that $1x = x$ and $-1x = -x$.*   ■

□ DO EXERCISE 3.

Only like terms may be combined. To combine like terms we add the numerical coefficients.

EXAMPLE 4   Combine like terms.

a. $4x + 2x + 3y$

$$4x + 2x + 3y = 4x + 2x + 3y = 6x + 3y$$

We may think of adding 4 apples, 2 apples, and 3 oranges.

b. $x + 2y - 3x$

$$
\begin{aligned}
x + 2y - 3x &= x - 3x + 2y &&\text{Rearranging terms} \\
&= x + (-3x) + 2y &&\text{Subtraction rule} \\
&= -2x + 2y &&\text{Recall that } x = 1x
\end{aligned}
$$

c. $1 - 5 + 2y$

$$1 - 5 + 2y = 1 + (-5) + 2y = -4 + 2y$$

d. $6x^2 + 4x + 3x^2$

$$6x^2 + 4x + 3x^2 = 6x^2 + 3x^2 + 4x = 9x^2 + 4x$$

Distributive law of multiplication over subtraction $a(b - c) = ab - ac$

e. $x + 4 - 3x - 5$

$$x + 4 - 3x - 5 = x - 3x + 4 - 5 \qquad \text{Rearranging terms}$$
$$= x + (-3x) + 4 + (-5) \qquad \text{Subtraction rule}$$
$$= -2x - 1$$

f. $-7x + 8 - 3x - 4$

$$-7x + 8 - 3x - 4 = -7x - 3x + 8 - 4$$
$$= -7x + (-3x) + 8 + (-4)$$
$$= -10x + 4 \qquad \blacksquare$$

□ **DO EXERCISE 4.**

3 The Distributive Laws

Recall there is a distributive law of multiplication over addition.

$$a(b + c) = ab + ac \qquad \text{Distributive law}$$

There is also a **distributive law of multiplication over subtraction**.

$$a(b - c) = ab - ac \qquad \text{Distributive law}$$

EXAMPLE 5 Multiply.

a. $3(x + 4) = 3x + 3(4) = 3x + 12$

b. $a(y - 3) = ay - a(3) = ay - 3a$

c. $-2(x + y - 4) = -2x + (-2)y - (-2)(4) = -2x - 2y + 8 \qquad \blacksquare$

□ **DO EXERCISE 5.**

We often use the fact that finding the opposite of an expression is the same as multiplying the expression by -1.

EXAMPLE 6 Multiply.

a. $-(x + 4) = -1(x + 4) = -1x + (-1)(4) = -x - 4$ *Recall that* $-1x = -x$

b. $-(3x - y + 4) = -1(3x - y + 4) = -1(3x) - (-1)(y) + (-1)(4)$
$$= -3x + y - 4 \qquad \blacksquare$$

□ **DO EXERCISE 6.**

□ **Exercise 4**   Combine like terms.

a. $2x + 2 + 3x$

b. $y - 3y - 4$

c. $-3x - 4 + 2x + 2$

d. $-8a^2 + 7a - 3a^2$

e. $5y - 8 - 9y - 4$

f. $7 - 3x - 9 - 6x$

□ **Exercise 5**   Multiply.

a. $5(x + 2)$

b. $-3(x + 4)$

c. $a(x + 3y - z)$

d. $-5(x - z - 3)$

□ **Exercise 6**   Multiply.

a. $-(y + 4)$

b. $-(x - 2)$

c. $-(x + 3y - 2z)$

d. $-(a - 6b - 3c)$

Grouping symbols   symbols which show which operation to do first. Examples of grouping symbols are parentheses ( ), brackets [ ], and braces { }.

4 Grouping Symbols

Grouping symbols tell us which operation to do first. Common grouping symbols are parentheses ( ), brackets [ ], and braces { }.

> To remove grouping symbols, work out the innermost grouping symbols first. Combine terms within grouping symbols before removing the next set of grouping symbols.

Remember that a negative sign in front of a grouping symbol indicates that we are to multiply all the terms in the grouping symbol by -1. A plus sign in front of a grouping symbol indicates that we are to keep the same signs that are inside the grouping symbol. Remember that a term with no sign is understood to be positive.

EXAMPLE 7   Simplify, by writing with the least number of symbols.

a. $[3x - (2x + 1)]$

$$[3x - (2x + 1)] = [3x - (+2x + 1)] \qquad 2x \text{ means } +2x$$
$$= [3x - 2x - 1] \qquad \text{Removing the innermost grouping symbols}$$
$$= [x - 1] \qquad \text{Combining like terms}$$
$$= x - 1 \qquad \text{The sign in front of the brackets is plus}$$

b. $[4x + (x - 3)]$

$$[4x + (x - 3) = [4x + (+x - 3)] \qquad x \text{ means } +x$$
$$= [4x + x - 3]$$
$$= [5x - 3]$$
$$= 5x - 3$$

c. $[5 - (3x - 2)]$

$$[5 - (3x - 2)] = [5 - (+3x - 2)] \qquad 3x \text{ means } +3x$$
$$= [5 - 3x + 2] \qquad \text{Removing the innermost grouping symbols}$$
$$= [5 + 2 - 3x] \qquad \text{Rearranging terms}$$
$$= [7 - 3x]$$
$$= 7 - 3x$$

© 1989 by Prentice Hall

d. $\{2x - [4 + (3x + 2)]\}$

$\{2x - [4 + (3x + 2)]\} = \{2x - [4 + 3x + 2]\}$ Removing the innermost grouping symbols

$= \{2x - [4 + 2 + 3x]\}$ Rearranging terms

$= \{2x - [6 + 3x]\}$ Combining like terms

$= \{2x - 6 - 3x\}$ Removing the innermost grouping symbols

$= \{2x - 3x - 6\}$ Rearranging terms

$= \{-x - 6\} = -x - 6$

e. $2[4 + (3x - 2)]$

$2[4 + (3x - 2)] = 2[4 + 3x - 2]$

$= 2[4 - 2 + 3x]$ Rearranging terms

$= 2[2 + 3x]$

$= 4 + 6x$ Using a distributive law

f. $-\{2[x - 3(3x + 4)]\}$

$-\{2[x - 3(3x + 4)]\} = -\{2[x - 9x - 12]\}$ Using a distributive law to remove the innermost parentheses
Notice that -3 is treated as -3

$= -\{2[-8x - 12]\}$

$= -\{-16x - 24\}$ Using a distributive law

$= 16x + 24$ ■

□ **DO EXERCISE 7.**

□ **Exercise 7** Simplify, by writing with the least number of symbols.

a. $3x - (2x + 4)$

b. $2x + (3x - 4)$

c. $5 - [x + (3 - x)]$

d. $\{2 - [x + 4 - (3x + 2)]\}$

e. $-\{3x - [8 - x + (2x + 1)]\}$

f. $2[3 - (2x + 2)]$

g. $-2\{x - 2(x + 4)\}$

h. $-8(x + 2) + 2(x + 4)$

i. $5(x + 3) - [7 + (x - 3)]$

j. $-[(x + 2) - 5] + (4 + x)$

Answers to Exercises —————————————————————————

1. a. $x, 3$ **b.** y **c.** $x, -5$ **d.** $x^2, 3x, -4$ **e.** $\dfrac{x}{2}, -3$

f. $y, 3$

2. a. $1, -4, 5, -3$, respectively   **b.** $-1, 3, 1, 2$, respectively

3. a. $6a, 3a$ **b.** $12xy, -3xy$ **c.** $9, -6$ **d.** $4y^2, -3y^2$

4. a. $5x + 2$ **b.** $-2y - 4$ **c.** $-x - 2$ **d.** $-11a^2 + 7a$

e. $-4y - 12$ **f.** $-2 - 9x$

5. a. $5x + 10$ **b.** $-3x - 12$ **c.** $ax + 3ay - az$

d. $-5x + 5z + 15$

6. a. $-y - 4$ **b.** $-x + 2$ **c.** $-x - 3y + 2z$

d. $-a + 6b + 3c$

7. a. $x - 4$ **b.** $5x - 4$ **c.** 2 **d.** $2x$ **e.** $-2x + 9$

f. $2 - 4x$ **g.** $2x + 16$ **h.** $-6x - 8$ **i.** $4x + 11$ **j.** 7

PROBLEM SET 2.6

A. *Find the terms in each expression.*

1. $7x - y$

2. $10x - 3y$

3. $-3x + 2y + 2$

4. $-3y + 10z + 4$

5. $x + 4$

6. $x - 2$

7. $\dfrac{x}{2} + 3$

8. $\dfrac{y}{4} - x$

9. $x - y$

10. $x + 8$

B. *Find the coefficient of each term.*

11. $6x^2 - x + 3$

12. $x^3 - 3x + 8$

13. $\dfrac{y}{4} - 7$

14. $\dfrac{x}{3} - 9$

C. *Identify the like terms.*

15. $8a - 3b + 4a$

16. $9x - 3y + 7x$

17. $3 - 2z + 9$

18. $8y + 3 - 7$

D. *Combine like terms.*

19. $7x + 3x - 2$

20. $8 - 4y - 2y$

21. $9a + 3 - a$

22. $4t - 6 + t$

23. $3 - 2s + 1$

24. $7 + 5z + 6$

25. $6b + 2 - 3b + 8$

26. $7 + 3q + 4 - 5q$

27. $4a^2 - 3a^2 + a$

28. $b - 3b^2 + 6b^2$

29. $9x + y + 6 - 4x - 15 - y$

30. $12a + 3b - 10 + 9 - 2a - 3b$

31. $1.7x + y - 0.2x - 2.5y$

32. $7.1a + 3.4b - 2.1a - 3.1b$

E. *Multiply.*

33. $6(x - 1)$

34. $4(x - 5)$

35. $-3(y - 1)$

36. $-7(y - 5)$

37. $-2(6 - t)$

38. $-4(12 - n)$

39. $-3(a + 4b)$

40. $-2(r + 3s)$

41. $6(-r - 3s + 2)$

42. $4(-2x - 7y + 2)$

43. $-(3a + 4)$

44. $-(2a + 3)$

45. $-(3x - 3y + 6z)$ **46.** $-(4a - 2b + 5c)$ **47.** $-(7r + 6s + 4)$ **48.** $-(6x + 5y + 2)$

49. $-(3a + 4b - 7)$ **50.** $-(5u + 3v - 3)$ **51.** $-(-2x - 4y - 17)$ **52.** $-(-7x - 8y - 25)$

F. *Simplify, by writing with the least number of symbols.*

53. $8x - (3x + 2)$ **54.** $6m - (3m + 8)$ **55.** $3y + (4y - 7)$ **56.** $11a + (2a - 4)$

57. $3x + 6x - (2x + 6)$ **58.** $5b + 3a - (2a + 7)$ **59.** $4y - 6x - (8x - 5y)$ **60.** $7b - 5a - (6a + 2b)$

61. $2y - v - (3y - v + w)$ **62.** $r - 2s - (5r - 3s + 2t)$ **63.** $7 - 3(x + 4)$ **64.** $3 - 8(x + 1)$

65. $[10 - 4(a - 2)]$ **66.** $[15 - 3(b - 6)]$ **67.** $\{6x - [3 + (4 + x)]\}$ **68.** $\{5x - [7 + (3 + 2x)]\}$

69. $2\{3 + [(x + 8) - 2x]\}$ **70.** $3\{1 + [(2a + 4) - 3a]\}$ **71.** $-\{3[x - 2(x + 7)]\}$ **72.** $-\{4[x - 3(x - 1)]\}$

73. A term in one variable is a _____ or the product of a number and a variable raised to a power.

74. The numerical _____ of a variable in a term is the number by which the variable is multiplied.

75. If there is no numerical coefficient shown multiplying a variable, it is understood to be _____.

76. Like terms are terms that have the same variable(s) and the variable(s) of each term must have the same _____.

77. To remove grouping symbols, work out the _____ grouping symbols first.

Checkup

The following problems provide a review of some of Section 1.3. These problems will help you with the next section. Write in exponential notation.

78. $2 \cdot 2 \cdot 2$

79. $5 \cdot 5 \cdot 5 \cdot 5$

Expand.

80. a^5

81. b^3

What is the meaning of the following?

82. 4^0 **83.** $a^0, \quad a \neq 0$ **84.** 7^1 **85.** a^1

2.7 EXPONENTS

Recall the meaning of exponent from Section 1.3. For example,

$$3 \cdot 3 \cdot 3 \cdot 3 = 3^4$$

where 3 is the base and 4 is the exponent. Exponents may be negative and there are rules for multiplying and dividing exponential numbers and raising them to powers.

◼ Negative Exponents

Consider the pattern in the following:

$$3^2 = 3 \cdot 3$$
$$3^1 = 3$$
$$3^0 = 1$$

Notice that in each case, we divided the right side of the preceding equation by 3. Continuing the pattern,

$$3^{-1} = \frac{1}{3}$$

$$3^{-2} = \frac{1}{3} \div 3 = \frac{1}{3} \cdot \frac{1}{3} = \frac{1}{3^2}$$

For any positive integer n, and any nonzero real number a,

$$a^{-n} = \frac{1}{a^n}$$

EXAMPLE 1 Rewrite with positive exponents.

a. 2^{-3}

$$2^{-3} = \frac{1}{2^3}$$

Do not confuse the negative exponent with the sign of the result.

$$2(-3) = -6$$

but

$$2^{-3} = \frac{1}{2^3}$$

b. x^{-4}

$$x^{-4} = \frac{1}{x^4}, \quad x \neq 0$$

c. $3y^{-5}$

$$3y^{-5} = 3 \cdot \frac{1}{y^5} = \frac{3}{y^5}, \quad y \neq 0 \quad ◼$$

☐ **DO EXERCISE 1.**

OBJECTIVES

1 Rewrite a number with or without negative exponents

2 Multiply numbers with the same base by adding the exponents

3 Divide numbers with the same base by subtracting the exponents

4 Raise an exponential number to a power

☐ **Exercise 1** Rewrite with positive exponents. Assume that all variables represent nonzero real numbers.

a. 2^{-3}

b. 3^{-4}

c. x^{-5}

d. y^{-4}

e. $-7x^{-2}$

f. $8y^{-4}$

□ **Exercise 2** Rewrite with negative exponents. Assume that all variables represent nonzero real numbers.

a. $\dfrac{1}{2^4}$

b. $\dfrac{1}{7^5}$

c. $\dfrac{1}{x^6}$

d. $\dfrac{1}{b^5}$

e. $\dfrac{9}{x}$

f. $\dfrac{15}{y^2}$

We may reverse the rule.

EXAMPLE 2 Rewrite with negative exponents.

a. $\dfrac{1}{2^5}$

$$\frac{1}{2^5} = 2^{-5}$$

b. $\dfrac{1}{x^3}$

$$\frac{1}{x^3} = x^{-3}, \quad x \neq 0$$

c. $\dfrac{10}{x^4}$

$$\frac{10}{x^4} = 10x^{-4}, \quad x \neq 0$$

Notice that when an exponential number is moved from the numerator to the denominator or from the denominator to the numerator, the sign on the exponent is changed. ∎

□ **DO EXERCISE 2.**

2 Multiplication

Recall that

$$a^4 \cdot a^2 \quad \text{means} \quad (a \cdot a \cdot a \cdot a)(a \cdot a) = a^6$$

If we have a large number of factors, this method is too long.

To multiply two exponential numbers with the same base, keep the base and add the exponents. For any integers m and n, and any nonzero real number a,

$$a^m \cdot a^n = a^{m+n}$$

EXAMPLE 3 Multiply and write with positive exponents. Assume that all variables represent nonzero real numbers.

a. $2^3 \cdot 2^2 = 2^{3+2} = 2^5$

b. $(-4)^3 \cdot (-4)^2 = (-4)^{3+2} = (-4)^5$

c. $x^5 \cdot x^{-3} = x^{5+(-3)} = x^2$

d. $y^{-4} \cdot y^{-2} = y^{-4+(-2)} = y^{-6} = \dfrac{1}{y^6}$

Remember that $a = a^1$.

e. $a \cdot a^5 = a^{1+5} = a^6$

We may not use the multiplication rule to multiply numbers such as $5^2 \cdot 4^3$ since the bases are different. Also, *recall that $a^0 = 1$, $a \neq 0$.*

f. $a^{-4} \cdot a^4 = a^{-4+4} = a^0 = 1$

g. $(-3)^2 \cdot (-3)^{-2} = (-3)^{2+(-2)} = (-3)^0 = 1$ ∎

□ **DO EXERCISE 3.**

3 Division

$$\frac{a^4}{a^2} \quad \text{means} \quad \frac{a \cdot a \cdot a \cdot a}{a \cdot a} = a^2$$

> The rule is: When dividing with like bases, keep the base and subtract the exponent in the denominator from the exponent in the numerator. For any integers m and n, and any nonzero real number a,
>
> $$\frac{a^m}{a^n} = a^{m-n}$$

EXAMPLE 4 Divide and write with positive exponents. Assume that all variables represent nonzero real numbers.

a. $\dfrac{3^4}{3^2} = 3^{4-2} = 3^2$

b. $\dfrac{7^6}{7^8} = 7^{6-8} = 7^{6+(-8)} = 7^{-2} = \dfrac{1}{7^2}$

c. $\dfrac{2^{-3}}{2^{-1}} = 2^{-3-(-1)} = 2^{-3+1} = 2^{-2} = \dfrac{1}{2^2}$

d. $\dfrac{b^5}{b^{-2}} = b^{5-(-2)} = b^{5+2} = b^7$

e. $\dfrac{x^{-5}}{x^{-4}} = x^{-5-(-4)} = x^{-5+4} = x^{-1} = \dfrac{1}{x}$

f. $\dfrac{y^{-3}}{y} = \dfrac{y^{-3}}{y^1} = y^{-3-1} = y^{-4} = \dfrac{1}{y^4}$ ∎

□ **DO EXERCISE 4.**

□ **Exercise 3** Multiply and write with positive exponents. Assume that all variables represent nonzero real numbers.

a. $3^3 \cdot 3^4$ b. $3^{-2} \cdot 3^7$

c. $6^{-3} \cdot 6^{-5}$ d. $8 \cdot 8^4$

e. $a^{-2} \cdot a^{-7}$ f. $b^{-6} \cdot b^{-5}$

g. $a^{-3} \cdot a^3$ h. $a^5 \cdot a^{-5}$

□ **Exercise 4** Divide and write with positive exponents. Assume that all variables represent nonzero real numbers.

a. $\dfrac{7^5}{7^3}$ b. $\dfrac{8^{-7}}{8^3}$

c. $\dfrac{4^{-2}}{4}$ d. $\dfrac{a^4}{a^{-3}}$

e. $\dfrac{z^{-6}}{z^{-6}}$ f. $\dfrac{a^{-2}}{a^{-7}}$

□ **Exercise 5** Simplify by raising to powers. Write with positive exponents. Assume that all variables represent nonzero real numbers.

a. $(4^{-2})^3$

b. $(5^{-3})^{-6}$

c. $(z^{-4})^3$

d. $(a^2)^{-6}$

e. $(2^2)^2$

f. $(-2)^4$

g. -2^4

h. $(-3)^2$

4 **Raising an Exponential Number to a Power**

$$(a^3)^2 \quad \text{means} \quad a^3 \cdot a^3 = (a \cdot a \cdot a)(a \cdot a \cdot a) = a^6$$

To raise an exponential number to a power, keep the base and multiply the exponents. For any nonzero real number a, and any integers m and n,

$$(a^m)^n = a^{mn}$$

EXAMPLE 5 Simplify by raising to powers. Write with positive exponents. Assume that all variables represent nonzero real numbers.

a. $(5^3)^2 = 5^{3(2)} = 5^6$

b. $(3^5)^{-2} = 3^{5(-2)} = 3^{-10} = \dfrac{1}{3^{10}}$

c. $(6^{-4})^{-3} = 6^{(-4)(-3)} = 6^{12}$

d. $(x^{-3})^3 = x^{-9} = \dfrac{1}{x^9}$

e. $(b^{-4})^{-2} = b^8$

f. $(a^3)^{-5} = a^{-15} = \dfrac{1}{a^{15}}$

g. $(2^2)^3 = 2^6 = 64$

h. $(-2)^2 = (-2)(-2) = 4$

i. $-2^2 = -(2)(2) = -4.$

Notice from Examples h and i that

$$(-2)^2 \neq -2^2 \qquad \blacksquare$$

□ **DO EXERCISE 5.**

Notice that

$$(a^4b^2)^3 = (a^4b^2) \cdot (a^4b^2) \cdot (a^4b^2)$$
$$= a^4 \cdot a^4 \cdot a^4 \cdot b^2 \cdot b^2 \cdot b^2$$
$$= (a^4)^3(b^2)^3 = a^{12}b^6$$

> A product of exponential numbers to a power is simplified by raising each number to the given power. For any nonzero real number a and any integers m and n,
>
> $$(ab)^m = a^m b^m$$

EXAMPLE 6 Simplify by raising to powers. Write with positive exponents. Assume that all variables represent nonzero real numbers.

a. $(ab)^4 = a^{1(4)}b^{1(4)} = a^4b^4$

b. $(a^2b^4)^3 = (a^2)^3(b^4)^3 = a^6b^{12}$

c. $(a^{-3}b^2)^2 = (a^{-3})^2(b^2)^2 = a^{-6}b^4 = \dfrac{1}{a^6} \cdot b^4 = \dfrac{b^4}{a^6}$

We must be careful to write each number to the given power.

d. $(2x^4)^3 = 2^3(x^4)^3 = 8x^{12}$

e. $(-2y^{-3})^2 = (-2)^2(y^{-3})^2 = 4y^{-6} = \dfrac{4}{y^6}$ ∎

☐ **DO EXERCISE 6.**

The following is a summary of the definitions and rules for exponents that we have discussed.

Definitions	Rules
$a^{-n} = \dfrac{1}{a^n}, \quad a \neq 0$	$a^m \cdot a^n = a^{m+n}$
$a^1 = a$	$\dfrac{a^m}{a^n} = a^{m-n}$
$a^0 = 1, \quad a \neq 0$	$(a^m)^n = a^{mn}$
	$(ab)^m = a^m b^m$

☐ **Exercise 6** Simplify by raising to powers. Write with positive exponents. Assume that all variables represent nonzero real numbers.

a. $(2xy)^3$

b. $(3ab)^2$

c. $(x^2y^3)^3$

d. $(a^{-2}b^3)^{-3}$

e. $(2y^3)^2$

f. $(-2x^4)^2$

g. $(3x^5)^2$

h. $(-3y^2)^2$

Answers to Exercises

1. a. $\dfrac{1}{2^3}$ b. $\dfrac{1}{3^4}$ c. $\dfrac{1}{x^5}$ d. $\dfrac{1}{y^4}$ e. $\dfrac{-7}{x^2}$ f. $\dfrac{8}{y^4}$

2. a. 2^{-4} b. 7^{-5} c. x^{-6} d. b^{-5} e. $9x^{-1}$ f. $15y^{-2}$

3. a. 3^7 b. 3^5 c. $\dfrac{1}{6^8}$ d. 8^5 e. $\dfrac{1}{a^9}$ f. $\dfrac{1}{b^{11}}$

   g. 1   h. 1

4. a. 7^2 b. $\dfrac{1}{8^{10}}$ c. $\dfrac{1}{4^3}$ d. a^7 e. 1   f. a^5

5. a. $\dfrac{1}{4^6}$ b. 5^{18} c. $\dfrac{1}{z^{12}}$ d. $\dfrac{1}{a^{12}}$ e. 16   f. 16

   g. -16 h. 9

6. a. $8x^3y^3$ b. $9a^2b^2$ c. x^6y^9 d. $\dfrac{a^6}{b^9}$ e. $4y^6$ f. $4x^8$

   g. $9x^{10}$ h. $9y^4$

PROBLEM SET 2.7

Assume that all variables represent nonzero real numbers.

A. *Rewrite with positive exponents.*

1. $3^{-2} = \dfrac{1}{3^2}$

2. $5^{-3} = \dfrac{1}{5^3}$

3. a^{-7} $\dfrac{1}{a^7}$

4. b^{-4} $\dfrac{1}{b^4}$

5. x^{-1} $\dfrac{1}{x}$

6. z^{-1} $\dfrac{1}{z}$

7. r^{-8} $\dfrac{1}{r^8}$

8. s^{-9} $\dfrac{1}{s^9}$

B. *Rewrite with negative exponents.*

9. $\dfrac{1}{3^2} = 3^{-2}$

10. $\dfrac{1}{4^5} = 4^{-5}$

11. $\dfrac{1}{a^5}$ a^{-5}

12. $\dfrac{1}{b^7}$

13. $\dfrac{1}{x^4}$ x^{-4}

14. $\dfrac{1}{y^7}$ y^{-7}

15. $\dfrac{1}{u}$ u^{-1}

16. $\dfrac{1}{v}$ v^{-1}

C. *Multiply and write with positive exponents.*

17. $2^5 \cdot 2^2$ 2^{7}

18. $4^3 \cdot 4^5$ 4^{8}

19. $7 \cdot 7^{-3}$ 7^{-2}

20. $5^{-4} \cdot 5$ 5^{-3}

21. $x^{-7} \cdot x^{-3}$

22. $y^{-3} \cdot y^{-2}$

23. $a^{-4} \cdot a^6$

24. $b^{-3} \cdot b^7$

25. $z^3 \cdot z^4$

26. $z^5 \cdot z^9$

27. $u^3 \cdot u^{-3}$

28. $v^{-4} \cdot v^4$

D. *Divide and write with positive exponents.*

29. $\dfrac{6^3}{6^2}$

30. $\dfrac{4^5}{4^3}$

31. $\dfrac{x^{-5}}{x^3}$

32. $\dfrac{z^{-4}}{z^2}$

33. $\dfrac{a^5}{a}$

34. $\dfrac{b^7}{b}$

35. $\dfrac{z^{-3}}{z^{-2}}$

36. $\dfrac{t^{-8}}{t^{-3}}$

37. $\dfrac{r^{-5}}{r^{-5}}$

38. $\dfrac{s^{-6}}{s^{-6}}$

39. $\dfrac{x^{-4}}{x^{-3}}$

40. $\dfrac{y^{-7}}{y^{-2}}$

E. *Simplify by raising to powers. Write with positive exponents.*

41. $(3^2)^3$ **42.** $(5^4)^2$ **43.** $(8^3)^{-2}$ **44.** $(2^{-3})^4$ **45.** $(a^{-2})^{-4}$ **46.** $(b^{-3})^{-5}$

47. $(3^2)^3$ **48.** $(4^3)^4$ **49.** $(-3)^2$ **50.** $(-2)^2$ **51.** -3^2 **52.** -2^2

53. $(-5)^2$ **54.** $(-4)^2$ **55.** $(x^3y^2)^3$ **56.** $(y^2z^2)^3$ **57.** $(a^{-2}b^4)^{-3}$ **58.** $(r^3s^{-4})^{-3}$

59. $(-2x^2)^2$ **60.** $(-3x^4)^2$ **61.** $(4y^4)^2$ **62.** $(3z^3)^3$ **63.** $(-2x^2)^3$ **64.** $(-3y^4)^3$

65. Look for a pattern in the following:

$$10^2 = 100$$
$$10^3 = 1000$$
$$10^4 = 10,000$$

What relationship does the exponent have to the number of zeros in the expanded form?

66. Use the result found in Problem 65 to find 10^6.

67. If an exponential number is moved from the numerator to the denominator, the sign on the _____ is changed.

68. To multiply two exponential numbers with the same base, keep the _____ and add the exponents.

69. To raise an exponential number to a power, keep the base and _____ the exponents.

Checkup

The following problems provide a review of some of Section 1.6. These problems will help you with the next section. Evaluate when $x = 9$ and $y = 8$.

70. $9x + 9y$

71. $7x + 8y$

Find the perimeter of the rectangles with the following dimensions.

72. Length: 9.3 cm; width: 7.2 cm

73. Length: 8 m; width: 6.3 m

Find the circumference of the circles with the given radii. Use 3.14 for π.

74. Radius: 7.8 ft

75. Radius: 18.4 in.

2.8 EVALUATIONS

1 We did some evaluations in Chapter 1 in the section on variables and geometric formulas. Now that we have studied negative numbers and exponents, we may do additional evaluations. Evaluations are used in the applications of mathematics to business, science, and other fields.

EXAMPLE 1 Evaluate when $x = -4$.

a. $3x + 4$

$$3x + 4 = 3(-4) + 4 = -12 + 4 = -8$$

Notice that when we substitute for x, we put the number substituted in parentheses.

b. $5 - 3x$

$$5 - 3x = 5 - 3(-4) = 5 + 12 = 17 \quad \blacksquare$$

☐ **DO EXERCISE 1.**

EXAMPLE 2 Evaluate $x^2 - 4x + 1$ for the given values of x.

a. $x = 3$

$$x^2 - 4x + 1 = (3)^2 - 4(3) + 1 = 9 - 12 + 1 = -3 + 1 = -2$$

b. $x = -3$

$$x^2 - 4x + 1 = (-3)^2 - 4(-3) + 1 = 9 + 12 + 1 = 22 \quad \blacksquare$$

☐ **DO EXERCISE 2.**

EXAMPLE 3 Evaluate $x^3 - x^2 + 5x - 4$ for the given values of x.

a. $x = -2$

Recall that $(-2)^3 = (-2)(-2)(-2) = 4(-2) = -8$. Hence

$$x^3 - x^2 + 5x - 4 = (-2)^3 - (-2)^2 + 5(-2) - 4$$
$$= -8 - 4 - 10 - 4 = -26$$

b. $x = 0$

$$x^3 - x^2 + 5x - 4 = (0)^3 - (0)^2 + 5(0) - 4$$
$$= 0 - 0 + 0 - 4 = -4 \quad \blacksquare$$

☐ **DO EXERCISE 3.**

OBJECTIVE

1 *Evaluate an expression for given values of the variables*

☐ **Exercise 1** Evaluate when $x = -2$.

a. $2x - 1$ **b.** $7 - x$

☐ **Exercise 2** Evaluate $x^2 + x - 3$ for the given values of x.

a. $x = 2$ **b.** $x = -2$

c. $x = 4$ **d.** $x = -4$

☐ **Exercise 3** Evaluate $-x^3 + 2x + 5$ for the given values of x.

a. $x = 2$ **b.** $x = -2$

c. $x = 0$ **d.** $x = 1$

□ **Exercise 4** Evaluate $(x - 4)/y$ for the given values of x and y.

a. $x = 10, y = 3$

b. $x = -5, y = 3$

□ **Exercise 5** Evaluate $3x^2y + 2xyz$ for the given values of x, y, and z.

a. $x = -1, y = 5, z = -2$

b. $x = 2, y = -3, z = 2$

EXAMPLE 4 Evaluate $(a - 2)/b$ when $a = -6$ and $b = 4$.

$$\frac{a - 2}{b} = \frac{(-6) - 2}{4} = \frac{-8}{4} = -2 \quad \blacksquare$$

□ **DO EXERCISE 4.**

EXAMPLE 5 Evaluate $xy^2 - 2xyz$ when $x = 2$, $y = -1$, and $z = 4$.

$$xy^2 - 2xyz = 2(-1)^2 - 2(2)(-1)(4)$$
$$= 2(1) - 2(2)(-1)(4)$$
$$= 2 - 4(-4)$$
$$= 2 + 16 = 18 \quad \blacksquare$$

□ **DO EXERCISE 5.**

Some of the ways in which evaluations are used in applied problems are the following.

EXAMPLE 6

a. Find the temperature in degrees Fahrenheit when the Celsius temperature is 20 degrees if the formula for converting from Celsius to Fahrenheit is $F = \frac{9}{5}C + 32$.

$$F = \frac{9}{5}C + 32 \qquad \text{Formula}$$

$$= \frac{9}{5}(20) + 32 \qquad \text{Substituting 20 for } C$$

$$= 36 + 32 = 68$$

The temperature is 68 degrees.

b. Find the retail price R of a coat if the cost C is \$40 and the percent markup r is 35%. Use the formula $R = C + rC$.

$$R = C + rC$$
$$= 40 + (35\%)(40)$$
$$= 40 + 0.35(40)$$
$$= 40 + 14 = 54$$

The price is \$54.

c. If an automobile traveling at rate r hits a tree, the force of the impact is the same as the force with which it would hit the ground when falling from a building s feet high, where

$$s = 0.034r^2$$

If a car traveling at a rate r of 50 miles per hour hits a tree, this is the same as pushing it from how high a building?

$$s = 0.034r^2$$
$$= 0.034(50)^2 \qquad \text{Substituting 50 for } r$$
$$= 0.034(2500) = 85$$

This is the same as pushing it from a building 85 feet high. ∎

☐ **DO EXERCISE 6.**

☐ **Exercise 6**

a. Find the temperature in degrees Celsius when the Fahrenheit temperature is 50 degrees if the formula for converting from Fahrenheit to Celsius is

$$C = \frac{5}{9}(F - 32) \quad (Hint: F = 50)$$

b. Find the power W delivered to a load by a current i of 2 amperes if the power in watts is given by

$$W = 110i - 22i^2$$

c. Find the efficiency E of a machine if it requires an input I of 200 horsepower to run a dynamo that generates an output O of only 180 horsepower of electricity. Efficiency E is given by the formula

$$E = \frac{O}{I}$$

(*continued*)

□ **Exercise 6** *continued*

d. The height h in feet of water in a reservoir may be determined by the velocity v with which water is escaping from a horizontal pipe of diameter d and length L. Find the height of water in the reservoir if the diameter of the pipe is 6 inches, the length of the pipe is 50 feet, the velocity of water is 4 feet per second, and the formula is given by

$$h = \frac{L(4v^2 + 5v - 2)}{100d}$$

Note This formula is designed to use length in feet and diameter in inches.

Answers to Exercises

1. a. -5 b. 9
2. a. 3   b. -1 c. 17   d. 9
3. a. 1   b. 9   c. 5   d. 6
4. a. 2   b. -3
5. a. 35   b. -60

6. a. 10 degrees   b. 132 watts   c. $\dfrac{9}{10}$ d. 6.833 feet

PROBLEM SET 2.8

A. *Evaluate when* $x = -3$.

1. $3x + 4$

2. $2x + 6$

3. $5x - 2$

4. $7x - 1$

5. $3 - 2x$

6. $4 - 8x$

B. *Evaluate* $-x^2 + 3x - 4$ *for the given values of* x.

7. $x = 1$

8. $x = 3$

9. $x = -1$

10. $x = -3$

C. *Evaluate* $2x^2 - x + 3$ *for the given values of* x.

11. $x = 0$

12. $x = 5$

13. $x = -4$

14. $x = -1$

D. *Evaluate* $-x^3 + x^2 + 1$ *for the given values of* x.

15. $x = 0$

16. $x = 1$

17. $x = 2$

18. $x = -2$

19. $x = -1$

20. $x = 3$

E. *Given* $x = -2$, $y = 2$, *and* $z = 3$, *evaluate the expression.*

21. $\dfrac{x - 3}{y}$

22. $\dfrac{x - 5}{y}$

23. $2xy^2 - xyz$

24. $-x^2y + 2xyz$

25. $\dfrac{7 - x}{z}$

26. $\dfrac{8 - y}{x}$

27. $xz - 2y^2$

28. $yz - 2x^2$

29. Find the interest I on a savings account of principal P of $2000 with a rate r of 7% invested for a time t of 1 year. Use $I = Prt$.

30. Find the distance d traveled by a car moving at a rate r of 54 miles per hour for a time t of 4 hours. Use $d = rt$.

31. Find the current I in amperes needed to run a refrigerator with a voltage E of 110 volts and a resistance R of 20 ohms. Use $I = E/R$.

32. Find the volume V of a gas in cubic feet at a temperature T of 200 kelvin at a pressure P of 20 pounds per square inch. Use $V = 2T/P$.

33. Find the height s above the ground after a time t of 8 seconds of a rocket launched at a velocity v of 153 feet per second. Use $s = vt - 16t^2$.

34. What is the weight w (in pounds) of a man of height h of 65 inches if height and weight are related by the formula $w = \frac{11}{2}h - 220$?

Checkup

The following problems provide a review of some of Section 1.7. Find the area of the following. Use 3.14 for π.

35. A square with side 4.5 in.

36. A rectangle with width 7 cm and length 11 cm.

37. A parallelogram with base 12 ft and height 6.8 ft.

38. A triangle with base 9 m and height 12 m.

39. A circle with diameter 17 in.

40. Find the lateral area of a right circular cylinder if its height is 9 cm and the radius of its base is 6 cm.

Find the volume of the following. Use 3.14 for π.

41. A rectangular solid with width 3 in., length 7 in., and height 5.4 in.

42. A right circular cylinder with height 10 m and diameter of the base 16 m.

CHAPTER 2   ADDITIONAL EXERCISES (OPTIONAL)

Section 2.1

Find the opposite.

1. $\dfrac{3}{4}$

2. -7.84

Simplify.

3. $-(-9.3)$

4. $-\left(\dfrac{5}{8}\right)$

Replace the ? < or >.

5. $-8 \, ? \, -5$

6. $0 \, ? \, -4$

7. $-7.9 \, ? \, -3.4$

8. $\dfrac{3}{5} \, ? \, \dfrac{5}{7}$

Find the following.

9. $\left|-\dfrac{11}{5}\right|$

10. $|-7.45|$

11. $-|-4.2|$

12. $-\left|\dfrac{3}{8}\right|$

Section 2.2

Add.

13. $-5 + 9$

14. $7 + (-4)$

15. $-6 + (-3)$

16. $-2 + (-8)$

17. $-9 + 6$

18. $-8 + (-9)$

19. $6 + (-8) + (5) + (-2)$

20. $(-7) + (-3) + 8 + (-1)$

Section 2.3

Subtract.

21. $7 - 9$ **22.** $6 - 8$ **23.** $-8 - 7$ **24.** $-9 - 7$

25. $3 - (-4)$ **26.** $5 - (-2)$ **27.** $70 - (-40) - 25$ **28.** $85 - 73 - (-22)$

Section 2.4

Multiply.

29. $(-7)8$ **30.** $9(-9)$ **31.** $(-7)(-4)(2)$ **32.** $(-8)(9)(-3)$

Divide.

33. $-8 \div 16$ **34.** $-10 \div (-25)$

Calculate.

35. $\dfrac{36 - 2 \cdot 3^2}{9}$ **36.** $8(-2) - 3(4)$

Section 2.5

Add.

37. $-\dfrac{7}{8} + \left(-\dfrac{3}{2}\right)$ **38.** $-7.1 + (-4.8)$ **39.** $-9.4 + 6.3$ $-\dfrac{8}{5} + \dfrac{3}{4}$

Subtract.

41. $-\dfrac{7}{8} - \left(-\dfrac{4}{5}\right)$ **42.** $-3.9 - (-4.8)$ **43.** $-16.2 - 13.8$ **44.** $-\dfrac{7}{10} - \dfrac{8}{5}$

Multiply.

45. $(-4.8)(-3.7)$ **46.** $\left(-\dfrac{3}{4}\right)\left(-\dfrac{8}{7}\right)$

Divide.

47. $-\dfrac{17}{3} \div \dfrac{5}{6}$ **48.** $7.2 \div (-2.4)$

Find the reciprocal.

49. 12 **50.** -15

Find three other equivalent fractions by changing signs.

51. $\dfrac{3}{4}$ **52.** $-\dfrac{9}{5}$

Section 2.6

Find the terms in each expression.

53. $\dfrac{2}{3}x^2 + x - 4$ **54.** $5x^2 - \dfrac{1}{4}x + 2$

Find the coefficient of each term.

55. $x - \dfrac{1}{4}$

56. $-x + 3$

57. $5x^2 - 2x + 3$

58. $7x^2 + 4x - 2$

Simplify.

59. $-\{3x - 2(4x - 5) - 6\}$

60. $4(3x - 4) - [2(4x - 5)]$

Section 2.7

Assume that all variables represent positive numbers.
Multiply and write with positive exponents.

61. $3^2 \cdot 3^4$

62. $6^5 \cdot 6^{-9}$

63. $x^{2n} \cdot x^{-3n}$

64. $y^{-4n} \cdot y^{-2n}$

Divide and write with positive exponents.

65. $\dfrac{x^7}{x}$

66. $\dfrac{x^{-3}}{x^{-5}}$

67. $\dfrac{y^n}{y^{5n}}$

68. $\dfrac{x^{7n}}{x^{3n}}$

Simplify by raising to powers. Write with positive exponents.

69. $(x^{4n})^2$

70. $(y^{-3n})^4$

Section 2.8

Evaluate the following expressions for the given value of x.

71. $x^3 + x - 3, \quad x = -2$

72. $-x^3 + 3x + 4, \quad x = -1$

73. $-x^2 - x + 5, \quad x = 3$

74. $x^2 - 4x + 2, \quad x = 2$

CHAPTER 2   PRACTICE TEST

1. Replace the ? with < or >: -5 ? -3.

1.	_____

Find the following.

2. $|8.5|$ **3.** $-|-3|$

2.	_____
3.	_____
4.	_____

Add.

4. $-11 + 8$ **5.** $-3 + (-2.5)$ **6.** $-\dfrac{3}{4} + \dfrac{1}{6}$

5.	_____
6.	_____

Subtract.

7. $-8 - (-2)$ **8.** $-3.4 - 6.8$ **9.** $-\dfrac{7}{8} - \dfrac{3}{4}$

7.	_____
8.	_____
9.	_____

Multiply.

10. $-7(-4)$ **11.** $\dfrac{1}{2}\left(-\dfrac{2}{3}\right)$

10.	_____
11.	_____
12.	_____

Divide.

12. $\dfrac{20}{-5}$ **13.** $-\dfrac{2}{7} \div -\dfrac{4}{3}$

13.	_____
14.	_____

14. Find the reciprocal: -3.

15.	_____

15. List the terms of the polynomial: $x^2 - 6x + 4$

16. _____

16. Combine like terms: $7a - 3 - 6a - 8$.

17. _____

17. Multiply: $-2(3x - 3)$.

Simplify.

18. _____

18. $8a - (3 - 2a)$

19. _____

19. $2(3x + 4) - 3[2 - (x - 2)]$

For Problems 20–24, assume that all variables represent nonzero real numbers.

20. _____

20. Write with a negative exponent: $\dfrac{1}{3^4}$.

21. _____

21. Write with a positive exponent: 5^{-2}.

22. _____

22. Multiply: $y^{-3} \cdot y^5$.

23. _____

23. Divide: $\dfrac{b}{b^{-2}}$.

24. _____

24. Raise to the power and write with a positive exponent: $(x^3)^{-5}$.

25. _____

25. Evaluate $-x^2 + 2x - 1$ for $x = -2$.

26. _____

26. Find the distance d traveled by a truck moving at a rate r of 50 miles per hour for a time t of 6 hours. Use $d = rt$.

CUMULATIVE REVIEW   CHAPTERS 1 AND 2

1. Change to a decimal and round to three places: $\frac{11}{3}$.

2. Change to a fraction: 0.004.

3. Evaluate: 2^4.

4. Change to a decimal: $\frac{1}{2}\%$.

5. Change to percent: 0.3.

Find the area of each figure.

6.

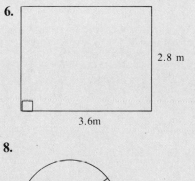

2.8 m

3.6m

7.

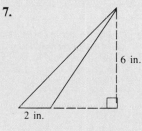

6 in.

2 in.

8.

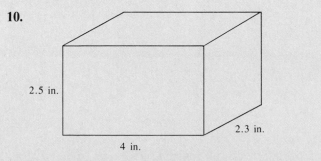

16 cm

9. Find the circumference of the circle.

2.5 ft

Find the volume of each figure.

10.

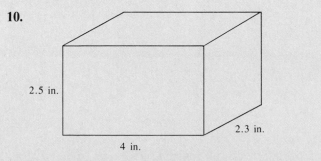

2.5 in.

4 in.

2.3 in.

11.

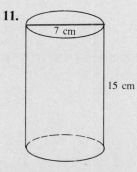

7 cm

15 cm

12. Replace the ? with $<$ or $>$: -2.5 ? -3.4.

13. Find the following: $-|8|$.

Do the operations indicated.

14. $\dfrac{-3(-4+5)+3}{8}$

15. $\dfrac{5(-2-4)-2(3)}{-6}$

16. Find the reciprocal of $\dfrac{-7}{9}$.

Combine like terms.

17. $3a - 4b - 7a - 3b$

18. $-2x + 4y - 7x + 8y$

Simplify by removing grouping symbols and combining like terms.

19. $7x - (3y - 2x) - 8y$

20. $-3\{2 + (3 - 9a) + 4a\} - 6$

Do the operation indicated and write with positive exponents. Assume that all variables represent nonzero real numbers.

21. $6x^4 \cdot 3x^{-5}$

22. $\dfrac{a^7}{a^{10}}$

23. $(3x^4)^2$

24. $(2b^3)^{-4}$

25. Evaluate $-3x^2 + 5x - 2$ for $x = -3$

26. Find the simple interest on a 1-year loan of $1500 if the interest rate is 9%. Use $I = Prt$.

Linear Equations and Inequalities in One Variable

3.1 LINEAR EQUATIONS

1 True, False, and Conditional Equations

An **equation** is a statement that two mathematical expressions are equal. Some equations are always true, some are always false, and some are sometimes true and sometimes false, depending on the value assigned to the variable. We call the latter equations **conditional**.

EXAMPLE 1 Which of the equations are true, false, or conditional?

a. $5 + 3 = 10$ is false

b. $2 + 6 = 8$ is true

c. $x - 4 = 9$ is conditional; we have not assigned a value to the variable x

d. $2(x + 3) = 2x + 6$ is true for any value of the variable x ■

□ DO EXERCISE 1.

2 Linear Equations in One Variable

A **linear equation** in one variable is an equation with an exponent of 1 on the variable in any term. Terms containing only numbers are sometimes considered to contain the variable $x^0 = 1$, for example, $3 = 3x^0 = 3(1) = 3$. For purposes of the definition of a linear equation, we will assume that terms containing only numbers do not include a variable.

EXAMPLE 2 Which of the following are linear equations?

a. $x + 5 = 9$ Linear; the exponent on x is understood to be 1

b. $x^2 - 3 = 7$ Not linear; the exponent on x is 2

c. $4 - y = 8$ Linear; the exponent on y is 1

d. $\dfrac{2}{3}x + 5 = \dfrac{1}{5}x$ Linear; the exponent on each x is 1

e. $x^2 + 3x + 4 = 0$ Not linear; there is an exponent other than 1 on the variable ■

□ DO EXERCISE 2.

□ **Exercise 1** Which of the equations are true, false, or conditional?

a. $7 + 1 = 8$ **b.** $x + 5 = 8$

c. $9 + 5 = 12$ **d.** $x - 3 = 1$

e. $3(x + 1) = 3x + 3$

f. $2x - 4 = 2(x - 2)$

□ **Exercise 2** Which of the following are linear equations?

a. $x - 7 = 15$

b. $9 - 4 = y$

c. $3 + y^2 = 9$

d. $\dfrac{3}{4}x = \dfrac{1}{6}x + \dfrac{1}{16}$

117

Linear equation in one variable an equation with an exponent of *one* on the variable in any term.

A solution of a linear equation a number that gives a true equation when it is substituted for the variable.

□ **Exercise 3** Are the values of the variables solutions for the given equations?

a. $x + 7 = 14$, $x = 7$

b. $9 - z = 3$, $z = 5$

c. $y - 4 = -8$, $y = -4$

d. $5 - x = 8$, $x = -3$

e. $y + \dfrac{2}{3} = \dfrac{4}{3}$, $y = \dfrac{2}{3}$

f. $x + 1.7 = 3$, $x = 1.3$

3 **Solutions**

A **solution of a linear equation** is a number that gives us a true equation when we substitute it for the variable. We say that the number *satisfies* or *is a root of* the equation.

EXAMPLE 3 Are the values of the variables solutions for the given equations?

a. $x + 6 = 8$, $x = 2$

$$\begin{array}{c|c} (2) + 6 & 8 \\ 8 & 8 \end{array} \quad \text{Yes}$$

Remember that when we substitute the number, we put it in parentheses.

b. $9 - y = 5$, $y = 4$

$$\begin{array}{c|c} 9 - (4) & 5 \\ 5 & 5 \end{array} \quad \text{Yes}$$

c. $7 + z = 12$, $z = 3$

$$\begin{array}{c|c} 7 + (3) & 12 \\ 10 & 12 \end{array} \quad \text{No}$$

d. $4 - x = 6$, $x = -2$

$$\begin{array}{c|c} 4 - (-2) & 6 \\ 6 & 6 \end{array} \quad \text{Yes} \quad \blacksquare$$

□ **DO EXERCISE 3.**

We will use our knowledge of linear equations in Chapter 4 to solve applied problems.

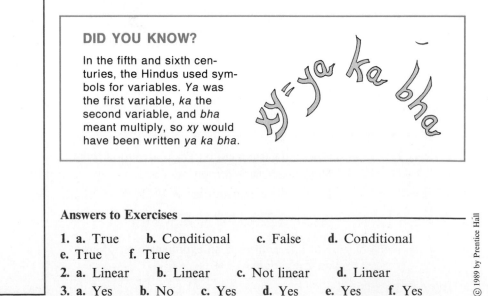

DID YOU KNOW?

In the fifth and sixth centuries, the Hindus used symbols for variables. *Ya* was the first variable, *ka* the second variable, and *bha* meant multiply, so *xy* would have been written *ya ka bha*.

Answers to Exercises _____

1. a. True   **b.** Conditional   **c.** False   **d.** Conditional
e. True   **f.** True
2. a. Linear   **b.** Linear   **c.** Not linear   **d.** Linear
3. a. Yes   **b.** No   **c.** Yes   **d.** Yes   **e.** Yes   **f.** Yes

PROBLEM SET 3.1

A. *Which of the equations are true, which are false, which conditional?*

1. $3 + 2 = 5$ T

2. $7 - 4 = 3$ T

3. $x - 7 = 2$ C

4. $y + 3 = 8$ C

5. $8 - 4 = 5$ F

6. $9 + 4 = 12$ F

7. $16 - 7 = 9$

8. $18 - 9 = 9$

9. $x + 2 = 3$

10. $y - 7 = 9$

11. $5(x - 1) = 5x - 5$

12. $6x + 12 = 6(x + 2)$

B. *Which of the following are linear equations?*

13. $y + 4 = 12$

14. $x - 7 = 8$

15. $9 - z = 14$

16. $15 - y = 10$

17. $x^2 - 3 = 7$

18. $5 - y^2 = 2$

19. $\frac{1}{3}z + \frac{2}{5} = \frac{1}{8}z$

20. $7 = \frac{1}{2}y + \frac{4}{9}$

21. $8 + x = 12$

22. $7 - y = 15$

23. $x^2 + x + 7 = 0$

24. $z^2 + z = 0$

25. $3 + x^2 = 4$

26. $y^2 + 7 = 12$

C. *Are the values of the variables solutions for the given equations?*

27. $x - 6 = 10, x = 16$

28. $y + 4 = 8, y = 4$

29. $z - 7 = 8, z = 14$

30. $9 - x = 6, x = 2$

31. $\dfrac{3}{5} + y = \dfrac{6}{5}$, $y = \dfrac{3}{5}$

32. $z - \dfrac{5}{7} = \dfrac{3}{7}$, $z = \dfrac{8}{7}$

33. $9 - x = 11$, $x = -2$

34. $10 - y = 12$, $y = -2$

35. $z - 3 = -10$, $z = -7$

36. $x - 8 = -11$, $x = -3$

37. $8 - y = 11$, $y = -4$

38. $7 - z = 15$, $z = -9$

39. Equations that are true or false depending on the value assigned to the variable are called _____ .

40. A _____ equation in one variable is an equation with an exponent of 1 on the variable in any term.

41. A number that gives us a true equation when we substitute it for the variable in an equation is called a _____ or _____ of the equation.

Checkup

The following problems review some of Section 2.1. Replace the ? with < or >.

42. $-6 \,?\, -1$

43. $-9 \,?\, -15$

44. $9 \,?\, -2$

45. $-4 \,?\, 3$

Find the following.

46. $|-9|$

47. $-|8|$

48. $-|2.5|$

49. $-|-3.6|$

50. $\left|\dfrac{4}{5}\right|$

51. $\left|-\dfrac{3}{7}\right|$

Addition property of equality If $a = b$ then $a + c = b + c$ where a, b, and c are real numbers.

3.2   THE ADDITION PROPERTY OF EQUALITY

1 OBJECTIVE

1 *Solve equations using the addition property of equality*

1 When we find the solution of an equation, we say that we solve it. We may solve some equations by adding the same number to both sides of the equation.

> **Addition Property of Equality**
>
> $$\text{If } a = b$$
> $$\text{then } a + c = b + c$$
>
> where a, b, and c are real numbers.

We may think of the equation as balancing. If we add the same number to both sides of the equation, it will still balance.

Our objective in solving an equation is to get the variable alone on one side of the equation.

EXAMPLE 1 Solve.

a.
$$
\begin{aligned}
x - 6 &= 9 \\
+\,6 &= +6 \qquad \text{Adding 6 to each side of the equation} \\
\hline
x &= 15 \qquad \text{Remember that a number plus its opposite is zero}
\end{aligned}
$$

We may check the answer by substituting 15 for x in the original equation.

Check $x - 6 = 9$

$$
\begin{array}{c|c}
(15) - 6 & 9 \\
9 & 9
\end{array}
$$

b.
$$
\begin{aligned}
-7 + y &= 4 \\
+7 &= +7 \qquad \text{Adding 7 to each side} \\
\hline
y &= 11
\end{aligned}
$$

Check $-7 + y = 4$

$$
\begin{array}{c|c}
-7 + (11) & 4 \\
4 & 4
\end{array}
$$

c.
$$
\begin{aligned}
x - \frac{1}{2} &= -\frac{3}{4} \\
+\frac{1}{2} &= +\frac{1}{2} \\
\hline
x \phantom{-\frac{1}{2}} &= -\frac{1}{4}
\end{aligned}
$$

□ **Exercise 1** Solve.

a. $x - 4 = 8$

b. $y - 7 = 10$

c. $-6 + x = 4$

d. $-\dfrac{1}{3} + y = \dfrac{5}{6}$

□ **Exercise 2** Solve.

a. $x + 8 = 12$

b. $y + 3 = -2$

c. $\dfrac{1}{5} + x = \dfrac{1}{4}$

d. $6 + y = 4$

Check $x - \dfrac{1}{2} = -\dfrac{3}{4}$

$$
\begin{array}{c|c}
-\dfrac{1}{4} - \dfrac{1}{2} & -\dfrac{3}{4} \\
\hline
-\dfrac{3}{4} & -\dfrac{3}{4} \quad \blacksquare
\end{array}
$$

□ **DO EXERCISE 1.**

We may also solve an equation by subtracting the same number from both sides of the equation because we are adding a negative number to both sides.

EXAMPLE 2 Solve.

a.
$$
\begin{array}{rl}
x + 5 = & -9 \\
-5 = & -5 \qquad \text{Adding } -5 \text{ to both sides} \\
\hline
x \quad = & -14
\end{array}
$$

Check $x + 5 = -9$

$$
\begin{array}{c|c}
(-14) + 5 & -9 \\
-9 & -9
\end{array}
$$

b.
$$
\begin{array}{rl}
\dfrac{1}{2} + y = & 2 \\
-\dfrac{1}{2} \quad\; = & -\dfrac{1}{2} \\
\hline
y = & \dfrac{3}{2}
\end{array}
$$

Check $\dfrac{1}{2} + y = 2$

$$
\begin{array}{c|c}
\dfrac{1}{2} + \dfrac{3}{2} & 2 \\
2 & 2 \quad \blacksquare
\end{array}
$$

□ **DO EXERCISE 2.**

Answers to Exercises ——————————————

1. a. 12   **b.** 17   **c.** 10   **d.** $\dfrac{7}{6}$

2. a. 4   **b.** -5 **c.** $\dfrac{1}{20}$ **d.** -2

PROBLEM SET 3.2

Solve and check.

1. $x - 4 - 7$

2. $x - 3 = 9$

3. $x - 2 = -4$

4. $x - 1 = -3$

5. $-2 + y = 3$

6. $-4 + y = 8$

7. $-5 + x = -7$

8. $-6 + x = -1$

9. $x - \dfrac{1}{3} = \dfrac{2}{3}$

10. $x - \dfrac{1}{5} = \dfrac{2}{5}$

11. $x - \dfrac{1}{3} = \dfrac{1}{4}$

12. $x - \dfrac{1}{6} = \dfrac{1}{8}$

13. $y - 2.1 = 3.4$

14. $y - 1.1 = 2.7$

15. $7.8 + x = 4$

16. $2.9 + x = 3$

17. $x + 7 = 10$

18. $x + 9 = 14$

19. $-11 = y + 3$

20. $-7 = y + 4$

21. $-9 + y = -12$

22. $-8 + y = -15$

23. $x + \dfrac{1}{4} = \dfrac{1}{2}$

24. $x + \dfrac{1}{6} = \dfrac{1}{12}$

25. $-2.1 = y - 3.5$

26. $-1.2 = y - 7.1$

27. $33 + y = 38.9$

28. $56.5 + y = 54$

29. $3.7 = 3.7 + x$

30. $6.1 = 6.1 + x$

31. Some equations may be solved by adding the _____ number to both sides of the equation.

32. The same number may be subtracted from _____ sides of an equation.

Checkup

The following problems provide a review of some of Section 2.5. Some of these problems will help you with the next section.
Find the reciprocal.

33. $\dfrac{7}{8}$

34. $\dfrac{-11}{4}$

35. 7

36. 3

37. -2

38. -9

Find three other equivalent fractions by changing signs.

39. $\dfrac{1}{4}$

40. $\dfrac{9}{8}$

41. $\dfrac{-3}{5}$

42. $\dfrac{7}{-9}$

43. $-\dfrac{5}{2}$

44. $-\dfrac{5}{6}$

Multiplication property of equality If $a = b$ then $ac = bc$ where a, b, and c are real numbers.

3.3 THE MULTIPLICATION PROPERTY OF EQUALITY

■ We may also solve an equation by multiplying both sides of the equation by the same nonzero number.

Multiplication Property of Equality
If $a = b$
then $ac = bc$

where a, b, and c are real numbers.

EXAMPLE 1 Solve: $\dfrac{x}{3} = 5$.

$$3 \cdot \frac{x}{3} = 5 \cdot 3$$

$$1x = 15$$

$$x = 15$$

Check $\dfrac{x}{3} = 5$

$$\begin{array}{c|c} \dfrac{(15)}{3} & 5 \\ \hline 5 & 5 \end{array} \quad \blacksquare$$

Remember that our objective is to get the variable x alone on one side of the equation.

□ DO EXERCISE 1.

OBJECTIVE

■ *Solve equations using the multiplication property of equality*

□ **Exercise 1**   Solve.

a. $\dfrac{x}{6} = 4$

b. $\dfrac{y}{3} = 7$

We may divide both sides of an equation by the same number since this is the same as multiplying both sides of the equation by the reciprocal of the number. If $a = b$,

$$a \cdot \frac{1}{c} = b \cdot \frac{1}{c} \qquad c \neq 0$$

This is the same as

$$\frac{a}{c} = \frac{b}{c} \qquad c \neq 0$$

where a, b, and c are real numbers.

EXAMPLE 2 Solve.

a. $4x = 8$

$$\frac{4x}{4} = \frac{8}{4}$$

$$x = 2$$

Check $4x = 8$

$4(2)$	8
8	8

If the coefficient of x is a fraction, it is easier to multiply both sides of the equation by the reciprocal of this number.

b. $\dfrac{4}{3}x = 5$

$$\frac{3}{4} \cdot \frac{4}{3}x = 5 \cdot \frac{3}{4}$$

$$x = \frac{15}{4}$$

Check $\dfrac{4}{3}x = 5$

$\dfrac{4}{3}\left(\dfrac{15}{4}\right)$	5
5	5

Remember that we may get the variable alone on either side of the equation. This solves the equation.

c. $4.4 = 2.2y$

$$\frac{4.4}{2.2} = \frac{2.2y}{2.2}$$

$$2 = y$$

Check $4.4 = 2.2y$

4.4	2.2(2)
4.4	4.4 ■

□ DO EXERCISE 2.

We cannot solve an equation $ax = c$ when $a = 0$ using the methods of this section because we would have to divide both sides of the equation by zero. *We may not divide by zero.* The division c/a is defined to be a number b such that $c = ab$.

Suppose that we could divide by zero. Then $c/0 = b$ and $c = 0 \cdot b = 0$. Then it must be true that $c = 0$ and we have the division $0/0$.

Suppose that $0/0 = b$. Then $0 = 0 \cdot b$ and b can be any number. $0/0 = 6$ because $0 = 0 \cdot 6$ and $0/0 = -9$ because $0 = 0 \cdot (-9)$. Both 6 and -9 and other numbers work. But a solution to a problem must be a unique number. Therefore,

---

Do not divide by zero.

It is possible, however, to divide zero by a number other than zero.

EXAMPLE 3 Divide.

a. $\dfrac{5}{0}$

Not possible.

b. $\dfrac{0}{3} = 0$ Because $0 = 3 \cdot 0$ ■

□ DO EXERCISE 3.

□ **Exercise 2**   Solve.

a. $5x = 20$

b. $\dfrac{5}{3}x = 8$

c. $2.5y = 10$

d. $35 = 7y$

□ **Exercise 3**   Divide.

a. $\dfrac{8}{0}$ Not possible

b. $\dfrac{0}{7} = 0$

c. $\dfrac{-5}{0}$

d. $\dfrac{0}{-8}$

PROBLEM SET 3.3

A. *Solve and check.*

1. $8x = 24$

2. $5x = 15$

3. $4x = -16$

4. $6y = -30$

5. $-2y = -8$

6. $-3x = -15$

7. $\dfrac{2}{3}x = 9$

8. $\dfrac{3}{5}x = 5$

9. $-\dfrac{1}{5}x = -10$

10. $-\dfrac{5}{7}x = -1$

11. $\dfrac{7}{3}y = -\dfrac{1}{3}$

12. $-\dfrac{8}{5}y = -\dfrac{7}{8}$

13. $1.3z = 3.9$

14. $5.1z = 10.2$

15. $3.2y = -12.8$

16. $7.1y = -21.3$

17. $-7.7 = -1.1x$

18. $-16.8 = -4.2x$

19. $-\dfrac{8}{5} = \dfrac{5}{4}x$

20. $-\dfrac{9}{7} = \dfrac{7}{3}x$

21. $-\dfrac{3}{4} = -\dfrac{3}{2}y$

22. $-\dfrac{7}{8} = -\dfrac{7}{16}y$

23. $\dfrac{3}{2}x = 10.2$

24. $\dfrac{2}{3}x = 15.4$

B. *Divide.*

25. $\dfrac{9}{0}$
26. $\dfrac{7}{0}$
27. $\dfrac{0}{5}$
28. $\dfrac{0}{3}$

29. $\dfrac{-8}{0}$
30. $\dfrac{-14}{0}$
31. $\dfrac{0}{-3}$
32. $\dfrac{0}{-15}$

33. Sometimes an equation is solved by multiplying both sides of the equation by the same _____ number.

34. To solve an equation, isolate the _____ on one side of the equation.

35. Some equations may be solved by dividing both sides of the equation by the _____ non-zero number.

36. A fraction may be _____ times both sides of an equation.

37. Division by zero is _____.

38. Zero divided by a number other than zero is _____.

Checkup

The following problems provide a review of some of Section 2.4. Calculate.

39. $\dfrac{15 - 2 \cdot 8}{-1}$
40. $\dfrac{2 + 2 \cdot 7}{8}$
41. $4 - 3(2)^2$

42. $7 + 2(3)^3$
43. $\dfrac{5 - 5(3)^2}{-2 + 4(-2)}$
44. $\dfrac{4^2 + 4(5)}{-1 - 5}$

3.4 THE ADDITION AND MULTIPLICATION PROPERTIES OF EQUALITY

1 To solve most equations, it is necessary to use both the addition and the multiplication properties of equality. When both properties are used, it is generally easier to use the addition property first.

EXAMPLE 1 Solve.

a.
$$5x - 4 = 11$$
$$\underline{+ 4 = +4} \qquad \text{Addition property}$$
$$5x \quad = 15$$

$$\frac{5x}{5} = \frac{15}{5} \qquad \text{Dividing both sides of the equation by 5}$$

$$x = 3$$

Check

$5x - 4 = 11$	
$5(3) - 4$	11
$15 - 4$	11
11	11

b.
$$4x + 4 - 12$$
$$\underline{- 4 = -4}$$
$$4x \quad = 8$$

$$\frac{4x}{4} = \frac{8}{4}$$

$$x = 2$$

Check

$4x + 4 = 12$	
$4(2) + 4$	12
$8 + 4$	12
12	12	■

☐ **DO EXERCISE 1.**

☐ **Exercise 1** Solve.

a. $3x - 7 = 20$

b. $5x + 4 = 24$

c. $\dfrac{2}{3}x - 4 = 2$

d. $\dfrac{1}{2}y + 2 = 7$

□ **Exercise 2** Solve.

a. $3 - 4x = 15$

b. $-x - 8 = 15$

c. $-4y + 8 = -16$

d. $-6 + y = -9$

To solve an equation, the sign of the coefficient of the variable must be positive 1 in the last step. Hence it may be necessary to divide both sides of the equation by a negative number as a final step in solving the equation.

EXAMPLE 2 Solve.

a.
$$\begin{aligned} -6x + 7 &= 25 \\ -7 &= -7 \qquad \text{Addition property} \\ \hline -6x &= 18 \end{aligned}$$

$$\frac{-6x}{-6} = \frac{18}{-6} \qquad \text{Dividing both sides of the equation by } -6$$

$$x = -3$$

Check
$$\begin{array}{c|c} \multicolumn{2}{c}{-6x + 7 = 25} \\ \hline -6(-3) + 7 & 25 \\ 18 + 7 & 25 \\ 25 & 25 \end{array}$$

b.
$$\begin{aligned} 14 - x &= 25 \\ -14 &= -14 \\ \hline -x &= 11 \end{aligned}$$

$$\frac{-1x}{-1} = \frac{11}{-1} \qquad \text{Recall that } -x = -1x$$

$$x = -11$$

Check
$$\begin{array}{c|c} \multicolumn{2}{c}{14 - x = 25} \\ \hline 14 - (-11) & 25 \\ 14 + 11 & 25 \\ 25 & 25 \end{array} \qquad \blacksquare$$

□ **DO EXERCISE 2.**

If there are variables and numbers on both sides of the equation, our first objective is to get the variables on one side of the equation and the numbers on the other side. We use the addition property as many times as necessary to do this. Finally, we use the multiplication property to get the variable with a coefficient of positive 1 alone on one side of the equation. This solves the equation.

EXAMPLE 3 Solve.

a.

$$5x + 4 = 3x - 2$$
$$\underline{ - 4 = - 4} \qquad \text{Addition property}$$
$$5x = 3x - 6$$
$$\underline{-3x = -3x} \qquad \text{Addition property}$$
$$2x = - 6$$

$$\frac{2x}{2} = \frac{-6}{2} \qquad \text{Dividing both sides of the equation by 2}$$

$$x = -3$$

Check $\quad 5x + 4 = 3x - 2$

$5(-3) + 4$	$3(-3) - 2$
$-15 + 4$	$-9 - 2$
-11	-11

We combine terms whenever possible to make our work easier.

b. $12 = 4x + 2x$

$$12 = 6x \qquad \text{Combining like terms}$$

$$\frac{12}{6} = \frac{6x}{6}$$

$$2 = x$$

Check $\quad 12 = 4x + 2x$

12	$4(2) + 2(2)$
12	$8 + 4$
12	12

c. $3x - 2 + 4x = 7 - x + 15$

$$3x + 4x - 2 = -x + 7 + 15 \qquad \begin{array}{l}\text{Rearranging terms using the}\\ \text{commutative law of addition}\end{array}$$

$$7x - 2 = -x + 22 \qquad \text{Combining like terms}$$
$$\underline{ + 2 = + 2}$$
$$7x = -x + 24$$
$$\underline{+x = +x}$$
$$8x = 24$$

$$x = \frac{24}{8}$$

$$x = 3$$

Check $\quad 3x - 2 + 4x = 7 - x + 15$

$3(3) - 2 + 4(3)$	$7 - (3) + 15$
$9 - 2 + 12$	$7 - 3 + 15$
$7 + 12$	$4 + 15$
19	19 ∎

□ **DO EXERCISE 3.**

□ **Exercise 3** Solve.

a. $6x + 3 = 5x - 5$

b. $-2x - 3 = -5x + 6$

c. $2x + 3x = 35$

d. $15 = 7x - 4x$

e. $2y + 6y - 5 = 3y - 9 + 3$

f. $3y + 4 - 2y - 7 = 4y + 3$

Answers to Exercises

1. **a.** 9   **b.** 4   **c.** 9   **d.** 10
2. **a.** −3   **b.** −23   **c.** 6   **d.** −3
3. **a.** −8   **b.** 3   **c.** 7   **d.** 5   **e.** $-\dfrac{1}{5}$ **f.** −2

PROBLEM SET 3.4

Solve and check.

1. $4x + 3 = 19$ **2.** $2x + 5 = -1$ **3.** $3x - 8 = 4$

4. $4y + 8 = 16$ **5.** $6z + 12 = -12$ **6.** $2z - 6 = 6$

7. $-3x + 8 = -13$ **8.** $-4x + 3 = -9$ **9.** $2 = 4 - x$

10. $-2 - x = 5$ **11.** $-7 - 2y = 3$ **12.** $-4 - 3y = 5$

13. $2x + 0.6 = 2.4$ **14.** $4x - 0.2 = -4.6$ **15.** $3 - 2x = 2.8$

16. $5 - 5x = 10.5$ **17.** $4y + 8 = 2y - 6$ **18.** $2y + 9 = 5y - 6$

19. $3z - 4 = 6z - 19$ **20.** $y + 1 = 16 - 4y$ **21.** $2y - 1 = 4 + y$

22. $6x - 8 - 4x = 18$ **23.** $2x - 7 = x + 1$ **24.** $3y - 8 = y + 6$

25. $2x + 3x + 3 = 38$ **26.** $10y - 7y = 15$ **27.** $12y - 14y = -18$

28. $6x + 2x = 48$

29. $8.4y - 4.4y = 16$

30. $2.2x + 3.8x = 54$

31. $-6x - 18 - 6 = -9x - 33 - 6$

32. $10y + 6 - 3 = 12y - 18 + 15$

33. $2x - 5 + 6x = 3x - 9 + 3$

34. $3x - 6 + 4x = x + 9 - x$

35. $4z - 4 + z = 6z + 20 - 4z$

36. $5z - 7 + z = 7z + 21 - 5z$

37. $2.1y + 45.2 = 3.2 - 8.4y$

38. $3.4x + 16 - 3.24x = 0.8x - 0.64 + 16$

39. When both the addition and multiplication properties are used to solve an equation, it is generally easier to use the _____ property first.

40. To solve an equation, the sign of the coefficient of the variable must be _____ in the last step.

41. If there are variables and numbers on both sides of the equation, the first objective in solving the equation is to get the _____ on one side of the equation and the numbers on the other side.

Checkup

The following problems provide a review of some of Section 2.6. These problems will help you with the next section. Simplify.

42. $7x - (3x + 4)$

43. $5 - (2x + 2)$

44. $3x - 2(2x + 5)$

45. $7x - 3(2x - 4)$

46. $7 + 3(2x - 3)$

47. $6 + 2(5x - 7)$

48. $6x - 4(x - 3)$

49. $7x - 2(6x - 5)$

3.5 EQUATIONS WITH PARENTHESES

1 Recall from Chapter 2 that we remove parentheses by using the distributive laws.

EXAMPLE 1 Multiply.

a. $5(2x - 3) = 5(2x) + 5(-3) = 10x - 15$

b. $-(4x + 7) = -1(4x + 7) = -4x - 7$ ■

□ **Exercise 1** Multiply.

a. $3(x + 5)$

□ **DO EXERCISE 1.**

Many equations with parentheses may be solved by first removing the parentheses.

b. $-4(x - 3)$

EXAMPLE 2 Solve.

a.
$$3x + 9 = -3(2x + 3)$$
$$\underline{3x + 9 = -6x - 9} \quad \text{Using a distributive law}$$
$$\underline{ -9 = -9}$$
$$\underline{3x = -6x - 18}$$
$$\underline{+6x = +6x}$$
$$9x = -18$$

$$x = -\frac{18}{9}$$

$$x = -2$$

c. $-(x + 9)$

d. $-(x - 6)$

Check
$$\begin{array}{c|c} 3x + 9 = -3(2x + 3) \\ \hline 3(-2) + 9 & -3[2(-2) + 3] \\ -6 + 9 & -3[-4 + 3] \\ 3 & -3(-1) \\ 3 & 3 \end{array}$$

b.
$$3(y - 4) - (2y - 3) = -4$$
$$3y - 12 - 2y + 3 = -4$$
$$3y - 2y - 12 + 3 = -4 \quad \text{Rearranging terms}$$
$$y - 9 = -4 \quad \text{Combining like terms}$$
$$\underline{+9 = +9}$$
$$y = 5$$

□ **Exercise 2** Solve.

a. $4x + 2 = -2(x + 2)$

b. $8 - 2(3y - 4) = 2y$

c. $6x - (3x + 8) = 16$

Check
$$\begin{array}{c|c} 3(y - 4) - (2y - 3) = -4 \\ \hline 3[(5) - 4] - [2(5) - 3)] & -4 \\ 3[5 - 4] - [10 - 3] & -4 \\ 3(1) - (7) & -4 \\ 3 - 7 & -4 \\ -4 & -4 \end{array}$$ ■

d. $13 - (2z + 2) = 2(z + 2) + 3z$

□ **DO EXERCISE 2.**

If an equation contains fractions, it is usually easier to multiply both sides of the equation by the lowest common denominator of the fractions to remove them.

EXAMPLE 3 Solve.

a. $\dfrac{1}{4}(2x + 1) = \dfrac{3}{2}x - \dfrac{5}{2}$

The LCD is 4. Multiply both sides of the equation by 4.

$$4 \cdot \dfrac{1}{4}(2x + 1) = 4\left(\dfrac{3}{2}x - \dfrac{5}{2}\right)$$

$$4 \cdot \dfrac{1}{4}(2x + 1) = 4 \cdot \dfrac{3}{2}x - 4 \cdot \dfrac{5}{2} \qquad \text{Multiplying all terms by 4}$$

$$\begin{array}{rl}
2x + 1 = & 6x - 10 \qquad \text{Simplifying} \\
\underline{-2x \quad\; = -2x} & \qquad\qquad\;\; \text{Addition property} \\
1 = & 4x - 10 \\
\underline{+10 = \qquad + 10} & \qquad\qquad\;\; \text{Addition property} \\
11 = & 4x
\end{array}$$

$$\dfrac{11}{4} = x$$

The number $\frac{11}{4}$ checks, so it is the solution.

b. $\dfrac{3}{8}(3x - 4) - 2 = \dfrac{1}{2}(x - 2)$

The LCD is 8. Multiply both sides of the equation by the LCD.

$$8\left[\dfrac{3}{8}(3x - 4) - 2\right] = 8\left[\dfrac{1}{2}(x - 2)\right]$$

$$8 \cdot \dfrac{3}{8}(3x - 4) - 8 \cdot 2 = 8 \cdot \dfrac{1}{2}(x - 2) \qquad \text{Using a distributive law}$$

$$
\begin{array}{lll}
3(3x - 4) - 16 = 4(x - 2) & \text{Simplifying} \\
9x - 12 - 16 = 4x - 8 & \\
9x - 28 \quad\;\; = \quad 4x - 8 & \\
\underline{-4x \qquad\qquad = -4x} & \text{Addition property} \\
5x - 28 \quad\;\; = \qquad -8 & \\
\underline{+28 \qquad\;\; = \qquad +28} & \text{Addition property} \\
5x \qquad\quad\; = \qquad\;\; 20 &
\end{array}
$$

$$x = 4$$

The number 4 checks, so it is the solution. ■

☐ **DO EXERCISE 3.**

The following are steps in solving linear equations in one variable.

1. Multiply on both sides of the equation to remove fractions.
2. Remove parentheses.
3. Combine like terms on each side of the equation, if possible.
4. Use the addition property of equality to get all the terms with variables on one side of the equation and all the other terms on the other side.
5. Combine like terms, if possible.
6. Use the multiplication property of equality to solve for the variable.

☐ **Exercise 3**   Solve.

a. $\dfrac{1}{3}\left(2x - \dfrac{1}{2}\right) = -\dfrac{1}{2}x + \dfrac{7}{6}$

b. $\dfrac{1}{4}(2y + 1) - 3 = -\dfrac{1}{2}(y - 2)$

Answers to Exercises _____

1. **a.** $3x + 15$ **b.** $-4x + 12$ **c.** $-x - 9$ **d.** $-x + 6$

2. **a.** -1 **b.** 2   **c.** 8   **d.** 1

3. **a.** $\dfrac{8}{7}$ **b.** $\dfrac{15}{4}$

PROBLEM SET 3.5

A. Multiply.

1. $4(x - 3)$

2. $6(x + 5)$

3. $-(3x - 2)$

4. $-(5x - 4)$

5. $-3(x + 4)$

6. $-7(x - 1)$

B. Solve and check.

7. $2(y - 1) = -6$

8. $4(3y + 2) = 20$

9. $3(2x - 3) = 15$

10. $4(2x - 3) = 28$

11. $2(5x + 5) - 20 = 10$

12. $5(3x - 4) + 5 = -15$

13. $3 - (x - 4) = 5$

14. $4(x - 2) - 3x = 3$

15. $5z - (2z + 8) = 16$

16. $6z - (3z + 8) = 16$

17. $5(2x - 1) + 3x = 4$

18. $3(x + 4) = 2x + 15$

19. $2(2y - 1) = -3(y + 3)$

20. $3(3x + 5) = 2(5x + 5)$

21. $5(x + 4) = 7(x - 2)$

22. $5(z + 4) = 3(z - 2)$

23. $5(2x - 1) = 4(2x + 1) + 7$

24. $5(4y + 3) = 6(3y + 2) - 1$

25. $3(z - 6) + 2 = 4(z + 2) - 21$

26. $7(2x + 6) = 9(x + 3) + 5$

27. $3(y - 1) + 18 = 5(2y + 3)$

28. $4(3x - 2) = 7(x - 1) - 1$

29. $\dfrac{1}{5}(x + 2) - \dfrac{3}{5} = \dfrac{1}{10}x$

30. $\dfrac{1}{8}(7y - 2) - \dfrac{1}{16} = \dfrac{1}{4}y$

31. $\dfrac{1}{3}(4x - 7) - 5 = \dfrac{1}{4}(3x - 10)$

32. $\dfrac{1}{5}(x - 1) = \dfrac{1}{2}(x + 3) + 2$

33. Parentheses are often removed by using the _____ laws.

34. To solve equations with parentheses, the first step is usually to remove the _____.

Checkup

The following problems provide a review of some of Sections 3.3 and 3.4. These problems will help you with the next section.

35. $2x = 3x - 4$

36. $5x = 2x + 9$

37. $8.1 = 2.7x$

38. $6.5 = 0.5y$

39. $6 = 4 + 5y$

40. $9 = 3 - 2x$

41. $15 = -\dfrac{5}{7}x$

42. $8 = \dfrac{4y}{3}$

3.6 FORMULAS

1 Formulas are used extensively in applications of algebra to business, the sciences, and other fields. We studied geometric formulas in Chapter 1. One example of a formula is the one for the perimeter P of a triangle with sides of length a, b, and c.

$$P = a + b + c$$

We may know the perimeter and the lengths of two of the sides of a triangle. We can find a formula for the length of the other side by using the addition property of equality.

EXAMPLE 1 Solve for c in the formula $P = a + b + c$.

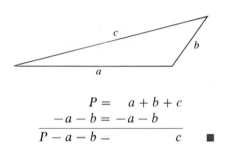

$$
\begin{array}{rl}
P = & a + b + c \\
-a - b = & -a - b \\
\hline
P - a - b - & \qquad c
\end{array}
$$ ■

Since a and b are *added* to the right side of the equation, we *subtract* a and b from both sides to remove them from the right side. This gives us a formula for the length of the side c, in terms of the perimeter and the other two sides, a and b.

☐ **DO EXERCISE 1.**

OBJECTIVE

1 *Use the addition or multiplication properties of equality or both properties to solve a formula for a specific variable.*

☐ **Exercise 1** Solve.

a. $P = a + b + c$ for b

b. $P = a + b + c$ for a

□ **Exercise 2**   Solve.

a. $d = rt$ for r

b. $I = Prt$ for t

□ **Exercise 3**   Solve.

a. $P = 2L + 2W$ for L

b. $y = 3x + 4$ for x

We may rearrange some other formulas by using the multiplication property of equality.

EXAMPLE 2   Solve for the given variable.

a. $d = rt$ for t (a distance formula)

Our objective is to get t alone on one side of the equation. Since r has been *multiplied* by t, we *divide* by r to remove it.

$$\frac{d}{r} = \frac{rt}{r}$$ Remember that dividing by r is the same as multiplying by $\frac{1}{r}$

$$\frac{d}{r} = t$$

b. $I = Prt$ for P (an interest formula)

$$\frac{I}{rt} = \frac{Prt}{rt}$$

$$\frac{I}{rt} = P$$ ■

□ **DO EXERCISE 2.**

It may also be necessary to use both the addition and the multiplication properties of equality.

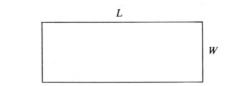

EXAMPLE 3   Solve $P = 2L + 2W$ for W.

$$
\begin{array}{rl}
P = & 2L + 2W \\
-2L = & -2L \\
\hline
P - 2L = & 2W
\end{array}
$$
Using the addition property

$$\frac{P - 2L}{2} = \frac{2W}{2}$$
Using the multiplication property since dividing by 2 is the same as multiplying by $\frac{1}{2}$

$$\frac{P - 2L}{2} = W$$ ■

□ **DO EXERCISE 3.**

If a formula contains a fraction, first multiply each side of the formula by the denominator of the fraction to remove the fraction. Then solve the problem.

EXAMPLE 4 Solve.

a. $A = \dfrac{1}{2} bh$ for b This is the formula for the area of a triangle.

$$2A = 2\left(\frac{1}{2} bh\right) \qquad \text{Multiply both sides by 2 to remove the fraction}$$

$$2A = bh$$

$$\frac{2A}{h} = \frac{bh}{h} \qquad \text{Dividing both sides by } h$$

$$\frac{2A}{h} = b$$

b. $y = \dfrac{x + z}{4}$ for x

$$4(y) = \left(\frac{x + z}{4}\right)4 \qquad \text{Multiplying by 4 to remove the fraction}$$

$$4y = x + z$$
$$\underline{-z = \quad -z}$$
$$4y - z = x \qquad \blacksquare$$

We could have worked the above example as follows.

$$y = \frac{x + z}{4}$$

$$4(y) = \left(\frac{x + z}{4}\right)4$$

$$4y = x + z$$

$$4y - z = x \qquad \text{We mentally added } -z \text{ to both sides of the equation}$$

□ **DO EXERCISE 4.**

□ **Exercise 4**   Solve.

a. $A = \dfrac{1}{2} bh$ for h

b. $A = \dfrac{a + b + c}{3}$ for c

Answers to Exercises _____

1. **a.** $b = P - a - c$ **b.** $a = P - b - c$

2. **a.** $r = \dfrac{d}{t}$ **b.** $t = \dfrac{I}{Pr}$

3. **a.** $L = \dfrac{P - 2W}{2}$ **b.** $x = \dfrac{y - 4}{3}$

4. **a.** $h = \dfrac{2A}{b}$ **b.** $c = 3A - a - b$

PROBLEM SET 3.6

A. *Solve for x.*

1. $y = x + 3$ **2.** $y = x - 4$ **3.** $y = x + z - 6$ **4.** $y = x - z + 8$

5. $-x = z + 3$ **6.** $-x = y + 7$ **7.** $y - x + 8 = 0$ **8.** $z - x + 5 = 0$

B. *Solve.*

9. $V = LWH$ for H
(volume of a rectangular solid)

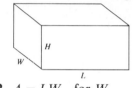

10. $V = LWH$ for L

11. $C = 2\pi r$ for r
(circumference of a circle)

12. $A = LW$ for W
(area of a rectangle)

13. $z = xy$ for y

14. $y = 3x$ for x

15. $A = 4\pi r^2$ for r^2
(surface area of a sphere)

16. $V - \pi r^2 h$ for r^2
(volume of a right circular cylinder)

17. $A = p + prt$ for t
(amount of a loan)

18. $L = a + dn - d$ for n
(arithmetic progression formula)

19. $S = 2\pi rh + 2\pi r^2$ for h
(surface area of a right circular cylinder)

20. $S = vt - 16t^2$ for v
(height above ground at time t, of an object launched with velocity v)

C. *Solve for x.*

21. $y = 6x - 2$ **22.** $y = 8x + 4$ **23.** $3x + z = 10$ **24.** $7x - y = 14$

25. $y + ax = b$ **26.** $z + bx = c$ **27.** $4y - x = 3$ **28.** $7z - x = 10$

29. $ay - bx = c$ **30.** $cy - dx = a$

D. *Solve.*

31. $A = \frac{1}{3} Bh$ for h **32.** $A = \frac{1}{3} Bh$ for B **33.** $y = \frac{1}{4} xz$ for x **34.** $x = \frac{1}{6} yz$ for z

35. $F = \frac{9}{5} C + 32$ for C **36.** $A = \frac{x + y}{2}$ for x **37.** $A = \frac{x + y + z}{3}$ for z **38.** $A = \frac{r}{2L}$ for L

39. $k = \frac{mv^2}{2g}$ for g **40.** $R = \frac{E}{I}$ for E **41.** $s = \frac{a}{1 - r}$ for a **42.** $s = \frac{a}{1 - r}$ for r

43. A relationship between certain variables may be called a _____.

44. If a formula contains a fraction, in order to solve it for a given variable we may _____ both sides of the equation by the denominator of the fraction to eliminate the denominator.

Checkup

The following problems provide a review of some of Sections 2.7 and 3.2. Some of these problems will help you with the next section. Assume that variables represent nonzero real numbers.
Write with positive exponents.

45. 3^{-2} **46.** 7^{-1} **47.** a^{-4} **48.** x^{-8}

Write with negative exponents.

49. $\frac{1}{7^3}$ **50.** $\frac{1}{6^5}$ **51.** $\frac{1}{x^4}$ **52.** $\frac{1}{b^{10}}$

Solve.

53. $x - \frac{1}{3} = \frac{1}{9}$ **54.** $y + \frac{2}{5} = \frac{3}{4}$

Inequality   a statement that two mathematical expressions are not equal; generally symbolized by $<$, $>$, $\leq$, or $\geq$. (An inequality usually has many solutions.)

Addition property of inequality   The same number may be added to both sides of an inequality. If $a < b$ then $a + c < b + c$.

3.7   THE ADDITION PROPERTY OF INEQUALITY

OBJECTIVE

1 *Solve inequalities using the addition property of inequality and graph the solutions on the number line*

Recall from Chapter 2 that $a > b$ means that a is to the right of b on the number line for any numbers a and b. We say that a is greater than b. We also stated in Chapter 2 that for any numbers a and b, $b < a$ means that b is to the left of a on the number line. In this case, we say that b is less than a.

"Greater than or equal to" is written $\geq$. "Less than or equal to" is written $\leq$.

$a > b$	means	a is greater than b
$a < b$	means	a is less than b
$a \geq b$	means	a is greater than or equal to b
$a \leq b$	means	a is less than or equal to b

An **inequality** is like an equation with the equal sign replaced by $<$, $>$, $\leq$, or $\geq$. We use the same techniques for solving inequalities as we used for solving equations except that when we multiply or divide each side of an inequality by a negative number, we must reverse the inequality sign. We explain the multiplication property of inequality in Section 3.8. *An inequality usually has many solutions.*

1 The Addition Property

We used the addition property of equality to solve some equations. There is also an addition property of inequality.

Addition Property of Inequality

The same number may be added to both sides of an inequality.

$$\text{If } a < b$$

$$\text{then } a + c < b + c$$

The property is also true for $\leq$, $>$, and $\geq$.

EXAMPLE 1   Solve and graph.

a. $x - 4 < 2$
$\underline{+ 4 = +4}$
$x < 6$

The solutions are all numbers less than 6. The solutions to inequalities in one variable may be graphed on the number line. We have graphed $x < 6$ below.

Notice that we use "○" because the equality sign is not included; x may not be equal to 6.

149

□ **Exercise 1** Solve and graph.

a. $x - 3 > 2$

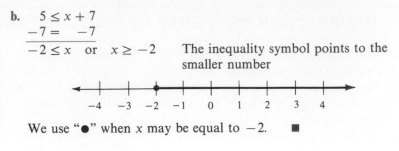

b. $x - 4 \le -6$

c. $-4 + x \ge -5$

d. $-4 > x - 7$

□ **Exercise 2** Solve and graph.

a. $4x + 2 > 3x + 3$

b. $5x - 7 \le 4x - 8$

---

b.
$$5 \le x + 7$$
$$\underline{-7 = \quad -7}$$
$$-2 \le x \quad \text{or} \quad x \ge -2 \qquad \text{The inequality symbol points to the smaller number}$$

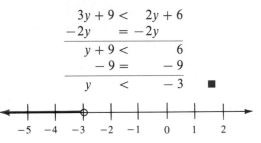

We use "●" when x may be equal to -2. ∎

□ **DO EXERCISE 1.**

We may use the addition property of inequality more than once. Notice that our objective is to get the variable alone on one side of the inequality.

EXAMPLE 2 Solve and graph.
$$3y + 9 < 2y + 6$$
$$\underline{-2y \quad = -2y}$$
$$y + 9 < \quad 6$$
$$\underline{-9 = \quad -9}$$
$$y \quad < \quad -3 \quad ∎$$

Notice that the same exercise may be worked by mentally adding a term to each side of the inequality.

$$3y + 9 < 2y + 6$$

$$3y - 2y + 9 < 6 \qquad \text{Adding } -2y \text{ to each side of the inequality}$$

$$y + 9 < 6 \qquad \text{Combining terms}$$

$$y < 6 - 9 \qquad \text{Adding } -9 \text{ to each side of the inequality}$$

$$y < -3$$

□ **DO EXERCISE 2.**

Answers to Exercises _____

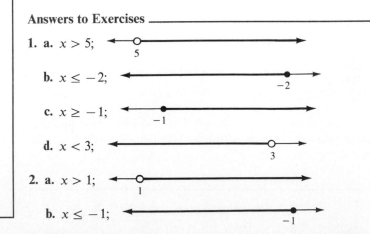

1. **a.** $x > 5$;

   b. $x \le -2$;

   c. $x \ge -1$;

   d. $x < 3$;

2. **a.** $x > 1$;

   b. $x \le -1$;

PROBLEM SET 3.7

Solve and graph.

1. $x - 1 < 4$

2. $y - 3 < 2$

3. $x + 4 \leq -1$

4. $x + 2 \leq -2$

5. $y - 6 > -12$

6. $x - 8 > -10$

7. $8 \leq y + 3$

8. $6 \leq x + 2$

9. $2y + 3 > y + 5$

10. $3x + 5 > 2x - 6$

11. $5y - 2 \leq 4y - 5$

12. $3x - 9 \leq 2x + 11$

13. $4y + 1 \geq 3y + 6$

14. $12x + 2 \geq 11x + 5$

15. $-10 < -6 + y$

16. $-11 < -8 + x$

17. $x + \dfrac{1}{2} \geq 1$

18. $y - \dfrac{1}{4} \geq 2$

19. $x + 3 < 3$

20. $y - 2 < -2$

21. $2(x + 1) \geq x + 3$

22. $3(y - 1) \leq 2y - 4$

23. $x - 2.5 \leq 0.5$

24. $y + 3.1 \geq 4.1$

25. $0.3x + 2 < -0.7x + 1$ $x < -1;$

26. $0.4y + 3 > -0.6y + 4$ $y > 1;$

27. If a is less than b, then a is to the _____ of b on the number line.

28. An inequality usually has _____ solutions.

29. The _____ number may be added to both sides of an inequality.

30. In an inequality, the inequality sign points to the _____ number.

31. In the graph of the inequality $x \leq 4$, the "•" means that _____ is included in the solutions.

32. In the graph of the inequality, $x > -2$, the "○" means that -2 is _____ _____ in the solutions.

Checkup

The following problems provide a review of some of Sections 2.8 and 3.3. Some of these problems will help you with the next section.

33. Find the amount paid back on a loan P of $3000 with a rate r of 8% interest invested for a time t of 2 years. Use the formula $A = P + Prt$.

34. Find the time t it takes a car to travel a distance d of 250 miles at a rate r of 75 miles per hour. Use the formula $t = d/r$.

Solve.

35. $\dfrac{x}{5} = 7$

36. $\dfrac{y}{-3} = 4$

37. $9y = 14$

38. $8x = -72$

Divide.

39. $\dfrac{0}{4}$

40. $\dfrac{7}{0}$

41. $\dfrac{-6}{0}$

42. $\dfrac{0}{-5}$

Multiplication property of inequality Both sides of an inequality may be multiplied by the same *positive* number. If both sides of an inequality are multiplied by the same *negative* number the inequality sign must be reversed.

3.8 THE MULTIPLICATION PROPERTY OF INEQUALITY

1 Using the Multiplication Property

Notice that
$$7 > 5 \quad \text{is a true statement}$$
$$3(7) > 3(5) \quad \text{Multiplying both sides by positive 3}$$
$$21 > 15 \quad \text{is a true inequality}$$

Consider
$$3 < 6 \quad \text{is a true statement}$$
$$-2(3) > -2(6) \quad \text{Multiplying both sides by } -2 \text{ and reversing the inequality sign}$$
$$-6 > -12 \quad \text{is a true inequality}$$

Multiplication Property of Inequality

Both sides of an inequality may be multiplied by the same *positive* number without reversing the inequality sign. If both sides of an inequality are multiplied by the same *negative* number, the inequality sign must be reversed.

EXAMPLE 1 Solve.

a. $\dfrac{x}{4} > 5$

$$4\left(\frac{x}{4}\right) > 4(5)$$

$$x > 20$$

b. $\dfrac{y}{-2} < 3$

$$-2\left(\frac{y}{-2}\right) > -2(3) \quad \text{Reversing the inequality sign}$$

$$y > -6$$

We may do a partial check. All numbers greater than -6 are solutions. Try $y = -2$.

$$\frac{y}{-2} < 3$$

$$\frac{-2}{-2} < 3$$

$$1 < 3 \quad \text{True} \quad \blacksquare$$

□ **DO EXERCISE 1.**

□ **Exercise 1** Solve.

a. $\dfrac{x}{3} > 2$

b. $\dfrac{y}{-2} \leq 4$

c. $\dfrac{x}{3} \geq -3$

d. $\dfrac{y}{-5} < -1$

□ **Exercise 2** Solve.

□ **Exercise 2** Solve.

a. $6x \geq 30$

b. $-4x < 32$

c. $3y > -11$

d. $-5y \leq -25$

Remember that dividing both sides of an inequality by a number is the same as multiplying both sides by the reciprocal of the number. Hence we may divide both sides of an inequality by the same positive number. However, if we *divide both sides of the inequality* by the same *negative number*, we must *reverse the inequality sign* in order to get an equivalent inequality.

EXAMPLE 2 Solve.

a. $3x > 15$

$$\frac{3x}{3} > \frac{15}{3}$$

$$x > 5$$

b. $-7x < 28$

$$\frac{-7x}{-7} > \frac{28}{-7} \qquad \text{Dividing by } -7 \text{ and reversing the inequality sign}$$

$$x > -4 \qquad ■$$

□ **DO EXERCISE 2.**

2 **Using Both the Addition and Multiplication Properties**

To solve some inequalities, we must use both the addition and the multiplication properties. It is easier to use the addition property first, combine terms, and then use the multiplication property to get the variable alone on one side of the inequality.

EXAMPLE 3 Solve.

a. $3x - 7 < 5$

$$3x < 5 + 7 \qquad \text{Adding 7 to both sides of the inequality}$$

$$3x < 12$$

$$x < \frac{12}{3}$$

$$x < 4$$

b. $5y + 7 > y - 9$

   $5y - y + 7 > -9$ Adding $-y$ to both sides of the inequality

   $4y + 7 > -9$ Combining terms

   $4y > -9 - 7$ Adding -7 to both sides of the inequality

   $4y > -16$

   $y > -\dfrac{16}{4}$

   $y > -4$

c. $2x - 8 < 4x + 3$

   $2x - 4x - 8 < 3$

   $-2x - 8 < 3$

   $-2x < 3 + 8$

   $-2x < 11$

   $x > \dfrac{11}{-2}$ Dividing by -2 and reversing the inequality sign

or $x > \dfrac{-11}{2}$ Changing two signs on the fraction   ■

□ **Exercise 3**   Solve.

a. $6 - 3y < -9$

b. $5x + 9 > 4x + 7$

c. $12 + 5y \le 8y + 16$

d. $8x + 20 \ge 12x + 16$

Steps in Solving a Linear Inequality in One Variable

1. Use the addition property of inequality to get the variables on one side of the inequality and all other terms on the other side.
2. Use the multiplication property of inequality to get the variable alone on one side. This solves the inequality.

□ **DO EXERCISE 3.**

Answers to Exercises

1. a. $x > 6$ **b.** $y \geq -8$ **c.** $x \geq -9$ **d.** $y > 5$

2. a. $x \geq 5$ **b.** $x > -8$ **c.** $y > -\dfrac{11}{3}$ **d.** $y \geq 5$

3. a. $y > 5$ **b.** $x > -2$ **c.** $y \geq -\dfrac{4}{3}$ **d.** $x \leq 1$

PROBLEM SET 3.8

Solve.

1. $\dfrac{x}{4} < 7$

2. $\dfrac{x}{5} > 5$

3. $\dfrac{y}{-3} > 2$

4. $\dfrac{y}{-8} < 1$

5. $\dfrac{x}{3} \le -7$

6. $\dfrac{x}{2} \le -5$

7. $\dfrac{x}{-3} \ge -5$

8. $\dfrac{x}{-4} \ge -1$

9. $3x < 12$

10. $5x < 10$

11. $-7x \ge 14$

12. $-6x \ge 24$

13. $-4x \le -12$

14. $-3x \le -27$

15. $4x > 32$

16. $5x > 50$

17. $-\dfrac{1}{3}y \ge 3$

18. $-\dfrac{1}{2}y \ge 2$

19. $-3y \le -10$

20. $-6x \le -9$

21. $3x + 4 > 7$

22. $2x - 8 < 4$

23. $-x + 2 \le 4$

24. $-x - 3 < 7$

25. $4y - 2 < 3y + 4$

26. $9y + 6 > 8y + 2$

27. $8y - 4 < 2y + 8$

28. $7y + 3 > 2y - 2$ **29.** $-3x - 4 > -7 + x$ **30.** $-7x + 3 > -8 + 2x$

31. $4y + 4 < 6y + 4$ **32.** $4y - 2 < 3y - 2$

33. Both sides of an inequality may be multiplied by the same _____ number without reversing the inequality sign.

34. If both sides of an inequality are multiplied by the same negative number, we must reverse the _____ _____.

35. If both the addition and multiplication properties are used to solve an inequality, it is usually easier to use the _____ property first.

Checkup

The following problems provide a review of some of Section 3.1. Which of the following are linear equations?

36. $x - 3 = 4$ **37.** $y^2 - 2 = 8$

38. $x^2 = 7$ **39.** $y + 5 = 7$

40. $\dfrac{1}{2}x + 2 = \dfrac{3}{4}$ **41.** $\dfrac{1}{4}x^2 - 3 = 5$

Are the values of the variables solutions for the given equations?

42. $\dfrac{2}{5} + x = \dfrac{7}{15}, \quad x = \dfrac{1}{15}$ **43.** $y - 5 = -\dfrac{11}{3}, \quad y = \dfrac{2}{3}$

CHAPTER 3   ADDITIONAL EXERCISES (OPTIONAL)

Section 3.1

Determine whether the equations are true, false, or conditional.

1. $87 - 54 = 23$

2. $126 - 97 = 29$

3. $4x - 19 = 48$

4. $5y + 33 = 79$

5. $75(x - 2) = 75x - 150$

6. $83x + 249 = 83(x + 3)$

Are the following equations linear equations?

7. $\dfrac{1}{3}x + 4 = 7$

8. $\dfrac{1}{5}x^2 - x = 7$

9. $5x^2 - \dfrac{7}{8}x + 3 = 0$

10. $\dfrac{y}{2} - 3 = 8y$

Section 3.2

Solve.

11. $x - 6 = 17$

12. $x + 7 = 9$

13. $7 + y = 6$

14. $8 + x = 5$

15. $x - \dfrac{7}{8} = \dfrac{3}{5}$

16. $y - \dfrac{3}{4} = \dfrac{5}{12}$

17. $x + 0.4 = 3$

18. $y + 0.8 = 5$

Section 3.3

Solve.

19. $\dfrac{3}{5}x = 10$

20. $-\dfrac{3}{7}x = -2$

21. $-8y = 72$

22. $7x = -49$

23. $-0.8y = 4$

24. $0.7x = 49$

Divide, if possible.

25. $\dfrac{0}{4.3}$

26. $\dfrac{9.8}{0}$

27. $\dfrac{0.5}{0}$

28. $\dfrac{0}{0.7}$

Section 3.4

Solve.

29. $4x + 7 = 31$

30. $3y - 6 = 6$

31. $8y + 9 = 9$

32. $7y + 3 = 12$

33. $\frac{3}{5}x + 2 = 14$ **34.** $\frac{8}{7}y - 3 = -11$ **35.** $\frac{3}{2}y - 4 = \frac{1}{6}$ **36.** $\frac{3}{4}x + 2 = \frac{1}{12}$

Section 3.5

Solve.

37. $2(3y - 1) = -7$ **38.** $4(2y + 3) = 5$ **39.** $4 - (x + 3) = 8$ **40.** $6 - (3y - 2) = 2$

41. $\frac{2}{3}\left(x - \frac{1}{4}\right) = \frac{3}{4}x - \frac{1}{4}$ **42.** $\frac{1}{5}(x + 2) = \frac{1}{2}\left(x - \frac{1}{5}\right)$

43. $\frac{1}{4}(8y + 4) - 17 = -\frac{1}{2}(4y - 8)$ **44.** $\frac{2}{3}(6x + 24) - 40 = -\frac{1}{2}(12x - 72)$

Section 3.6

Solve.

45. $V = xyz$ for x **46.** $P = \dfrac{a + b + c}{3}$ for a **47.** $A = \dfrac{h}{2}(b + c)$ for b

48. $A = \dfrac{h}{2}(b + c)$ for c **49.** $x + 3y = 4$ for y **50.** $4x + 5y = 10$ for y

Section 3.7

Solve and graph.

51. $3x - 2 < 2x - 4$ **52.** $5x + 7 > 4x + 8$

53. $\frac{3}{5}x + 2 \geq -\frac{7}{5}x$ **54.** $3 + \frac{9}{4}x < \frac{5}{4}x$

Section 3.8

Solve and graph.

55. $-9x < 27$ **56.** $16x \geq -64$

57. $x - \frac{1}{2} \leq \frac{1}{4}x + 1$ **58.** $x - \frac{1}{2} \geq \frac{3}{4}x + 1$

CHAPTER 3   PRACTICE TEST

1. Which of the following is a linear equation: (a) $8 - x^3 = 0$, (b) $y + 3 = 7$, or (c) $x^2 + 2 = 4$?

1. _____

2. Which of the following is a solution of $9 - x = 11$: (a) $x = 2$, (b) $x = -2$, or (c) $x = 4$?

2. _____

Solve.

3. $y - 8 = -4$

3. _____

4. $7.1 + x = 12.2$

4. _____

5. $-2x = -8$

5. _____

6. $-\dfrac{3y}{7} = \dfrac{7}{5}$

6. _____

Divide.

7. $\dfrac{0}{9}$

7. _____

8. $\dfrac{-3}{0}$

8. _____

Solve.

9. _____

9. $-3x + 4 = 5x + 6$

10. _____

10. $-0.2y + 7 = 1.8y - 1$

11. _____

11. $5(2y - 7) = 5$

12. _____

12. $2(2x - 3) = 4(2x + 2)$

13. _____

13. $5x + y = 4$ for x

14. _____

14. $R = \dfrac{E}{I}$ for I

Solve and graph.

15. _____

15. $x + 5 \geq 2$ ⟵————————————⟶

16. _____

16. $4x + 4 > 3x + 1$ ⟵————————————⟶

Solve.

17. _____

17. $-5y \leq -10$

18. _____

18. $6y + 3 > 2y - 4$

4 Applied Problems

4.1 INTRODUCTION TO APPLIED PROBLEMS

A major application of algebra is to the solution of applied problems in the working world. Many word problems are done in a similar way and we illustrate some of the different types of problems in this chapter. We must also learn to translate phrases into symbols, and some examples are shown below. Notice that we already know how to translate the arithmetic statements. The algebraic phrases are similar except that instead of a specific number 5, we are using an unknown number x.

■ Translating Phrases into Symbols

EXAMPLE 1 Translate the phrases into symbols.

Addition

Arithmetic Phrase	Symbols	Algebraic Phrase	Symbols
Five plus three	$5 + 3$	A number plus three	$x + 3$
Three more than five	$5 + 3$	Three more than a number	$x + 3$
Five increased by three	$5 + 3$	A number increased by three	$x + 3$
The sum of five and three	$5 + 3$	The sum of a number and three	$x + 3$

■

□ DO EXERCISE 1.

OBJECTIVES

1 *Translate phrases into symbols*

2 *Translate word problems to equations*

□ **Exercise 1** Translate the phrases into symbols.

a. A number plus one

b. Two more than a number

c. A number increased by six

d. The sum of a number and seven

□ **Exercise 2** Translate the phrases into symbols.

a. A number decreased by eight

b. The difference between a number minus four

c. Subtract six from a number

d. Subtract ten from a number

□ **Exercise 3** Write symbols for the phrases.

a. Seven times a number

b. The product of a number and fifteen

c. Three-fourths of a number

d. Double a number

□ **Exercise 4** Write symbols for the phrase "a number divided by eight."

EXAMPLE 2 Translate the phrases into symbols.

Subtraction

Arithmetic Phrase	Symbols	Algebraic Phrase	Symbols
Five decreased by two	$5 - 2$	A number decreased by two	$x - 2$
Five minus two	$5 - 2$	A number minus two	$x - 2$
Two less than five	$5 - 2$	Two less than a number	$x - 2$
Subtract two from five	$5 - 2$	Subtract two from a number	$x - 2$

Notice that when we subtract two from five, the five is written first. Similarly, when a number is subtracted from x, the x comes first. ■

□ **DO EXERCISE 2.**

EXAMPLE 3 Write symbols for the phrases.

Multiplication

Arithmetic Phrase	Symbols	Algebraic Phrase	Symbols
Two times five	$2(5)$	Two times a number	$2x$
The product of two and five	$2(5)$	The product of two and a number	$2x$
Double five	$2(5)$	Double a number	$2x$
One-fourth of five	$\frac{1}{4}(5)$	One-fourth of a number	$\frac{1}{4}x$

Notice that in the phrase "one-fourth of five," "of" indicates multiplication. ■

□ **DO EXERCISE 3.**

EXAMPLE 4 Write symbols for the phrase.

Division

Arithmetic Phrase	Symbol	Algebraic Phrase	Symbol
Five divided by six	$\frac{5}{6}$	A number divided by six	$\frac{x}{6}$

■

□ **DO EXERCISE 4.**

2 Translating Word Problems to Equations

> The word "is" and the term "the result is" translate to equals, =.

EXAMPLE 5 Translate to an equation. Do not solve.

a. A number minus 2 is 8.
$$x - 2 = 8$$

b. A number divided by 12 is 3.
$$\frac{x}{12} = 3 \quad \blacksquare$$

☐ **DO EXERCISE 5.**

Often, one of the operations is combined with another in a word problem.

EXAMPLE 6 Translate to an equation.

a. A number divided by 6, increased by 8, is 12.
$$\frac{x}{6} + 8 = 12$$

b. Three times a number, decreased by 2, is the same as (equals) the product of 4 and the number.
$$3x - 2 = 4x \quad \blacksquare$$

☐ **DO EXERCISE 6.**

We must use parentheses when necessary.

EXAMPLE 7 Translate to an equation.

a. Three times a number, minus 2, is 9.
$$3x - 2 = 9$$

We multiply 3 times a number and then subtract 2. This equals 9.

b. Three times the result of subtracting 2 from a number is 9.
$$3(x - 2) = 9$$

We show the result of subtracting 2 from a number and multiply this quantity by 3. This equals 9. $\quad \blacksquare$

☐ **DO EXERCISE 7.**

☐ **Exercise 5** Translate to an equation. Do not solve.

a. A number plus 3 is 11.

b. A number divided by 8 is 21.

☐ **Exercise 6** Translate to an equation.

a. Double a number and decrease it by 3. The result is 7.

b. A number divided by 2 is the same as the number minus 3.

☐ **Exercise 7** Translate to an equation.

a. Four times a number, minus 20, is the product of the number and 3.

b. Four times the result of subtracting a number from 20 is the product of the number and 3.

c. Five, plus 6 times a number, equals 8.

d. If the sum of 5 and a number is multiplied by 6, the result is 8.

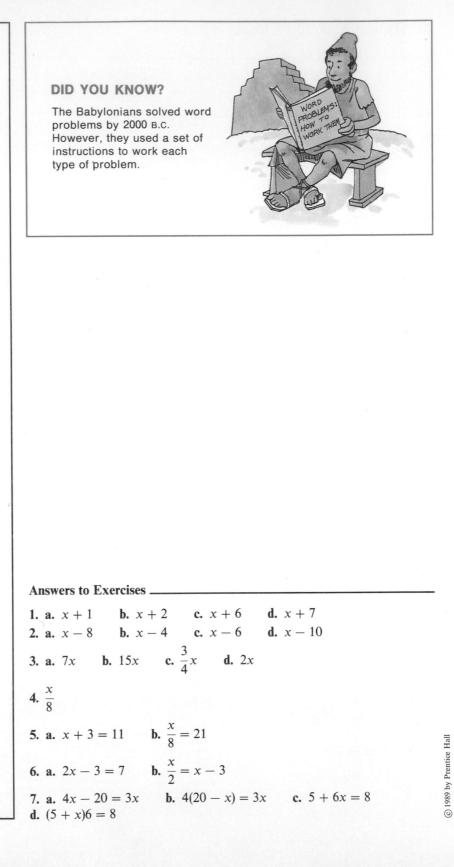

DID YOU KNOW?

The Babylonians solved word problems by 2000 B.C. However, they used a set of instructions to work each type of problem.

Answers to Exercises

1. a. $x + 1$ **b.** $x + 2$ **c.** $x + 6$ **d.** $x + 7$

2. a. $x - 8$ **b.** $x - 4$ **c.** $x - 6$ **d.** $x - 10$

3. a. $7x$ **b.** $15x$ **c.** $\frac{3}{4}x$ **d.** $2x$

4. $\frac{x}{8}$

5. a. $x + 3 = 11$ **b.** $\frac{x}{8} = 21$

6. a. $2x - 3 = 7$ **b.** $\frac{x}{2} = x - 3$

7. a. $4x - 20 = 3x$ **b.** $4(20 - x) = 3x$ **c.** $5 + 6x = 8$
d. $(5 + x)6 = 8$

PROBLEM SET 4.1

A. *Write symbols for the phrases.*

1. A number plus 6

2. A number plus 8

3. The sum of a number and 4

4. The sum of a number and -2

5. A number increased by 20

6. A number increased by 50

7. Four more than a number

8. Ten more than a number

9. A number less -2

10. A number less 14

11. A number decreased by 7

12. A number decreased by 16

13. A number subtracted from 15

14. A number subtracted from 25

15. The product of a number and 4

16. The product of a number and 7

17. $\frac{3}{7}$ of a number

18. $\frac{7}{8}$ of a number

19. A number divided by 9

20. A number divided by 15

21. A number divided by 10

22. A number divided by 4

B. *Translate to equations. Do not solve.*

23. The sum of a number and 5 is 24.

24. One-half of a number is -8.

25. Thirty-six divided by a number is -3.

26. Twice a number, minus 3, is -14.

27. A number plus 4 equals the number divided by 6.

28. Two more than a number is the same as 3 times the number, minus 8.

29. Five minus twice a number is the same as 4 times the number, plus 7.

30. If we subtract 4 from a number and then divide by 3, the result will be 7.

31. Three minus twice a number is equal to 5 times the number, plus 6.

32. Take 7 minus twice a number and divide the result by 4. The answer is 6.

33. Double the result of subtracting a number from 3. The result is 5 times the number, plus 6.

34. Take 2 times the result of subtracting a number from 7 and divide the result by 4. The answer is 6.

35. The expression $x - 2$ may be read "subtract _____ from _____ _____."

36. "Two times a number" may also be read "the _____ of 2 and a number."

37. The word "is" translates to _____.

Checkup

The following problems provide a review of some of Section 3.5. These problems will help you with the next section. Solve.

38. $3(x - 4) = 6$

39. $7(x + 2) = -14$

40. $2(4x + 3) = -10$

41. $5(2x - 1) = 25$

42. $3(2x - 1) + 2x = -11$

43. $2z - (4z + 5) = 1$

44. $6y - (2y - 3) = -13$

45. $3x - 2(4x + 3) = -11$

4.2 SOLVING APPLIED PROBLEMS

1

There are several steps that are helpful in solving word problems.

1. Read the problem carefully, several times.
2. Make a drawing, if applicable.
3. Find the questions to be answered and write variables for the unknowns.
4. Separate sentences into phrases to write an equation. Sometimes we may need to remember a formula that is not given.
5. Solve the equation using the techniques of Chapter 3.
6. Be sure that you have answered the questions and check your answer in the original word problem.

□ Exercise 1 Four times a number, decreased by 2, is 18. What is the number?

EXAMPLE 1 Twice a number, plus 3, is 7. What is the number?

Variable Let $x =$ the number

Equation $2x + 3 = 7$

Solve $\quad\quad 2x = 4$

$\quad\quad\quad\quad x = 2$

The number is 2.

Check Twice 2 plus 3 is 7. The answer checks. ■

□ DO EXERCISE 1.

□ Exercise 2 If we double a number and add 7, the result is the number divided by 4. What is the number?

EXAMPLE 2 Five times the result of subtracting 4 from a number is the same as the product of the number and 3, increased by 10. What is the number?

Variable Let $x =$ the number

Equation $5(x - 4) = 3x + 10$

Solve $\quad 5x - 20 = 3x + 10$

$\quad\quad 2x - 20 = 10$

$\quad\quad\quad\quad 2x = 30$

$\quad\quad\quad\quad\quad x = 15$

The number is 15.

Check The difference between 15 and 4 is 11. Five times 11 is 55.

This is the same as the product of 15 and 3 increased by 10, which is also 55. ■

□ DO EXERCISE 2.

□ **Exercise 3**   A man has 14 cookies to divide between himself, his wife, and 3 children. If he and his wife each get a double share, how many does each child receive?

EXAMPLE 3   Jim, Carlos, and Kim buy 16 candy bars. If Carlos buys twice as many as Jim, and Kim buys the same number as Jim, how many does Kim buy?

Variable   Let x = the number that Jim buys

Then $2x$ = the number that Carlos buys

and x = the number that Kim buys

Equation $x + 2x + x = 16$

Solve $4x = 16$

$x = 4$

Kim buys 4 candy bars.

Check Jim also buys 4 candy bars since he buys the same number as Kim.

Carlos buys twice as many as Jim, or $2(4) = 8$. The total number of candy bars that the boys buy is $4 + 4 + 8$, or 16. The answer checks. ■

□ **DO EXERCISE 3.**

EXAMPLE 4   John has 124 feet of fencing. He wishes to build a rectangular fence with the length 12 feet greater than the width. What should be the dimensions of the fence?

Drawing

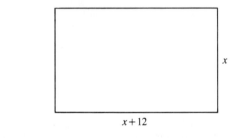

Variable   Let x = the width of the fence

Then $x + 12$ = the length of the fence

Equation Use the formula for the perimeter of a rectangle:

$$P = 2L + 2W$$

$$124 = 2(x + 12) + 2x$$

Consecutive integers integers which differ by one

Solve
$$124 = 2x + 24 + 2x$$
$$124 = 4x + 24$$
$$100 = 4x$$
$$25 = x$$
$$x + 12 = 37$$

The dimensions of the fence are 25 feet by 37 feet.

Check The perimeter of the fence is $2L + 2W$ or $2(37) + 2(25)$, which is 124, so the answers check. ■

□ **DO EXERCISE 4.**

If an integer is 1 greater than another integer, the integers are consecutive. For example, 3 and 4 are **consecutive integers**.

EXAMPLE 5 The sum of two consecutive integers is 47. What are the integers?

Variable   Let x = the first consecutive integer

Then $x + 1$ = the second consecutive integer

Equation $x + (x + 1) = 47$
$$x + x + 1 = 47$$
$$2x + 1 = 47$$
$$2x = 46$$
$$x = 23$$
$$x + 1 = 23 + 1 = 24$$

The integers are 23 and 24.

Check The sum of 23 and 24 is 47. The answer checks in the original problem. ■

□ **DO EXERCISE 5.**

□ **Exercise 4** The length of a rectangle is twice the width. The perimeter is 36 meters. Find the dimensions.

□ **Exercise 5** The sum of two consecutive integers is -23. Find the integers.

Answers to Exercises _____

1. $4x - 2 = 18$

$\qquad 4x = 20$

$\qquad\quad x = 5$

The number is 5.

2. $\qquad 2x + 7 = \dfrac{x}{4}$

$\quad 4(2x + 7) = \left(\dfrac{x}{4}\right)4$

$\qquad 8x + 28 = x$

$\qquad\qquad 7x = -28$

$\qquad\qquad\ x = -4$

The number is -4.

3. $2x + 2x + x + x + x = 14$

$\qquad\qquad\qquad\quad 7x = 14$

$\qquad\qquad\qquad\qquad x = 2$

Each child receives 2 cookies.

4. $\qquad P = 2L + 2W$

$\qquad 36 = 2(2W) + 2W$

$\quad -6W = -36$

$\qquad\ W = 6$

$\qquad 2W = 12$

The dimensions are 6 meters by 12 meters.

5. $x + x + 1 = -23$

$\qquad 2x + 1 = -23$

$\qquad\quad 2x = -24$

$\qquad\qquad x = -12$

$\qquad x + 1 = -11$

The integers are -12 and -11.

PROBLEM SET 4.2

1. If 8 is subtracted from 4 times a certain number, the result is 32. Find the number.

2. Six increased by 7 times a number is 27. What is the number?

3. Five times a number, plus 4, is the same as 3 times the number, minus 6. Find the number.

4. Twice a number, plus 5, is the same as 3 times the number, plus 7. What is the number?

5. A number divided by 2, plus $\frac{3}{4}$, is double the number. What is the number?

6. A number divided by 5 is the same as the number decreased by $\frac{2}{3}$. What is the number?

7. Three times a number, plus 5, is the same as 4 times the result of subtracting 4 from the number. What is the number?

8. A number minus 6 is the same as doubling the result of subtracting 5 from the number. What is the number?

9. In a triangle, one angle is twice as large as the smallest angle and the third angle is 20° larger than the smallest angle. The sum of the angles of a triangle is 180°. Find all three angles.

10. Find the number of degrees in each angle of a triangle if the second angle is 3 times the smallest and the third angle is twice the difference between the second angle and the smallest angle.

11. Karen, Joyce, and Linda buy 24 hot dogs. If Karen buys 4 fewer than Joyce and Linda buys twice as many as Karen, how many does each person buy?

12. Mrs. Worthington gave her children $60,000. If Jill received twice as much as Harry but Vicky only received one-half as much as Jill, how much did each receive?

13. The length of rectangle is 3 times the width. If the perimeter is 32 inches, find the length and width.

14. The length of a rectangle is 8 centimeters more than the width. The perimeter is 30 centimeters. Find the dimensions of the rectangle.

15. The width of a rectangle is one-third the length and the perimeter is 4 times the length minus 12 feet. Find the dimensions of the rectangle.

16. The perimeter of a triangle is 26 inches. If the longest side is twice the length of the shortest side and the third side is 4 inches less than twice the shortest side, find the length of each side of the triangle.

17. The sum of three consecutive integers is 108. What are the integers?

18. The sum of three consecutive integers is 84. Find the integers.

19. Two numbers are consecutive even integers if they are even and one number is 2 larger than the other. Find two consecutive even integers whose sum is 54.

20. The sum of two consecutive even integers is 114. Find the integers.

21. Two numbers are consecutive odd integers if they are odd and one number is 2 larger than the other. Find two consecutive odd integers whose sum is 112.

22. Find two consecutive odd integers whose sum is 64.

23. The first step in solving a word problem is to _____ the problem carefully, several times.

24. The last step in solving a word problem is to _____ your answer in the original word problem.

25. To solve a word problem, sometimes we use a _____ that is not given.

26. The formula for the _____ of a rectangle is $P = 2L + 2W$.

27. If an integer is 1 greater than another integer, the integers are _____.

Checkup

The following problems provide a review of some of Section 3.6. Solve.

28. $P = 4s$ for s

29. $V = k\pi r^3$ for r^3

30. $L = a + dn - d$ for a

31. $A = p + prt$ for r

32. $S = 2\pi rh$ for h

33. $V = \pi r^2 h$ for h

34. $y = 3x + 4$ for x

35. $2x - y = 3$ for y

Interest formula Interest = (Principal) · (rate) · (time) where Principal is the amount invested, rate is the percent interest per year and time is in years.

4.3 INTEREST PROBLEMS

1 We will discuss simple interest. In the present day, interest is frequently compounded and that formula is more complicated.

To solve simple interest problems, we must learn the following **interest formula**:

> , interest = principal · rate · time
>
> where the principal is the amount invested (in dollars), the rate is the percent interest, and the time is in years.

EXAMPLE 1 Lisa invested $800 at 8% for 2 years. How much interest did she earn?

Variable Let I = interest earned in 2 years

Equation Interest = principal · rate · time

$$I = Prt \quad \text{(in symbols)}$$

$$I = 800(0.08)(2)$$

Solve $\qquad I = \$128$

Lisa earned $128 interest.

Check The interest for 1 year is 0.08(800), or $64. The interest for 2 years is 2(64), or $128. ∎

☐ **DO EXERCISE 1.**

EXAMPLE 2 If $120 was received as interest for 1 year on a sum of money invested at 8% per year, how much was invested?

Variable Let P = amount invested

Equation $\qquad I = Prt$

$$120 = P(0.08)(1)$$

Solve $\qquad \dfrac{120}{0.08} = P$

$$\$1500 = P$$

The amount invested was $1500.

Check $1500 times 8% interest times 1 year is $120. ∎

☐ **DO EXERCISE 2.**

1 *Solve word problems about interest*

☐ **Exercise 1** How much interest did John earn on $500 invested at 9% for 4 years?

☐ **Exercise 2** An investment at 9% annual interest returned $576 in 4 years. How much was invested?

□ **Exercise 3** If $900 is borrowed at an interest rate of 9.5% per year, how much money must be paid back after 4 years?

EXAMPLE 3 James borrowed $700 at an interest rate of 10% per year for 2 years. He paid back the $700 and the amount of interest. What amount did he pay back?

$$\text{Total amount paid back} = \text{principal} + \text{interest}$$

Variable Let I = interest for 2 years

Equation $I = Prt$

$$I = 700(0.10)(2)$$

Solve $I = \$140$

$$\text{Total amount} = \text{principal} + \text{interest}$$
$$= \$700 + \$140$$
$$\text{Total amount} = \$840$$

James paid back $840.

Check The interest is $700 times 10% times 2 years, which is $140. This added to $700 gives $840. ■

□ **DO EXERCISE 3.**

EXAMPLE 4 If Jean adds $1000 more to her investment, her annual interest will be $340. The money is invested at 8.5%. How much was Jean's original investment?

Variable Let P = the amount of Jean's original investment

Then $P + 1000$ = the new amount invested

Equation $I = Prt$

$$340 = (P + 1000)(0.085)(1)$$

Solve $340 = P(0.085) + 1000(0.085)$

$$340 = 0.085P + 85$$

$$255 = 0.085P$$

$$\$3000 = P$$

□ **Exercise 4** Laura withdrew some money to buy stock from her account, which originally contained $6000. The account now yields $180 annually at 9% interest. How much did she withdraw?

Jean's original investment was $3000.

Check If Jean adds $1000 to the original investment of $3000, she has $4000 invested. $4000(0.085)(1) is $340. ■

□ **DO EXERCISE 4.**

EXAMPLE 5 The total annual interest on two sums of money is $700. How much money is invested at each rate if three times as much money is invested at 9% per year as is invested at 8% per year?

Variable Let P = the amount invested at 8%

Then $3P$ = the amount invested at 9%

Principal · rate · time =   interest

P	$\cdot\, 0.08 \cdot$	1	$= P(0.08)(1)$	Use these facts in the equation
$3P$	$\cdot\, 0.09 \cdot$	1	$= 3P(0.09)(1)$	

Equation Interest at 8% + interest at 9% = total interest

$$P(0.08)(1) + 3P(0.09)(1) = 700$$

Solve
$$0.08P + 0.27P = 700$$

$$8P + 27P = 70{,}000 \quad \text{Multiplying each term by 100}$$

$$35P = 70{,}000$$

$$P = 2000$$

$$3P = 6000$$

$2000 is invested at 8% and $6000 is invested at 9%.

Check The interest for 1 year at 8% is 0.08(2000) = $160. The interest for 1 year at 9% is 0.09(6000) = $540. $160 + $540 = $700. ■

□ **DO EXERCISE 5.**

EXAMPLE 6   If Jim withdraws $500 from his account, the new amount will give the same amount of interest each year at 10% as his original investment does at 9%. How much does he have invested at 9%?

Variable   Let P = amount invested at 9%

Then $P - 500$ = amount invested at 10%

Principal	· rate ·	time =	interest	Use these facts in
P	$\cdot\, 0.09 \cdot$	1 =	$P(0.09)(1)$	the equation
$(P - 500) \cdot 0.10 \cdot$		1 =	$(P - 500)(0.10)(1)$	

Interest at 9% = interest at 10%

$$P(0.09)(1) = (P - 500)(0.10)(1)$$

Solve
$$0.09P = (P - 500)(0.10)$$

$$9P = (P - 500)(10) \quad \text{Multiplying by 100}$$

$$9P = 10P - 5000$$

$$-P = -5000$$

$$P = 5000$$

Jim has $5000 invested at 9%.

Check Multiplying 5000(0.09)(1) gives $450 interest. This is the same as $(5000 - 500)(0.10)(1) = 4500(0.10) = 450. ■

□ **DO EXERCISE 6.**

□ **Exercise 5**   If a sum of money is invested at 9% per year and $4000 more than that amount is invested at 10% per year, how much is invested at each rate if the annual total interest is $780?

□ **Exercise 6**   If Julie adds $3000 to her account, the new amount will give the same amount of interest at 7% as the original amount does at 10%. How much does she have invested at 10%?

Answers to Exercises

1. $I = Prt$
$I = 500(0.09)(4)$
$I = 180$
John earned $180.

2. $I = Prt$
$576 = P(0.09)(4)$
$576 = 0.36P$
$1600 = P$
$1600 was invested.

3. $I = Prt$
$I = 900(0.095)(4)$
$I = \$342$
Total $= 900 + 342$
Total $= 1242$
After 4 years, $1242 must be paid back.

4. $I = Prt$
$180 = (6000 - x)(0.09)(1)$
$180 = 540 - 0.09x$
$-360 = -0.09x$
$-36,000 = -9x$
$4000 = x$
Laura withdrew $4000.

5. $Prt + Prt = \text{interest}$
$P(0.09)(1) + (P + 4000)(0.10)(1) = 780$
$0.09P + 0.10P + 400 = 780$
$0.19P = 380$
$19P = 38,000$
$P = 2000$
$P + 4000 = 6000$
$2000 is invested at 9%, $6000 is invested at 10%.

6. $Prt = Prt$
$(P + 3000)(0.07)(1) = P(0.10)(1)$
$0.07P + 210 = 0.10P$
$210 = 0.03P$
$21,000 = 3P$
$7000 = P$
Julie has $7000 invested at 10%.

PROBLEM SET 4.3

1. What is the interest earned on $2000 invested at 9% for 2 years?

2. If $3000 is invested for 5 years at 8.5% per year, what is the interest earned?

3. Kevin invested some money for 3 years at 8% interest. He received an interest payment of $1200. How much did he invest?

4. Debbie received $150 as interest for 1 year on money she invested at 7.5% per year. How much did she invest?

5. How much money must be returned after 5 years if $800 is borrowed at an interest rate of 11% per year?

6. Steve borrowed $1200 at an interest rate of 10% per year for 3 years. He paid back the $1200 and the interest. What amount did he pay back?

7. Joan withdrew some money to pay tuition from her account, which originally contained $1000. The account now yields $63 annually at 7% interest. How much did she withdraw?

8. If Ken adds $1500 to his investment, his annual interest will be $280. The money is invested at 8% annual interest. How much was Ken's original investment?

9. How much money is invested at each rate if the total annual interest on two sums of money is $460? Twice as much money is invested at 8% per year as is invested at 7% per year.

10. If a sum of money is invested at 7% per year and 3 times as much is invested at 8.5% per year, how much is invested at each rate if the total interest is $162.50?

11. If $400 is withdrawn from an account, the new amount will give the same interest each year at 9% as the original amount does at 7%. How much is invested at 7%?

12. If Susan adds $900 to her account, the new total will give the same annual interest at 8% as the original amount does at 11%. How much does she have invested at 11%?

13. Michael invested $6000, some money at 8% and the remainder at 9% per year. How much did he invest at each rate if his yearly interest was $525?

14. A total of $8000 is invested, some at 7% and the remainder at 10% per year. How much is invested at each rate if the total yearly interest is $725?

15. The formula for simple interest is Interest = _____ × rate × time.

16. The amount of a loan that must be paid back is the principal plus _____.

Checkup

The following problems provide a review of some of Section 3.7.
Solve and graph.

17. $4x + 3 > 2x - 1$

18. $7x - 8 < 3x + 4$

19. $5x - 1 \leq 3x - 5$

20. $6x - 1 \geq 2x - 5$

21. $6x + 2 < 7x + 4$

22. $3x + 4 > 4x - 1$

23. $x - 2.3 \leq 1.7$

24. $x + 3.8 \geq -1.2$

4.4 COIN AND MIXTURE PROBLEMS

OBJECTIVES

1 Solve coin problems

2 Solve mixture problems

1 Coin Problems

When we find the total value of a set of coins, we are working a "coin problem."

EXAMPLE 1

a. Carlos has 6 quarters and 9 nickels. What is the total value of the quarters and nickels that he has?

Equation $\left(\begin{array}{c}\text{Value of}\\\text{quarters}\end{array}\right) + \left(\begin{array}{c}\text{value of}\\\text{nickels}\end{array}\right) = \text{total value}$

$$0.25(6) \quad + \quad 0.05(9) \quad = \text{total value}$$

Solve $\qquad 1.50 \quad + \quad 0.45 \quad = \text{total value}$

$$1.95 \quad = \text{total value}$$

Carlos has $1.95.

We use the same word equation to solve problems where the number of items is unknown.

b. Kristin has $2.10 in dimes and nickels. She has a total of 25 coins. How many of each coin does she have?

$$25 = \text{total number of coins}$$

Variable $\qquad$ Let $x =$ number of dimes

$\qquad$ Then $25 - x =$ number of nickels

$$\left(\begin{array}{c}\text{Number}\\\text{of coins}\end{array}\right) \cdot \left(\begin{array}{c}\text{value of}\\\text{each}\end{array}\right) = \text{total value}$$

Dimes $\qquad x \quad \cdot \quad 0.10 \quad = 0.10x$

Nickels $\quad (25 - x) \quad \cdot \quad 0.05 \quad = 0.05(25 - x)$

Equation $\left(\begin{array}{c}\text{Value of}\\\text{dimes}\end{array}\right) + \left(\begin{array}{c}\text{value of}\\\text{nickels}\end{array}\right) = \left(\begin{array}{c}\text{total value}\\\text{of all the coins}\end{array}\right)$

$$0.10x \quad + 0.05(25 - x) = 2.10$$

Solve $\qquad 10x \quad + \quad 5(25 - x) \quad = 210 \qquad \text{Multiplying by 100}$

$$10x \quad + \quad 125 - 5x \quad = 210$$

$$5x + 125 \quad = 210$$

$$5x = \quad 85$$

$$x = \quad 17$$

$$25 - x = \quad 8$$

She has 17 dimes and 8 nickels.

Check Eight coins plus 17 coins is 25 coins. $0.10(17) + 0.05(8) = 1.70 + 0.40 = 2.10$. ∎

□ **Exercise 1** A bank contains nickels and quarters. If the value of the coins is $1.50 and there are 18 coins in the bank, how many are there of each kind?

□ **DO EXERCISE 1.**

2 Mixture Problems

The techniques used in solving mixture problems may be used in laboratories, where various strengths of solutions are required. To work mixture problems, we need to understand the meaning of percent of a quantity.

EXAMPLE 2   Heidi has 50 cookies. Sally ate 10% of them. How many did Sally eat?

Variable Let x = the number of cookies Sally ate

Equation $x = 10\%(50)$

Solve $x = 0.10(50)$

$x = 5$

Sally ate 5 cookies.

Check Ten percent of 50 cookies is 5 cookies.   ■

□ **DO EXERCISE 2.**

EXAMPLE 3   Five hundred liters of a solution is 35% pure alcohol. How many liters of pure alcohol are in the solution?

Variable Let x = liters of pure alcohol in the solution

Drawing

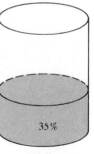

35%

500 liters

Equation $x = 35\%(500)$

Solve $x = 0.35(500)$

$x = 175$

The solution contains 175 liters of pure alcohol.

Check Thirty-five percent of 500 liters of solution is 175 liters of pure alcohol.   ■

□ **DO EXERCISE 3.**

EXAMPLE 4 *x* liters of solution is 18% alcohol. How much pure alcohol is in the solution?

Variable Let *y* = liters of pure alcohol in the solution

Drawing

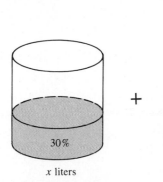 +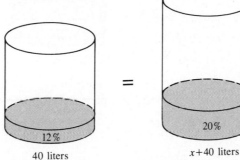

x liters   40 liters   *x*+40 liters

Equation $y = 18\%(x)$

Solve $y = 0.18x$

The solution contains $0.18x$ liters of pure alcohol. ∎

☐ **DO EXERCISE 4.**

Now we apply the foregoing techniques to the solution of mixture problems.

EXAMPLE 5 Wendy wants to make a 20% acid solution by mixing 40 liters of a 12% solution with a 30% solution. How many liters of the 30% solution should she use?

Variable   Let *x* = liters of 30% solution

           40 = liters of 12% solution

     Then *x* + 40 = total liters of 20% solution

$$\left(\begin{array}{c}\text{Percent of acid}\\\text{in solution}\end{array}\right) \cdot \left(\begin{array}{c}\text{liters of}\\\text{solution}\end{array}\right) = \left(\begin{array}{c}\text{liters of}\\\text{pure acid}\end{array}\right)$$

    30   ·   *x*   = 0.30(*x*)   Use these facts in the equation.

    12   ·   40   = 0.12(40)

    20   · (*x* + 40) = 0.20(*x* + 40)

The liters of pure acid in the solutions that are mixed must equal the liters of pure acid in the final solution.

☐ **Exercise 4** Solve.

a. *x* gallons of solution is 45% pure acid. How much pure acid is in the solution?

b. How many grams of pure salt is in *x* grams of 30% salt solution?

□ **Exercise 5** How many liters of 5% acid solution must be added to 30 liters of a 10% solution to make an 8% solution?

Drawing

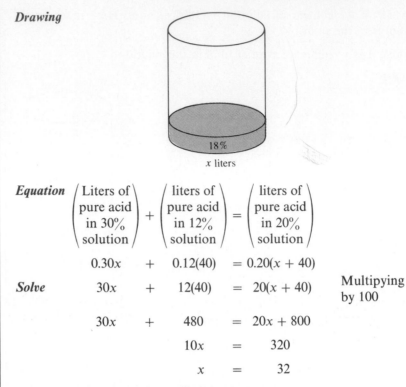

18%

x liters

Equation

$$\begin{pmatrix} \text{Liters of} \\ \text{pure acid} \\ \text{in 30\%} \\ \text{solution} \end{pmatrix} + \begin{pmatrix} \text{liters of} \\ \text{pure acid} \\ \text{in 12\%} \\ \text{solution} \end{pmatrix} = \begin{pmatrix} \text{liters of} \\ \text{pure acid} \\ \text{in 20\%} \\ \text{solution} \end{pmatrix}$$

$$0.30x + 0.12(40) = 0.20(x + 40)$$

Solve

$$30x + 12(40) = 20(x + 40) \qquad \text{Multipying by 100}$$

$$30x + 480 = 20x + 800$$

$$10x = 320$$

$$x = 32$$

Wendy should use 32 liters of 30% solution.

Check The number of liters of final 20% solution is $40 + x = 40 + 32 = 72$. The liters of pure acid in the 40 liters of 12% solution is $0.12(40) = 4.8$. The liters of pure acid in the 32 liters of 30% solution is $0.30(32) = 9.6$. The total liters of pure acid mixed together is $4.8 + 9.6 = 14.4$. This is the same as the liters of pure acid in the final 20% solution, which is $0.20(72) = 14.4$. ■

□ **DO EXERCISE 5.**

Answers to Exercises _____

1. $0.05x + 0.25(18 - x) = 1.50$
$5x + 25(18 - x) = 150$
$5x + 450 - 25x = 150$
$-20x + 450 = 150$
$-20x = -300$
$x = 15$
$18 - x = 3$
There are 15 nickels and 3 quarters in the bank.

2. 5

3. 30 gal

4. a. $0.45x$ **b.** $0.30x$

5. $0.05x + 0.10(30) = 0.08(30 + x)$
$5x + 10(30) = 8(30 + x)$
$5x + 300 = 240 + 8x$
$-3x = -60$
$x = 20$
20 liters of 5% acid solution must be added.

PROBLEM SET 4.4

1. A collection of 21 dimes and nickels is worth $1.40. How many of each type of coin are in the collection?

2. Juan has 14 coins with a total value of $1.90. The coins are nickels and quarters. How many of each coin does he have?

3. Susan has 30 dimes and quarters. If the value of the coins is $4.20, how many of each coin does she have?

4. A collection of 25 nickels and dimes is worth $1.60. How many of each coin are in the collection?

5. Karen has 3 times as many nickels as dimes. If the total value of the coins is $2.75, how many of each kind does she have?

6. A collection of dimes and quarters is worth $10.50. If there are 7 more quarters than dimes in the collection, how many of each are there?

7. Barbara had 30 granola bars. Her brother ate 15% of them. How many did he eat?

8. Joan completed 50 homework questions. Eighty percent of them were correct. How many were correct?

9. Find 45% of x.

10. What is 68% of x?

11. How many liters of a 40% solution of acid must Charlene add to 50 liters of a 15% solution to produce a 20% solution?

12. How many gallons of a 10% antifreeze solution must be added to 60 gallons of a 20% solution to produce a 16% solution?

13. Diane wanted to use 30 gallons of a 55% acid solution and a 90% acid solution to make a 65% acid solution. How many gallons of the 90% solution should be used?

14. Paul added 70% alcohol to 35 liters of a 30% alcohol solution to make a 45% solution. How much 70% alcohol solution did he add?

15. A 35% silver alloy is to be melted with a 65% silver alloy. How many kilograms of each should be used to get 20 kilograms of a 40% silver alloy?

16. A 35% copper alloy is to be melted with a 45% copper alloy. How many kilograms of each should be used to get 50 kilograms of a 40% copper alloy?

17. The value of 7 quarters is _____ .

18. Eight nickels have a value of _____ .

Checkup

The following problems provide a review of some of Section 3.8.
Solve.

19. $-x + 4 < 6$

20. $-y - 2 \geq 3$

21. $2x - 5 > 3$

22. $3x + 2 \leq -7$

23. $3x + 2 \leq 5x - 7$

24. $2x - 5 < 4x + 2$

25. $-2x + 3 \geq 6x + 1$

26. $3x + 2 > 5x - 3$

Distance formula distance = rate · time

4.5 DISTANCE PROBLEMS

OBJECTIVE

▪ 1 ▪ Solve distance problems

■1 If we travel at a constant rate of 50 miles per hour for 3 hours, we know that we have traveled 150 miles. The **distance formula** that we are using is

$$\text{rate} \cdot \text{time} = \text{distance}$$

This formula is also used to solve more complicated distance problems.

EXAMPLE 1 Maria drove from her home to Chicago at a rate of 55 miles per hour. If the trip took 4 hours, how far is it from Maria's home to Chicago?

Equation Rate · time = distance

Solve   55 ·   4 = distance

                         220 = distance

It is 220 miles from Maria's home to Chicago. ■

□ **DO EXERCISE 1.**

□ **Exercise 1** Tom hiked for 4 hours. If he walked at a rate of $3\frac{1}{2}$ miles per hour, how far did he hike?

Opposite Direction

EXAMPLE 2 Sally is traveling on her bicycle at 24 miles per hour. She met and passed Patricia, who is riding her bicycle at 20 miles per hour, in the opposite direction. In how many hours after they pass will they be 22 miles apart?

Variable Let $t =$ the time in hours when they will be 22 miles apart

	$\begin{pmatrix} \text{Rate in} \\ \text{miles per} \\ \text{hour} \end{pmatrix}$ ·	$\begin{pmatrix} \text{time in} \\ \text{hours} \end{pmatrix} =$	$\begin{pmatrix} \text{distance} \\ \text{in miles} \end{pmatrix}$
Sally	24 ·	t =	$24t$
Patricia	20 ·	t =	$20t$

Notice that Sally travels at 24 miles per hour for t hours, so she travels for a distance of $24t$ miles. Similarly, Patricia travels at 20 miles per hour for t hours, so she travels $20t$ miles. These distances must add up to 22 miles.

Drawing

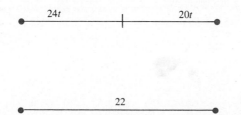

□ **Exercise 2** Jim and Joe start from camp and walk in opposite directions. If Jim walks 4 miles per hour and Joe walks 3 miles per hour, in how many hours will they be 18 miles apart?

Equation $24t + 20t = 22$

Solve $44t = 22$

$$t = \frac{22}{44} = \frac{1}{2}$$

They will be 22 miles apart in $\frac{1}{2}$ hour.

Check Sally and Patricia are 22 miles apart after $\frac{1}{2}$ hour. In $\frac{1}{2}$ hour Sally traveled $24(\frac{1}{2}) = 12$ miles and Patricia traveled $20(\frac{1}{2}) = 10$ miles. The sum of their distances is $12 + 10 = 22$, which is the required miles apart. ∎

□ **DO EXERCISE 2.**

Same Direction

EXAMPLE 3 David forgot his lunch. He is riding a school bus traveling 45 miles per hour. When the bus has traveled 4 miles from David's house, his father leaves home in his car to overtake the bus. He is traveling at 55 miles per hour. How long will it take David's father to overtake the bus?

Variable Let t = time in hours that it takes David's father to overtake the bus

$$\begin{pmatrix} \text{Rate in} \\ \text{miles per} \\ \text{hour} \end{pmatrix} \cdot \begin{pmatrix} \text{time in} \\ \text{hours} \end{pmatrix} = \begin{pmatrix} \text{distance} \\ \text{in miles} \end{pmatrix}$$

	Rate		time		distance
Bus	45	·	t	=	$45t$
Car	55	·	t	=	$55t$

Drawing

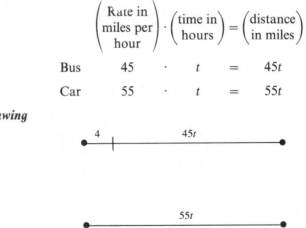

At the time that David's father leaves home, the bus has already traveled 4 miles and it will travel an additional distance of $45t$ miles before the car overtakes it. The car travels a distance of $55t$ miles. When the car overtakes the bus, they have traveled the same distance.

Equation $4 + 45t = 55t$

$$4 = 10t$$

$$\frac{4}{10} = t$$

$$\frac{2}{5} = t$$

It takes David's father $\frac{2}{5}$ of an hour to overtake the bus.

Check The bus traveled $4 + 45(\frac{2}{5}) = 4 + 18 = 22$ miles. David's father traveled $55(\frac{2}{5}) = 22$ miles, so the answer checks. ■

□ DO EXERCISE 3.

Round Trip

EXAMPLE 4 Kathy jogged to the store and then walked home. The round trip took 1 hour. If she jogged 5 miles per hour and walked 2 miles per hour, how far is it from her home to the store?

Variable   1 hour = total time

          Let t = time in hours that Kathy jogged

          Then $1 - t$ = time in hours that she walked

$$\left(\begin{array}{c}\text{Rate in}\\\text{miles per}\\\text{hour}\end{array}\right) \cdot \left(\begin{array}{c}\text{time in}\\\text{hours}\end{array}\right) = \left(\begin{array}{c}\text{distance}\\\text{in miles}\end{array}\right)$$

	Rate		time		distance
Jog	5	·	t	=	$5t$
Walk	2	·	$(1 - t)$	=	$2(1 - t)$

Drawing

$$\underset{5t}{\bullet\rule{4cm}{0.4pt}\bullet}$$

$$\underset{2(1-t)}{\bullet\rule{4cm}{0.4pt}\bullet}$$

The distances from her home to the store and from the store to her home are equal.

Equation $5t = 2(1 - t)$

          $5t = 2 - 2t$

          $7t = 2$

          $t = \dfrac{2}{7}$

The distance that she jogged, which is the distance from her home to the store, is $5t = 5(\frac{2}{7}) = \frac{10}{7} = 1\frac{3}{7}$ miles.

Check The distance that Kathy jogged is $1\frac{3}{7}$ miles. The distance she walked is $2(1 - t) = 2(1 - \frac{2}{7}) = 2(\frac{5}{7}) = 1\frac{3}{7}$ miles. Since the distances are the same, the answer checks. ■

□ DO EXERCISE 4.

□ **Exercise 3** After a sailboat has traveled 10 miles from Johnson Island, a motorboat leaves the island on the same route. If the sailboat travels 8 miles per hour and the motorboat travels 12 miles per hour, how long does it take the motorboat to overtake the sailboat?

□ **Exercise 4** Crystal Mountain ski lift carries a skier to the top of the lift at a rate of 120 feet per minute. She skis to the bottom on a path parallel to the lift at a rate of 280 feet per minute. How long is the lift if the round trip takes 30 minutes?

Answers to Exercises _____

1. 14 miles

2. $4t + 3t = 18$

   $7t = 18$

   $t = \dfrac{18}{7} = 2\dfrac{4}{7}$

   They will be 18 miles apart in $2\dfrac{4}{7}$ hours.

3. $10 + 8t = 12t$

   $-4t = -10$

   $t = \dfrac{10}{4} = 2\dfrac{1}{2}$

   The motorboat overtakes the sailboat in $2\dfrac{1}{2}$ hours.

4. $120t = 280(30 - t)$

   $120t = 8400 - 280t$

   $400t = 8400$

   $t = 21$

   distance $= 120t = 120(21) = 2520.$

   The lift is 2520 feet long.

PROBLEM SET 4.5

1. An Alpha Airlines plane flying due north at 420 miles per hour met and passed a Brit Air plane flying due south at 510 miles per hour. How many hours after passing each other will they be 2790 miles apart?

2. Bob and Bill leave the ice cream store at the same time going in opposite directions. If Bob walks 4 miles per hour and Bill walks 3 miles per hour, in how many hours will they be 10 miles apart?

3. Two boats left the dock at the same time. One boat traveled downstream at the rate of 10 miles per hour. The other boat went upstream at a rate of 8 miles per hour. How far apart were the boats after $\frac{1}{2}$ hour?

4. Peiheng and Karen left campus traveling in opposite directions. If Peiheng drove 35 miles per hour and Karen drove 30 miles per hour, how far apart were they after $1\frac{1}{2}$ hours?

5. Two cars leave Miami at the same time traveling in the same direction. If one car averages 48 miles per hour and the other car travels 54 miles per hour, in how many hours will they be 42 miles apart?

6. A passenger train is following a freight train on a parallel track. The passenger train is traveling at 115 kilometers per hour and the freight train is traveling at 75 kilometers per hour. If the passenger train is 100 kilometers behind the freight train, how long will it take the passenger train to overtake the freight train?

7. Jean leaves the starting point of a bicycle race traveling at a steady rate of 25 miles per hour. Twenty minutes (one-third of an hour) later Debra leaves the starting point going at 30 miles per hour. How long does it take Debra to catch up with Jean?

8. Mr. Carstairs leaves Phoenix traveling at an average rate of 40 miles per hour. Fifteen minutes (one-fourth of an hour) later his wife leaves Phoenix traveling 52 miles per hour. How long before Mrs. Carstairs passes her husband?

9. A pilot in a jet plane on a training mission leaves base flying due north at a speed of 900 kilometers per hour. He turns the plane and flies due south at a speed of 800 kilometers per hour. If the jet can remain in the air 5 hours, for how long was it flying north?

10. Karen walked to school at a rate of 3 miles per hour. She jogged home at 8 miles per hour. If the trip took $\frac{2}{3}$ of an hour, how far is it from her home to school?

11. A sailboat leaves its moorage and travels to Jones Island at a rate of 10 miles per hour. The return trip was made at a speed of 12 miles per hour. If the trip took $4\frac{2}{5}$ hours, how much time did it take to get to Jones Island?

12. The Jensen family went on a vacation to Yosemite National Park. Their camper averaged 46 miles per hour going to the park and 54 miles per hour returning home. How much time did they spend returning home if the entire trip took 10 hours?

13. The distance formula is distance = rate · _____ .

Checkup

The following problems provide a review of some of Section 2.6. These problems will help you with the next section. Identify the like terms.

14. $7x + 4 - 2x$

15. $3y - 2 + 5y$

16. $6 - 2a + 3$

17. $-8 + 5b + 4$

Combine like terms.

18. $9a + 6 - 15a$

19. $-3x - 4 + 7x$

20. $-3 + 4x - 9$

21. $8 - 7x + 9$

22. $7x + 3y - (5x + 8y)$

23. $-5a - 6b - (3a - 9b)$

24. $-\{2[x - 4(x - 3)]\}$

25. $\{-5 + [y - 2(3y - 4)]\}$

CHAPTER 4 ADDITIONAL EXERCISES (OPTIONAL)

Section 4.1

Write symbols for the phrases.

1. Five more than a number

2. A number increased by 5

3. The sum of a number and 8

4. The product of a number and 7

5. $\frac{3}{5}$ of a number

6. Five times a number

7. A number decreased by 4

8. A number less -5

9. A number subtracted from 8

10. A number divided by 4

11. $\frac{2}{3}$ of a number

12. A number subtracted from 4

Section 4.2

13. If 6 is added to the product of a number and 4, the result is twice the number, plus 3. Find the number.

14. A number minus 4 is the same as doubling the result of subtracting 8 from the number. What is the number?

15. The width of a rectangle is 4 less than the length. If the perimeter is 168 inches, find the length and the width.

16. The perimeter of a rectangle is 144 centimeters. If the length is 3 times the width, find the dimensions of the rectangle.

17. In a triangle, one angle is 3 times as large as the smallest angle and the third angle is 30° larger than the smallest angle. The sum of the angles is 180°. Find all three angles.

18. The sum of the angles of a triangle is 180°. If one angle is 15° larger than the smallest and the third angle is twice the sum of the other two angles, find the number of degrees in each angle.

19. The sum of three consecutive integers is 186. Find the integers.

20. Find two consecutive even integers whose sum is 94.

21. The sum of three consecutive odd integers is 105. Find the integers.

Section 4.3

22. Juan received $475 as interest for 1 year on money he invested at 9.5% per year. How much did he invest?

23. Linda invested some money for 4 years at 8% interest. She received interest of $1216. How much did she invest?

24. If $15,000 is borrowed at an interest rate of 10.5% per year, how much money must be paid back after 2 years?

25. If $15,000 is borrowed at an interest rate of 5.5% per year, how much money must be paid back after 2 years?

26. If a sum of money is invested at 6% per year and 3 times as much is invested at 9% per year, how much is invested at each rate if the total interest is $495?

27. Melissa invested $7500 for 1 year. Some of the money was invested at 8% interest and the remainder was invested at 6.5% interest. If Melissa earned $550.50 on her investments, how much did she invest at each rate?

28. Kim invested $5400, some money at 7.5% and the remainder at 6% per year. How much did he invest at each rate if his yearly interest was $363.75?

29. If $850 is added to an account, the new amount will give the same annual interest at 6% as the original amount did at 8.5%. How much is invested at 6%?

30. If $600 is withdrawn from an account, the new amount will give the same interest per year at 8% as the original amount did at 5.5%. How much is invested at 8%?

Section 4.4

31. Knute has 32 dimes and nickels worth $2.80. How many dimes does he have?

32. A collection of 72 dimes, nickels, and quarters is worth $13.20. If there are 3 times as many quarters as nickels, how many of each coin are in the

33. Maria has 57 coins with a total value of $7.05. If she has 6 more dimes than nickels and the rest of the coins are quarters, how many of each coin does she have?

34. How many quarts of a 30% salt solution must Sandra add to 40 quarts of a 10% solution to make a 25% solution?

35. Jeff added 60% alcohol to 5 gallons of a 40% alcohol solution to make a 45% solution. How much 60% alcohol solution did he add?

36. How many liters of 100% acid must be added to 60 liters of a 35% solution to make a 55% solution?

37. How many ounces of pure alcohol should be added to 10 ounces of a 20% alcohol solution to make a 45% solution?

38. Water is 0% salt solution. How much water should be added to a 70% salt solution to make 40 gallons of a 25% solution?

39. Water is 0% antifreeze solution. How much water should be added to a 65% antifreeze solution to make 12 gallons of a 50% solution?

Section 4.5

40. Two trains leave the station traveling in opposite directions. If one train traveled 50 miles per hour and the other train traveled 90 miles per hour, in how many hours will they be 700 miles apart?

41. Linda and Cheryl leave work and walk in opposite directions. If Linda walks 2.8 miles per hour and Cheryl walks 3.2 miles per hour, in how many hours will they be 4.5 miles apart?

42. Two cars leave Dallas at the same time traveling in opposite directions. If one car averages 58 miles per hour and the other car travels 64 miles per hour, in how many hours will they be 793 miles apart?

43. Two bicycles leave Chicago at the same time traveling in the same direction. If one bicycle goes 16 miles per hour and the other bicycle averages 24 miles per hour, in how many hours will they be 100 miles apart?

44. Gary leaves the starting point of a race 15 minutes before Jeff. If Jeff runs at 10 miles per hour and Gary runs at 8.4 miles per hour, how long does it take Jeff to catch up with Gary?

45. A plane leaves Chicago flying due east at 480 miles per hour. Forty minutes later another plane leaves Chicago flying due east at 530 miles per hour. How long before the second plane overtakes the first?

46. Cindy jogged to school at a rate of 7 miles per hour. She walked home at 2 miles per hour. If the entire trip took 2 hours, how long did it take her to get to school?

47. Cascade Pass ski lift carries a skier to the top of the lift at a rate of 140 feet per minute. He skis to the bottom on a path parallel to the lift at a rate of 260 feet per minute. How long is the lift if the round trip takes 20 minutes?

48. A sailboat left its moorage and traveled to San Juan Island at a rate of 8 miles per hour. The return trip was made at a speed of 10 miles per hour. If the entire trip took 5 hours and 45 minutes, how much time did it take to return from San Juan Island?

CHAPTER 4   PRACTICE TEST

Write symbols for the phrases.

1. A number subtracted from 8

1. _____

2. A number divided by 4, minus 7 times the number

2. _____

3. Four times a number, increased by 2, is the same as 3 times the result of subtracting 5 from a number. What is the number?

3. _____

4. A landscaper has 96 feet of fencing. He wishes to build a fence with length 8 feet greater than the width. What should be the dimensions of the fence?

4. _____

5. Joan borrowed $900 at an interest rate of 11% per year. How much must she pay back after 4 years?

5. _____

6. _____

7. _____

8. _____

9. _____

10. _____

6. Mrs. Westcott invested $6000, part at 8% and the remainder at 9%. How much did she invest at 8% if her yearly interest was $520?

7. How many liters of a 35% acid solution must be added to 50 liters of a 20% acid solution to make a 25% solution?

8. Anita has 22 dimes and quarters. If the value of the coins is $4.75, how many of each coin does she have?

9. A plane leaves Kennedy Airport traveling at 320 miles per hour. One hour later a plane leaves the airport traveling the same route at 400 miles per hour. How long does it take the second plane to overtake the first?

10. The Logan family visited relatives in St. Louis. They averaged 35 miles per hour going to St. Louis and 55 miles per hour returning. If the entire trip took 8 hours, how long did it take to get to St. Louis?

CUMULATIVE REVIEW   CHAPTERS 3 AND 4

1. Which of the following equations are linear?
   (a) $x^{-1} = 4$
   (b) $3x + 4 = 9$
   (c) $\dfrac{3}{4} y - 2 = \dfrac{5}{8} y$
   (d) $3x^3 + x^2 = 8$

2. Which of the following is a solution of $x + \frac{3}{8} = \frac{7}{5}$?
   (a) $\dfrac{41}{40}$
   (b) $\dfrac{71}{40}$
   (c) $-\dfrac{41}{40}$

Divide, if possible.

3. $\dfrac{0}{-8}$

4. $\dfrac{15}{0}$

Solve.

5. $7x - 4 = 9x + 12$

6. $\dfrac{2}{5} y + 3 = \dfrac{1}{3} y + \dfrac{7}{15}$

7. $5x - 3(2x - 5) = 4$

8. $4(0.3x + 1.8) = 3(1.4x - 1.84)$

Solve for the given variable.

9. $5x + 2y = 10$ for y

10. $3x - 4y = 7$ for y

11. $V = \dfrac{4}{3} \pi r^3$ for r^3

12. $S = \pi h(r + R)$ for R

Solve and graph.

13. $3x + 2 \geq x - 6$ $x \geq -4;$

14. $5y - 3 > 9y + 5$ $y < -2;$

15. The product of a number and 3 is the same as 16 plus the number. What is the number?

16. Find the number of degrees in each angle of a triangle if the second angle is 20° more than the first and the third angle is the sum of the first and second. The sum of the angles of a triangle is 180°.

17. The width of a rectangle is one-third the length plus 2 and the perimeter is 3 times the length plus 1. Find the dimensions of the rectangle.

18. The sum of three consecutive odd integers is 87. Find the integers.

19. Juan bought a car. He borrowed $12,000 at 12% interest for 4 years to pay for the car. How much interest did he pay?

20. If a sum of money was invested at 8% per year and one-half as much was invested at 6% per year, how much was invested at each rate if the total interest earned was $385?

21. Cheryl has one-fourth as many dimes as quarters. If the total value of the coins is $16.50, how many of each kind does she have?

22. How many liters of an 80% alcohol solution should be added to a 45% solution to get 30 liters of a 55% solution?

23. Ted and George start from home and walk in opposite directions. If Ted walks 2 miles per hour and Joe walks 3 miles per hour, in how many hours will they be 12 miles apart?

24. A pilot in a jet plane leaves base flying due east at a speed of 750 kilometers per hour. She turns the plane and flies due west at a speed of 850 kilometers per hour. If the jet remains in the air 6 hours, for how long was it flying west?

25. Karen Johnson leaves Dallas traveling at an average rate of 54 miles per hour. One-half hour later her husband leaves Dallas traveling at 65 miles per hour. How long does it take Mr. Johnson to catch up with his wife?

Polynomials

5.1 INTRODUCTION TO POLYNOMIALS

Operations with polynomials are very important in applications of mathematics. We begin our study of polynomials by reviewing terms. Recall that terms in an expression are separated by plus or minus signs.

1 Polynomials

> A **polynomial** in a variable is an expression whose terms contain only whole-number powers of the variable. (Remember that the whole numbers include zero.)

EXAMPLE 1   Are the following polynomials?

$x^4 + x^3 - 2x + 3$	Yes, recall that $3 = 3x^0$
$4x^2 - x^{-3} + 9$	No, the exponent -3 is not a whole number
$y + y^4$	Yes, because the exponents are whole numbers
$\dfrac{y}{4} + 7$	Yes, since $y/4 = 1/4y$
$\dfrac{3}{x} - 4$	No, $3/x = 3x^{-1}$; the exponent is not a whole number
5	Yes; $5 = 5x^0$
$4x^5 - 3x^2$	Yes ■

□ **DO EXERCISE 1.**

OBJECTIVES

1 *Identify polynomials*

2 *Identify trinomials, binomials, and monomials*

3 *Combine like terms*

4 *Find the degree of a term*

5 *Find the degree of a polynomial*

6 *Arrange polynomials in descending order*

□ **Exercise 1**   Which of the following are polynomials?

a. $3x^3 - x + 2$ **b.** $x^2 + 4$

c. $4x^{-1} + 8$ **d.** $\dfrac{3}{y} - 9$

e. $\dfrac{y}{4} - y^2$ **f.** 7

Trinomial   a polynomial with exactly 3 terms.
Binomial   a polynomial with exactly 2 terms.
Monomial   a polynomial with exactly 1 term.

□ **Exercise 2**   Are the following polynomials: trinomials, binomials, monomials—or none of these?

a. $x^3 - 1$

b. $x^4 - x + 3x^2 + 8$

c. 5

d. $x - 3 + x^3$

□ **Exercise 3**   Find the like terms and combine them.

a. $5x + 3x^2 - 7x$

b. $x^3 - 6x^2 + 5x^2 + 7x^3$

c. $6x + 5x^4 - 6x$

d. $x^2y^2 - x^3y^3 - x^2y^2$

2 **Special Polynomials**

Three types of polynomials are used more than others. **Trinomials** have exactly three terms, **binomials** have exactly two terms, and **monomials** have exactly one term.

EXAMPLE 2

Trinomials	*Binomials*	*Monomials*
$3x^2 + x - 2$	$x - 3$	$2x$
$x^5 + 4x^2 + x$	$y^2 + 8$	7
$x + 9 - x^3$	$7 - y^8$	$-5y^3$ ■

□ **DO EXERCISE 2.**

3 **Combining Like Terms**

Recall from Chapter 2 that like terms are terms that have exactly the same variable or variables and the variable(s) of each term must have the same exponents. Only like terms may be combined.

EXAMPLE 3   Find the like terms and combine them.

a. $3x^2 - 4x + 5x^2 = 3x^2 + 5x^2 - 4x$ Rearranging terms
$$= 8x^2 - 4x$$

b. $-6x^4 - 3x^2 - 4x^2 + 5x^4$
$$= -6x^4 + 5x^4 - 3x^2 - 4x^2$$ Rearranging terms
$$= -6x^4 + 5x^4 - 3x^2 + (-4x^2)$$
$$= -x^4 - 7x^2$$

Notice that $-1x^4 = -x^4$.

c. $4x^3 + 3y - 4x^3 = 4x^3 - 4x^3 + 3y = 0x^3 + 3y = 3y$

d. $x^2y + 2xy^2 + 3x^2y = x^2y + 3x^2y + 2xy^2 = 4x^2y + 2xy^2$ ■

□ **DO EXERCISE 3.**

Degree of a polynomial in one variable   the largest exponent on the variable of any of the terms.

Descending order   arranging the terms of a polynomial with the term with the largest exponent on the variable first, the term with the next largest exponent on the variable second and so on from left to right.

Degree of a term containing one variable   the exponent on the variable.

4 Degree of a Term

> The **degree of a term containing one variable** is the exponent on the variable.

EXAMPLE 4   Find the degree of each term of $4x^3 + 2x^2 - 3x + 8$.

Term	Degree	
$4x^3$	3	
$2x^2$	2	
$-3x$	1	Recall that $x = x^1$
8	0	Recall that $8 = 8x^0$ ∎

□ **DO EXERCISE 4.**

5 Degree of a Polynomial

> The **degree of a polynomial in one variable** is the largest exponent on the variable in any of the terms.

EXAMPLE 5   Find the degree of the polynomial $7x^6 - 2x^3 + 4x - 9$. The degree of the polynomial is 6 since 6 is the largest exponent on the variable in any of the terms.   ∎

□ **DO EXERCISE 5.**

6 Descending Order

When we work with polynomials we usually arrange them in **descending order**. We write the term with the largest exponent on the variable first, the term with the next-largest exponent on the variable second, and so on.

EXAMPLE 6   Arrange the polynomials in descending order.

a. $6 + x^4 - x + x^2 = x^4 + x^2 - x + 6$

b. $x^3 - 4x + 3x^5 - 7x^9 = -7x^9 + 3x^5 + x^3 - 4x$ ∎

□ **DO EXERCISE 6.**

□ **Exercise 4**   Find the degree of each term of the polynomial.

a. $5x^4 - 4x^2 - 8$

b. $y^3 + y - 1$

□ **Exercise 5**   Find the degree of the polynomial.

a. $x^4 - 2x + 6$ **b.** $y + 3$

□ **Exercise 6**   Arrange the polynomials in descending order.

a. $-3x + 8 - 6x^2$

b. $4 + x - 3x^3 + 8x^7$

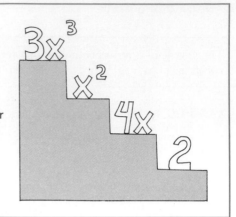
Answers to Exercises

1. (a), (b), (e), (f)
2. **a.** Binomial   **b.** None of these   **c.** Monomial   **d.** Trinomial
3. **a.** $-2x + 3x^2$ **b.** $8x^3 - x^2$ **c.** $5x^4$ **d.** $-x^3y^3$
4. **a.** 4, 2, and 0, respectively   **b.** 3, 1, and 0, respectively
5. **a.** 4   **b.** 1
6. **a.** $-6x^2 - 3x + 8$ **b.** $8x^7 - 3x^3 + x + 4$

PROBLEM SET 5.1

A. *Are the following polynomials?*

1. $x^4 - x^2 + 2$

2. $x^3 - x + 4$

3. $\dfrac{y}{8} - 9$

4. $\dfrac{x}{9} + 10$

5. $4x^{-2} + x$

6. $6x^4 + x^{-1}$

7. $\dfrac{5}{x} + 3$

8. $9 + \dfrac{6}{x^2}$

B. *Are the following polynomials: trinomials, binomials, monomials—or none of these?*

9. $x + 6$

10. $x^2 + 2x - 6$

11. $x^2 - x + 3$

12. $y - 4$

13. x

14. $y^7 - y^2 + y + 2$

15. $x^4 - x^3 + x^2 + 1$

16. y

17. 5

18. 9

C. *Find the like terms and combine them.*

19. $-6x + 4 + 3x$

20. $-9y - 7 - 8y$

21. $-4x^2 + 3x - 6x^2$

22. $4x^4 - x - 4x^4$

23. $3x^3 + x - 3x^3$

24. $-8y^2 + 7y + 6y^2$

25. $-x^2y + xy^2 - 3x^2y$

26. $yz^2 + 3yz^2 - y^2z$

27. $\dfrac{1}{2}x^2 - x + \dfrac{3}{4}x^2 + 2x$

28. $\dfrac{1}{5}x^3 + 4 - \dfrac{3}{10}x^3 + 2x - 8$

29. $0.3x + x^3 - 4 - 1.8x - 3x^3$

30. $x^4 - 1.9x^3 + 5 - 3x^4 - 8.7x^3$

D. *Find the degree of each term and the degree of the polynomial.*

31. $x^2 + x - 2$

32. $x^3 - 2x^2 + 4$

33. $x + 5$

34. $x^5 - x^4 + x$

E. *Arrange the polynomials in descending order.*

35. $7 + x - x^2$

36. $6 + x^3 + x^2$

37. $2x - 3x^5 + x^2 + 9x^7$

38. $4x^3 + 6x^5 - x + 3x^2$ **39.** $6x^7 - 9 + 4x^2 + 8x$ **40.** $9x^4 - 3x^2 + 5x + 7$

41. The height of an object thrown straight up in the air with a velocity of 32 feet per second, after t seconds, is given by the polynomial

$$h = -16t^2 + 32t$$

Find the height of the object after 2 seconds.

$$= -16(2) + 32(2)$$
$$-64 + 64$$
$$= 0.$$

42. The surface area of a sphere is given by the polynomial $4\pi r^2$. Find the surface area of a sphere with radius 3 centimeters. Use 3.14 for π.

43. A polynomial in a variable is an expression whose terms contain only _____ number powers of the variable.

44. A polynomial with exactly three terms is a _____.

45. The degree of a term containing one variable is the _____ on the variable.

46. The degree of a polynomial in one variable is the largest exponent on the variable in any of the _____.

Checkup

The following problems provide a review of some of Section 4.1.
Write symbols for the phrases.

47. A number minus 4

48. The product of a number and 8

49. A number subtracted from 6

50. One-half of a number

Translate to equations.

51. The product of 5 and a number, minus 6, is 4.

52. A number divided by 6, less 4, is 7.

53. Double the result of subtracting a number from 4. The answer is 9.

54. The product of 2 and the result of adding 5 to a number is 3.

5.2 ADDITION AND SUBTRACTION OF POLYNOMIALS

Addition and subtraction of polynomials is similar to addition and subtraction of signed numbers except that we must be careful to combine only *like* terms.

1 **Addition of Polynomials**

EXAMPLE 1 Add.

a. $6x^2 - 9x - 7$ and $-4x^2 + 3x - 2$

$(6x^2 - 9x - 7) + (-4x^2 + 3x - 2)$ A plus sign in front of a parentheses means keep the signs in the parentheses

$= 6x^2 - 9x - 7 - 4x^2 + 3x - 2$

$= 6x^2 - 4x^2 - 9x + 3x - 7 - 2$ Rearranging terms

$= 2x^2 - 6x - 9$ Combining like terms

b. $xy^2 - 3xy$ and $-7xy - 4xy^2$

$(xy^2 - 3xy) + (-7xy - 4xy^2)$

$= xy^2 - 3xy - 7xy - 4xy^2$ Removing the parentheses

$= xy^2 - 4xy^2 - 3xy - 7xy$ Rearranging terms

$= -3xy^2 - 10xy$ Combining like terms   ∎

☐ **DO EXERCISE 1.**

Polynomials may also be added by placing like terms in columns and adding them vertically, although polynomials are usually added horizontally.

EXAMPLE 2 Add.

a. $x^4 - 3x^2 - 7$ and $-2x^4 + x - 1$

$$
\begin{array}{r}
x^4 - 3x^2 - 7 \\
-2x^4 + x - 1 \\
\hline
-x^4 - 3x^2 + x - 8
\end{array}
$$

b. $3x^2 - 4xy + y^2$ and $4x^2 + 3xy - y^2$

$$
\begin{array}{r}
3x^2 - 4xy + y^2 \\
4x^2 + 3xy - y^2 \\
\hline
7x^2 - xy
\end{array}
$$ ∎

Notice that in the first expression in Example a, a space is left for the x term since there is an x term in the second expression. No space was left for an x^3 term since it does not appear in either expression.

☐ **DO EXERCISE 2.**

☐ **Exercise 1** Add.

a. $4x^2 + 3x - 2$ and $-2x^2 + 6x - 4$

b. $-6x^4 - 2x^2 + 3$ and $-3x^4 - 4x^2 - 7$

c. $7x^5 + 6x - 3$ and $3x^4 + x - 5$

d. $3xy - 4x^2y$ and $-5xy + 2x^2y$

☐ **Exercise 2** Add.

a. $x^3 - x + 8$ and $4x^3 + 2x^2 - 3x$

b. $2x^2 - xy + y^2$ and $-x^2 + xy + 3y^2$

a. $(-6x^2 + x - 7)$
    $- (-4x^2 + 5x - 7)$

b. $(7x^3 - x^2 + 8)$
    $- (6x^3 - 4x^2 - 3x)$

c. $(8y^5 - 9y^4 - y - 3)$
    $- (9y^6 + 7y^5 + 3y - 7)$

d. $(6y^2 - 4xy - x^2)$
    $- (7y^2 + 3xy - 4x^2)$

▣ Subtraction of Polynomials

Recall that we subtract a real number from another real number by adding the opposite of the second real number to the first real number: $a - b = a + (-b)$. We may use this same technique for polynomials.

EXAMPLE 3 Subtract.

a. $(3x^2 + 2x - 4) - (-4x^2 + 5x - 6)$

$\quad = (3x^2 + 2x - 4) + (+4x^2 - 5x + 6)$ Adding the opposite of the second polynomial

$\quad = 3x^2 + 2x - 4 + 4x^2 - 5x + 6$

$\quad = 3x^2 + 4x^2 + 2x - 5x - 4 + 6$ Rearranging terms

$\quad = 7x^2 - 3x + 2$

Notice that the opposite of the second polynomial is formed by changing all the signs on this polynomial.

We may make our work easier by remembering that a subtraction or negative sign in front of a parentheses means that we are to multiply each term in the expression by -1.

b. $(4x^5 - 3x^4 - x^2 + 5x) - (3\overset{\downarrow}{x}^5 + 6x^3 - x^2 - 8)$ The sign on $3x^5$ is plus

$\quad = (4x^5 - 3x^4 - x^2 + 5x) - 1(3x^5 + 6x^3 - x^2 - 8)$

$\quad = 4x^5 - 3x^4 - x^2 + 5x - 3x^5 - 6x^3 + x^2 + 8$ Using a distributive law

$\quad = 4x^5 - 3x^5 - 3x^4 - 6x^3 - x^2 + x^2 + 5x + 8$

$\quad = x^5 - 3x^4 - 6x^3 + 5x + 8$ ■

□ **DO EXERCISE 3.**

Vertical subtraction is used in long division of polynomials, which we will study later in this chapter. Vertical subtraction is similar to vertical addition except that we must remember to change the signs of the terms of the polynomial being subtracted. This is the same as multiplying this polynomial by -1.

EXAMPLE 4 Subtract.

a. $(7x^2 + 6x - 4) - (4x^2 - 3x + 5)$

$$\begin{array}{r} 7x^2 + 6x - 4 \\ -(4x^2 - 3x + 5) \\ \hline \end{array}$$

Now we change the signs of the polynomial being subtracted and add.

$$\begin{array}{r} 7x^2 + 6x - 4 \\ -4x^2 + 3x - 5 \\ \hline 3x^2 + 9x - 9 \end{array}$$

b. Subtract $3x^4 + 3x^2 + 7$ from $3x^4 - 2x^2 + 3$.

$$\begin{array}{rr} 3x^4 - 2x^2 + 3 & 3x^4 - 2x^2 + 3 \\ -(3x^4 + 3x^2 + 7) \quad = & -3x^4 - 3x^2 - 7 \\ \hline & -5x^2 - 4 \quad \blacksquare \end{array}$$

□ DO EXERCISE 4.

3 Addition and Subtraction

Addition and subtraction may be combined.

EXAMPLE 5 Add or subtract as indicated.

a. $(y^2 - y) + (3y^2 - 5) - (7y - 8)$
$= y^2 - y + 3y^2 - 5 - 7y + 8$ Removing parentheses
$= y^2 + 3y^2 - y - 7y - 5 + 8$ Rearranging terms
$= 4y^2 - 8y + 3$

b. $(x^2 + y^2) - (2x^2 - xy) + (xy + 3y^2)$
$= x^2 + y^2 - 2x^2 + xy + xy + 3y^2$
$= x^2 - 2x^2 + xy + xy + y^2 + 3y^2$ Rearranging terms
$= -x^2 + 2xy + 4y^2$ $\blacksquare$

□ DO EXERCISE 5.

□ **Exercise 4** Subtract.

a. $(4x^2 + 3x) - (4x^2 - 6x)$

b. $(7x - 4) - (7x + 8)$

c. $(9x - 3) - (9x - 3)$

d. Subtract $8x^4 - 7x^2 + 7$ from $8x^4 + 6x^2 + 3$.

□ **Exercise 5** Add or subtract as indicated.

a. $(3x^2 - 7x + 4) - (2x - 3)$
$+ (-3x^2 + 7x)$

b. $(3x^5 + x^3) + (4x^3 - 7)$
$- (-2x^5 - x^3)$

c. $(4x^2 - 2xy) + (y^2 - 3xy)$
$- (2x^2 + 3y^2)$

d. $(5xy + y^2) - (6xy - 5x^2)$
$+ (x^2 - 7y^2)$

Answers to Exercises _____

1. a. $2x^2 + 9x - 6$ **b.** $-9x^4 - 6x^2 - 4$ **c.** $7x^5 + 3x^4 + 7x - 8$
d. $-2xy - 2x^2y$

2. a. $5x^3 + 2x^2 - 4x + 8$ **b.** $x^2 + 4y^2$

3. a. $-2x^2 - 4x$ **b.** $x^3 + 3x^2 + 3x + 8$
c. $9y^6 + y^5 - 9y^4 - 4y + 4$ **d.** $-y^2 - 7xy + 3x^2$

4. a. $9x$ **b.** -12 **c.** 0 **d.** $13x^2 - 4$

5. a. $-2x + 7$ **b.** $5x^5 + 6x^3 - 7$ **c.** $2x^2 - 5xy - 2y^2$
d. $6x^2 - xy - 6y^2$

PROBLEM SET 5.2

A. *Add.*

1. $x^2 + 3x - 5$ and $3x^2 - 6x - 2$

2. $5x^3 + 6x - 4$ and $-6x^3 + 7x + 1$

3. $-2x^4 + x^2 - 5$ and $3x^4 - 2x^2 + 8$

4. $-5x^5 + 3x^2 - 1$ and $-x^5 + 2x^2 - 7$

5. $-8x^3 + 3x + 4$ and $-4x^2 + 6x - 7$

6. $7x^4 - 6x^2 - 3$ and $3x^4 + x - 5$

7. $(3x^2 + 6x - 2) + (7x^2 + 8)$

8. $(3x^3 - 2) + (x^2 - 4)$

9. $(6xy + x^2) + (y^2 - 5xy)$

10. $(y^3 - 2y) + (y^2 + y)$

11. $(x^2y^2 - 3xy + xy^2) + (3x^2y^2 - xy - xy^2)$

12. $(3yz^2 - 4yz + 7y^2z) + (-3yz^2 + 8yz + 5y^2z)$

13. $\left(\dfrac{1}{5}x^4 - x^2 + \dfrac{3}{8}\right) + \left(\dfrac{2}{3}x^4 - \dfrac{1}{2}x^2 + \dfrac{1}{16}\right)$

14. $\left(\dfrac{5}{8}x^2 - x + \dfrac{1}{6}\right) + \left(\dfrac{1}{4}x^2 - 2x - \dfrac{1}{12}\right)$

15. $(3.8x^3 + 1.4x^2 + 1.3) + (-1.6x^3 - 8.4x^2 - 1.1)$

16. $(-8.5x^2 - 0.3x - 1.7) + (3.4x^2 - 0.7x + 9.3)$

17. $\begin{aligned} x^2 + 3x - 2 \\ \underline{5x^2 - 2x + 8} \end{aligned}$

18. $\begin{aligned} 3x^3 - 4x^2 - 5 \\ \underline{-2x^3 + 6x^2 - 7} \end{aligned}$

19. $\begin{aligned} 8x^4 \qquad + 3x^2 + \ x \\ \underline{-6x^4 + x^3 \qquad - 2x} \end{aligned}$

20. $\begin{aligned} -4x^2 + 8 \\ \underline{5x^3 \qquad - 3} \end{aligned}$

B. *Subtract.*

21. $(-7x^2 - 3x - 2) - (-x^2 - 2x + 8)$

22. $(6x^2 + 4x - 1) - (3x^2 - 2x + 3)$

23. $(9x^3 - 4x^2 + 2) - (-8x^3 + x^2 - 3)$

24. $(-7x^4 + 3x - 3) - (2x^4 - 2x + 5)$

25. $(6x^5 - 8y^4 - y^2 + 4) - (-9x^5 + 3y^2 - 1)$

26. $(7y^4 - 8y^3 + 3) - (3y^3 - 7)$

27. $(7x^2 - 3xy + y^2) - (4x^2 + 6xy - y^2)$

28. $(-3x^2 + 6xy + 2y^2) - (4x^2 - 7xy + 5y^2)$

29. $\left(\dfrac{9}{5}x^3 + \dfrac{1}{4}x^2 - \dfrac{3}{8}\right) - \left(\dfrac{3}{5}x^3 + \dfrac{1}{2}x^2 - \dfrac{2}{3}\right)$

30. $\left(\dfrac{7}{3}x^3 - \dfrac{3}{2}x^2 + \dfrac{1}{7}\right) - \left(\dfrac{8}{3}x^3 - \dfrac{5}{6}x^2 - \dfrac{5}{2}\right)$

31. $(5.9x^3 - 0.02x + 0.3) - (7.1x^3 + 0.7x - 0.06)$

32. $(-3.1x^2 - 0.01x - 0.05) - (-2.1x^2 + 0.9x - 0.08)$

33. Subtract $z^2 + yz + y^2$ from $-z^2 + yz + y^2$.

34. Subtract $y^2 + yz$ from $-yz + z^2$.

C. *Subtract.*

35. $3x^2 + 6x$
$\quad\ \ x^2 - 7x$

36. $9x^3 - 4x$
$\quad\ \ 8x^3 - 7x$

37. $8x + 4$
$\quad\ 7x + 4$

38. $9x - 7$
$\quad\ 9x - 3$

39. $\quad 5x^3 + 6x^2 + 3$
$\quad -2x^3 \quad\quad\ - 3$

40. $\quad 3x^4 \quad\quad - 2$
$\quad -5x^4 + x^2$

41. $x^2 - 4$
$\quad x^2 + 4$

42. $y^2 - 8$
$\quad y^2 + 8$

43. $x^2 - 3$
$\quad x^2 - 3$

44. $z^2 + 7$
$\quad z^2 + 7$

D. *Add or subtract as indicated.*

45. $(2x^2 - 3x) + (5x^2 - 2) - (3x + 1)$

46. $(7x^3 - 2) + (3x^3 + x^2) + (x^2 + 5)$

47. $(8x^5 - 6x) - (7x^4 - 6x) + (3x^2 + 2x)$

48. $(8x^3 - 7x^4) - (8x^3 + 2x^2) + (6x^2 + 3x)$

49. $(9x^4 - x^2) - (6x^4 + 3) - (4x^2 + 7)$

50. $(7x^3 - 3x^2) - (8x^5 - 2) - (-3x^2 + 5)$

51. Polynomials may be added by placing _____ terms in columns and adding vertically.

52. The _____ of a polynomial is formed by changing all signs on the polynomial.

Checkup

The following problems review some of Sections 2.4 and 4.2, and some will help you with the next section.
Multiply.

53. $-2(7)$

54. $-8(-9)$

55. $-7(-8)$

56. $3(-6)$

57. If 6 is subtracted from three times a number, the result is 45. Find the number.

58. The sum of the angles of a triangle is $180°$. The second angle is three times as large as the smallest angle and the third angle is equal to the sum of the first and second angles. Find the third angle.

59. Hee Sook gave her friends $20,000. If Sunya received twice as much as Susan and Amy received as much as Susan, how much did Sunya receive?

60. The sum of two consecutive even integers is 186. Find the integers.

5.3 MULTIPLICATION OF POLYNOMIALS

OBJECTIVES

1 *Multiply monomials*

2 *Multiply a monomial and any polynomial*

3 *Multiply two polynomials*

4 *Multiply a binomial by a binomial using FOIL*

1 Multiplying Monomials

Recall from Section 2.7 that when we multiply two exponential numbers with the same base, we keep the base and add the exponents.

$$a^m \cdot b^m = a^{m+n}$$

To multiply two monomials multiply the coefficients and add the exponents on the identical variables.

EXAMPLE 1 Multiply.

a. $5(4x^2) = 5(4)(x^2) = 20x^2$

b. $3x(5x) = (3)(5)(x)(x) = 15x^2$

c. $(-6x^2y)(2x^3z^3) = (-6)(2)(x^2y)(x^3z^3) = -12x^5yz^3$

d. $-6ab(4a^2b^3) = (-6)(4)(ab)(a^2b^3) = -24a^3b^4$ Recall $ab = a^1b^1$ ■

□ **DO EXERCISE 1.**

2 Multiplying a Monomial and Any Polynomial

To multiply a monomial times any polynomial, multiply each term of the polynomial by the monomial. This is using a distributive law. Sometimes we apply it to more than two terms.

EXAMPLE 2 Multiply.

a. $4x(x^2 + 3) = 4x(x^2) + 4x(3)$ Using a distributive law
$$= 4x^3 + 12x$$

b. $-2y(y^2 + 4y - 7)$
$$= -2y(y^2) + (-2y)(4y) - (-2y)(7)$$ Distributive law for three terms
$$= -2y^3 - 8y^2 + 14y$$

Recall that a distributive law may be stated as $(b + c)a = ba + ca$.

c. $(x^2 - 2x + 3)3x = x^2(3x) - 2x(3x) + 3(3x) = 3x^3 - 6x^2 + 9x$ ■

□ **DO EXERCISE 2.**

□ **Exercise 1** Multiply.

a. $-7(3y)$

b. $-x(4x^2)$

c. $8xy^2(-2x^3y^5)$

d. $-x(-8x^6y^4)$

□ **Exercise 2** Multiply.

a. $-2x^2(x + 4)$

b. $3y(2y^2 - 4y + 8)$

c. $5y(y^3 - 7)$

d. $-x(-x^2 + 3x - 1)$

e. $(x - 3)(-4x^2)$

f. $(y^2 - 3y + 2)2y$

❸ Multiplying Two Polynomials

> Two polynomials are multiplied by multiplying each term of one polynomial by each term of the other polynomial. The distributive laws may be used more than once. Then combine like terms.

EXAMPLE 3 Multiply.

a.
$$\begin{aligned}
(x + 2)(x + 4) &= x(x + 4) + 2(x + 4) && \text{Using a distributive law} \\
&= x(x) + 4(x) + 2(x) + 2(4) && \text{Using a distributive law} \\
&= x^2 + 4x + 2x + 8 \\
&= x^2 + 6x + 8 && \text{Combining like terms}
\end{aligned}$$

b. $(x - 3)(x^2 + 4x - 1)$
$$\begin{aligned}
&= x(x^2 + 4x - 1) - 3(x^2 + 4x - 1) \\
&= x(x^2) + x(4x) - x(1) - 3(x^2) + (-3)(4x) - (-3)(1) \\
&= x^3 + 4x^2 - x - 3x^2 - 12x + 3 \\
&= x^3 + x^2 - 13x + 3 && \text{Combining like terms}
\end{aligned}$$

c. $(x^2 + 2x - 1)(x^2 - x + 1)$
$$\begin{aligned}
&= x^2(x^2 - x + 1) + 2x(x^2 - x + 1) - 1(x^2 - x + 1) \\
&= x^4 - x^3 + x^2 + 2x^3 - 2x^2 + 2x - x^2 + x - 1 \\
&= x^4 + x^3 - 2x^2 + 3x - 1 \quad ∎
\end{aligned}$$

□ **DO EXERCISE 3.**

4 Multiplying a Binomial by a Binomial

There is a special method for multiplying a binomial by a binomial that is often used. This method also helps us to understand factoring, which is discussed later in this chapter.

EXAMPLE 4 Multiply.

a. $(x + 4)(x + 3)$

We multiply the **f**irst terms of each binomial, $x \cdot x = x^2$
We multiply the **o**uter terms of each binomial, $x \cdot 3 = 3x$
We multiply the **i**nner terms of each binomial, $4 \cdot x = 4x$
We multiply the **l**ast terms of each binomial, $4 \cdot 3 = 12$
Then we add the results to get the answer: $x^2 + 3x + 4x + 12$ or $(x + 4)(x + 3) = x^2 + 7x + 12$

This method is called FOIL, which is an abbreviation for first, outer, inner, and last.

b. $(x + 5)(x - 2)$

Multiply the first terms, $x \cdot x \quad = x^2$
Multiply the outer terms, $x \cdot (-2) = -2x$
Multiply the inner terms, $5 \cdot x \quad = 5x$
Multiply the last terms, $5 \cdot (-2) \quad = -10$
Add the results: $(x + 5)(x - 2) = x^2 - 2x + 5x - 10 = x^2 + 3x - 10$
The following diagram for Example b makes the FOIL method shorter:

$$
\begin{array}{c}
O \\
I \\
(x + 5)(x - 2) \\
F \qquad L
\end{array}
$$

c. $(5x - 1)(3x - 4)$

$$
\begin{array}{c}
O \\
I \\
(5x - 1)(3x - 4) = 15x^2 - 20x - 3x + 4 = 15x^2 - 23x + 4 \quad \blacksquare \\
F \qquad L
\end{array}
$$

The method shown in Example 3 would give the same result.

□ **DO EXERCISE 4.**

□ **Exercise 4** Multiply.

a. $(x + 4)(x + 6)$

b. $(y + 3)(y + 1)$

c. $(2x - 7)(2x - 2)$

d. $(5x - 8)(x + 3)$

e. $(x + y)(x - y)$

f. $(3x + 4y)(2x - y)$

g. $(x - 5)(x + 5)$

h. $(y + 3)(y + 3)$

i. $(-3x + 1)(2x - 4)$

j. $(x^2 + 3)(x^2 + 5)$

PROBLEM SET 5.3

A. *Multiply.*

1. $3(-4x)$

2. $4(3x)$

3. $-5x(-4)$

4. $3y(2)$

5. $-4(2x^3)$

6. $-7(-5x^4)$

7. $x(7x^2)$

8. $x(-8x^3)$

9. $x^3(x^2)$

10. $x^4(-x)$

11. $3x^5y(-4xy^2)$

12. $-2xy(x^2y^3)$

13. $6x^3y^4(7xy^5)$

14. $-2xy^2(8x^3y)$

15. $-3x(-x^5y^4)$

16. $4x(x + 2)$

17. $3y(y^4 - 2)$

18. $-2(x^2 + 2x - 1)$

19. $4(y^3 - 2y^2 + 4)$

20. $x(x - 7)$

21. $x^4(-x + 2)$

22. $-2x(x^2 + 3x - 1)$

23. $(-y^2 - 3y + 4)5y$

24. $(x^2 - 9)6x^5$

B. *Multiply by multiplying each term of one polynomial by every term of the other polynomial.*

25. $(x - 7)(x - 1)$

26. $(x - 6)(x + 2)$

27. $(2y + 3)(y - 2)$

28. $(y - 6)(3y + 4)$

29. $(x - 3)(x^2 + 2x - 1)$

30. $(y + 2)(y^2 - 3y + 1)$

31. $(2x - 1)(x^2 + 3x - 4)$

32. $(3y + 2)(y^2 + y + 4)$

33. $(3z + 8)(z^4 - z + 5)$

34. $(7z - 1)(z^3 - z^2 + z)$

35. $(x^2 + x - 1)(x^2 - x + 1)$

36. $(2y^2 + y - 2)(y^2 + y + 1)$

37. $(y - 3)(y^3 - 2y + 4)$

C. *Multiply using FOIL*

38. $(x + 3)(x + 5)$ **39.** $(y + 4)(y - 1)$ **40.** $(x - 3)(x - 7)$ **41.** $(y - 3)(y - 6)$

42. $(3x + 4)(x - 8)$ **43.** $(6x - 2)(x - 1)$ **44.** $(3y + 2)(4y + 1)$ **45.** $(7y - 8)(y - 2)$

46. $(5x - 4)(2x - 5)$ **47.** $(6x - 6)(x - 2)$ **48.** $(x - y)(2x + y)$ **49.** $(5x - 6y)(x + y)$

50. $(8x - 1)(7x + 1)$ **51.** $(5y + 4)(2y + 3)$ **52.** $(4x + 8)(x + 3)$ **53.** $(x - 9)(x + 9)$

54. $(y - 4)(y + 4)$ **55.** $(x + 8)(x + 8)$ **56.** $(x + 7)(x + 7)$ **57.** $(x^4 + 2)(2x - 3)$

58. $(3x^2 - 1)(x^2 + 8)$ **59.** $(7x - 1)(x^3 - 3)$ **60.** $\left(x - \dfrac{1}{4}\right)\left(3x - \dfrac{1}{2}\right)$ **61.** $(2x - 0.4)(x + 0.2)$

62. To multiply two monomials, multiply the coefficients and _____ the exponents on the identical variables.

63. To multiply a monomial times any polynomial other than a monomial use a _____ law.

64. Two polynomials may be multiplied by multiplying each _____ of one polynomial by each term of the other polynomial.

65. A special method called FOIL may be used to multiply a _____ by a binomial.

Checkup

The following problems provide a review of some of Section 4.3.

66. Carlos received $180 interest on an amount of money invested at 8% interest for 3 years. How much did he invest?

67. Carrie withdrew some money, to pay for books, from her account, which originally contained $1500. The account now yields $103.50 annually at 7.5% interest. How much did she withdraw?

68. Jannel invested $4000, some money at 8% and the remainder at 9% per year. How much did she invest at 9% if her yearly interest was $355?

69. If $700 is withdrawn from an account, the new amount will give the same interest each year at 8% as the original amount does at 6%. How much was invested at 6%?

5.4 SPECIAL PRODUCTS OF BINOMIALS

When we multiply polynomials, we find their product. We may always multiply two binomials by the FOIL method, but there are special methods for certain binomials which allow us to do the multiplication more rapidly.

1 **Multiplying Two Binomials That Are Identical Except for the Sign between Them**

If we multiply using FOIL, we get the following results.

$$(x - 2)(x + 2) = x^2 + 2x - 2x - 4 = x^2 - 4$$

$$(2x + y)(2x - y) = 4x^2 - 2xy + 2xy - y^2 = 4x^2 - y^2$$

Notice that in each case the middle terms drop out. The answer is the square of the first term minus the square of the second term of either binomial. Hence

$$(A - B)(A + B) = A^2 - B^2$$

$A^2 - B^2$ is called "the difference of two squares."

EXAMPLE 1 Multiply.

$$(x - 3)(x + 3) = x^2 - 3^2 = x^2 - 9$$

$$(x - 4y)(x + 4y) = x^2 - (4y)^2 = x^2 - 16y^2 \quad \blacksquare$$

☐ **DO EXERCISE 1.**

2 **Multiplying Two Identical Binomials or Squaring a Binomial**

When we multiply using FOIL, we get the following results.

a. $(x + 3)^2 = (x + 3)(x + 3) = x^2 + 3x + 3x + 3^2 = x^2 + 6x + 9$

b. $(2x - 1)^2 = (2x - 1)(2x - 1) = 4x^2 - 2x - 2x + 1 = 4x^2 - 4x + 1$

Notice that the middle term of the answer is twice the product of the two terms of the original binomial. Hence a quick way to square a binomial is to square the first term, add twice the product of the two terms, and add the square of the last term.

$$(A + B)^2 = A^2 + 2AB + B^2$$

$$(A - B)^2 = A^2 + 2A(-B) + (-B)^2 = A^2 - 2AB + B^2$$

☐ **Exercise 1** Multiply. Do not use FOIL.

a. $(x - 5)(x + 5)$

b. $(2y + 4)(2y - 4)$

c. $(x + 1)(x - 1)$

d. $(3y - 6)(3y + 6)$

e. $(x + y)(x - y)$

f. $(x^2 - 3y)(x^2 + 3y)$

□ **Exercise 2**   Multiply. Do not use FOIL.

□ **Exercise 2**   Multiply. Do not use FOIL.

a. $(x + 9)^2$

b. $(x - 2)^2$

c. $(3x - 2)^2$

d. $(5x - 4)^2$

e. $(x - y)^2$

f. $(x + 4y)^2$

EXAMPLE 2   Multiply.

a. $(y - 4)^2 = y^2 - 2(y)(4) + 4^2 = y^2 - 8y + 16$

b. $(4x - 1)^2 = (4x)^2 - 2(4x)(1) + 1^2 = 16x^2 - 8x + 1$

c. $(x + 3y)^2 = x^2 + 2(x)(3y) + (3y)^2 = x^2 + 6xy + 9y^2$ ■

□ **DO EXERCISE 2.**

Notice that

$$(A + B)^2 \neq A^2 + B^2$$

For example, $(5 + 2)^2 \neq 5^2 + 2^2$, since $(5 + 2)^2 = 7^2 = 49$ and $5^2 + 2^2 = 25 + 4 = 29$.

Similarly,

$$(A - B)^2 \neq A^2 - B^2$$

For example, $(5 - 2)^2 \neq 5^2 - 2^2$, since $(5 - 2)^2 = 3^2 = 9$ and $5^2 - 2^2 = 25 - 4 = 21$.

Special Products of Binomials

$$(A - B)(A + B) = A^2 - B^2$$

$$(A + B)^2 = A^2 + 2AB + B^2$$

$$(A - B)^2 = A^2 - 2AB + B^2$$

DID YOU KNOW?

The mathematician Diophantus of Alexandria began using abbreviations or symbols when solving algebraic problems in the third century. Previously, mathematicians wrote out in their entirety all the words needed when solving problems.

Answers to Exercises

1. a. $x^2 - 25$ **b.** $4y^2 - 16$ **c.** $x^2 - 1$ **d.** $9y^2 - 36$
e. $x^2 - y^2$ **f.** $x^4 - 9y^2$
2. a. $x^2 + 18x + 81$ **b.** $x^2 - 4x + 4$ **c.** $9x^2 - 12x + 4$
d. $25x^2 - 40x + 16$ **e.** $x^2 - 2xy + y^2$ **f.** $x^2 + 8xy + 16y^2$

PROBLEM SET 5.4

Multiply. Do not use FOIL.

1. $(x - 4)(x + 4)$ **2.** $(y - 3)(y + 3)$ **3.** $(x + 2)(x - 2)$ **4.** $(y + 7)(y - 7)$

5. $(5 - x)(5 + x)$ **6.** $(8 - y)(8 + y)$ **7.** $(1 + y)(1 - y)$ **8.** $(9 + y)(9 - y)$

9. $(2x - 3)(2x + 3)$ **10.** $(5x - 1)(5x + 1)$ **11.** $(7x + 2)(7x - 2)$ **12.** $(6x + 1)(6x - 1)$

13. $(5x - 2y)(5x + 2y)$ **14.** $(5x - y)(5x + y)$ **15.** $(2x + 3y)(2x - 3y)$ **16.** $(x - 7y)(x + 7y)$

17. $(x^2 - 2)(x^2 + 2)$ **18.** $(y^4 + 3)(y^4 - 3)$ **19.** $(x + 2)^2$ **20.** $(y + 3)^2$

21. $(y - 4)^2$ **22.** $(x - 1)^2$ **23.** $(y - 7)^2$ **24.** $(x - 9)^2$

25. $(2x + 1)^2$ **26.** $(3x + 2)^2$ **27.** $(5x - 1)^2$ **28.** $(2y - 3)^2$

29. $(4y - 2)^2$ **30.** $(8y - 1)^2$ **31.** $(x + y)^2$ **32.** $(2x + y)^2$

33. $(5x - y)^2$ **34.** $(7x - 4y)^2$ **35.** $(8x - 5y)^2$ **36.** $(5x - 2y)^2$

37. $(x + 2)(x - 2)$ **38.** $(x + 4)(x - 4)$ **39.** $(x + 2)^2$ **40.** $(x - 4)^2$

41. $(5x - 8)^2$ **42.** $(3x + 2)^2$ **43.** $(3x - 2)(3x + 2)$ **44.** $(3x + 2)(3x - 2)$

45. $(6x - 1)(6x + 1)$ **46.** $(5y - 4)(5y + 4)$ **47.** $(6x - 1)^2$ **48.** $(5y + 4)^2$

49. $(3x - y)(3x + y)$ **50.** $(y - x)(y + x)$ **51.** $(3x - y)^2$ **52.** $(y + x)^2$

53. $(2x - 3y)^2$ **54.** $(3x - 4y)^2$ **55.** $\left(x - \dfrac{1}{4}\right)\left(x + \dfrac{1}{4}\right)$ **56.** $\left(x - \dfrac{3}{8}\right)^2$

57. $(0.4x - 3)^2$ **58.** $(x + 0.2)(x - 0.2)$

59. To multiply two binomials that are identical except for the sign between them, square the first term of either binomial and _____ the square of the second term.

60. To square a binomial, square the first term, add _____ the product of the two terms, and add the square of the last term.

Checkup

The following problems provide a review of some of Sections 2.7 and 4.4. Some of these problems will help you with the next section.
Divide and write with positive exponents.

61. $\dfrac{x^3}{x}$ **62.** $\dfrac{x^2}{x^2}$ **63.** $\dfrac{x^4}{x}$

64. Carrie has 35 dimes and quarters worth $7.55. How many dimes does she have?

65. Joe has twice as many dimes as nickels. If the total value of the coins is $5.75, how many dimes does he have?

66. How many liters of a 60% acid solution should be added to 30 liters of a 25% solution to make a 50% solution?

67. A 30% silver alloy is to be melted with a 55% silver alloy. How many kilograms of 55% alloy should be used to get 15 kilograms of a 40% silver alloy?

5.5 DIVISION OF POLYNOMIALS

1 Division of a Polynomial by a Monomial

We will use division of a polynomial by a monomial to help us factor in Section 5.6. We may add or subtract fractions that have polynomials in the numerator or denominator by methods similar to the addition or subtraction of rational numbers.

$$\frac{A + B}{C} = \frac{A}{C} + \frac{B}{C}$$

Since we can add the second expression to get the first expression when A, B, and C are polynomials.

Also,

$$\frac{A - B}{C} = \frac{A}{C} - \frac{B}{C}$$

Similarly,

$$\frac{4x^3 - 8x^2}{4x} = \frac{4x^3}{4x} - \frac{8x^2}{4x}$$

Now we may divide the coefficients and use the rule for dividing exponential numbers, $a^m/a^n = a^{m-n}$:

$$\frac{4x^3}{4x} - \frac{8x^2}{4x} = x^{3-1} - 2x^{2-1} = x^2 - 2x$$

> To divide a polynomial by a monomial, divide each term of the polynomial by the monomial.

EXAMPLE 1 Divide.

a. $\dfrac{x^3 + x^2}{x} = \dfrac{x^3}{x} + \dfrac{x^2}{x} = x^{3-1} + x^{2-1} = x^2 + x$

b. $\dfrac{5x^3 - 15x^2 + 10x}{5x} = \dfrac{5x^3}{5x} - \dfrac{15x^2}{5x} + \dfrac{10x}{5x} = x^2 - 3x + 2$

c. $\dfrac{8x^2 + 4x}{4x^2} = \dfrac{8x^2}{4x^2} + \dfrac{4x}{4x^2} = 2x^{2-2} + x^{1-2} = 2x^0 + x^{-1}$

Recall that $x^0 = 1$ and $x^{-1} = 1/x$. Hence

$$2x^0 + x^{-1} = 2 + \frac{1}{x} \qquad \text{and} \qquad \frac{8x^2 + 4x}{4x^2} = 2 + \frac{1}{x}$$

d. $\dfrac{4x^2 - 8x + 7}{x} = \dfrac{4x^2}{x} - \dfrac{8x}{x} + \dfrac{7}{x} = 4x - 8 + \dfrac{7}{x}$ ∎

Caution $\dfrac{B}{A + C} \neq \dfrac{B}{A} + \dfrac{B}{C}.$

☐ **DO EXERCISE 1.**

---

1 *Divide a polynomial by a monomial*

2 *Divide a polynomial by a polynomial other than a monomial*

☐ **Exercise 1** Divide.

a. $\dfrac{4x^4 - 8x^2}{4x}$

b. $\dfrac{6x^5 - 3x^3 + 9x^2}{3x^2}$

c. $\dfrac{4x^3 + 8x^2 - 4x}{4x}$

d. $\dfrac{7x^3 - 8x^2 + 3}{2}$

e. $\dfrac{9x^2 + 3x}{3x^2}$

f. $\dfrac{4x^3 + 8x + 5}{2x}$

2 Division of a Polynomial by a Polynomial Other Than a Monomial

We may divide a polynomial by a binomial or a trinomial. The procedure that we use is very much like the long division that we learned to do in arithmetic. We will show division in arithmetic along with division of polynomials so that you can see the similarity. In the division problem $64 \div 4 = 16$, 64 is called the dividend, 4 is the divisor, and 16 is the quotient.

<table>
<tr><td colspan="2" align="center">***Division in Arithmetic***</td><td colspan="2" align="center">***Division in Algebra***</td></tr>
<tr><td></td><td align="center">$\dfrac{441}{7}$</td><td></td><td align="center">$\dfrac{x^2 - 2x - 24}{x + 4}$</td></tr>
</table>

Step 1   Divide 7 into 44. The result is 6.

$$\begin{array}{r} 6 \\ 7\overline{)441} \end{array}$$

Divide x into x^2. The result is x.

$$\begin{array}{r} x \\ x+4\overline{)x^2 - 2x - 24} \end{array}$$

Step 2   Multiply $7(6) =$

Step 3   Subtract and bring down the 1.

$$\begin{array}{r} 6 \\ 7\overline{)441} \\ 42 \\ \hline 21 \end{array}$$

Multiply $x(x + 4) =$

Subtract and bring down the next term.

$$\begin{array}{r} x \\ x+4\overline{)x^2 - 2x - 24} \\ x^2 + 4x \\ \hline -6x - 24 \end{array}$$

The procedure begins to repeat at this point.

Step 4   Divide 7 into 21. The result is 3.

$$\begin{array}{r} 63 \\ 7\overline{)441} \\ 42 \\ \hline 21 \end{array}$$

Divide x into $-6x$. The result is -6.

$$\begin{array}{r} x - 6 \\ x+4\overline{)x^2 - 2x - 24} \\ x^2 + 4x \\ \hline -6x - 24 \end{array}$$

Step 5

$$\begin{array}{r} 63 \\ 7\overline{)441} \\ 42 \\ \hline 21 \\ 21 \\ \hline \end{array}$$

Multiply $7(3) =$

$$\begin{array}{r} x - 6 \\ x+4\overline{)x^2 - 2x - 24} \\ x^2 + 4x \\ \hline -6x - 24 \\ -6x - 24 \\ \hline 0 \end{array}$$

Change signs!

Multiply $-6(x + 4) =$ Change signs!

Hence $\dfrac{441}{7} = 63$ and $\dfrac{x^2 - 2x - 24}{x + 4} = x - 6.$

$$x^3 + 0x^2 + 0x + x + 1.$$

$$x^2 + x^2$$

EXAMPLE 2 Divide.

a. $\dfrac{x^2 + 4x - 14}{x + 6}$ Notice that when we divide by a binomial, we use long division.

1. Divide x^2 by x.
2. Multiply $x(x + 6) =$
3. Subtract and bring down -14.
4. Divide $-2x$ by x to get -2.
5. Multiply $-2(x + 6) =$
6. Subtract. The remainder is -2.

$$
\begin{array}{r}
x -2 \\
x + 6 \overline{)\, x^2 + 4x - 14} \\
\underline{x^2 + 6x} \qquad\qquad \text{Change} \\
-2x - 14 \qquad \text{signs!} \\
\underline{-2x - 12} \qquad \text{Change} \\
-2 \qquad \text{signs!}
\end{array}
$$

Recall that $\frac{9}{5} = 1\frac{4}{5}$. The remainder, 4, is placed over the divisor, 5, to give $\frac{4}{5}$. Similarly, $(x^2 + 4x - 14)/(x + 6) = x - 2 + (-2)/(x + 6)$ since -2 is the remainder and $x + 6$ is the divisor.

Hence $\dfrac{x^2 + 4x - 14}{x + 6} = x - 2 + \dfrac{-2}{x + 6}.$

b. $\dfrac{x^3 - 1}{x - 1}$ The x^2 and x terms are missing in the numerator. Use 0 as the coefficient for the missing terms.

$$
\begin{array}{r}
x^2 + x + 1 \\
x - 1 \overline{)\, x^3 + 0x^2 + 0x - 1} \\
\underline{x^3 - x^2} \\
x^2 + 0x \\
\underline{x^2 - x} \\
x - 1 \\
\underline{x - 1} \\
0
\end{array}
$$

$$\dfrac{x^3 - 1}{x - 1} = x^2 + x + 1 \quad \blacksquare$$

□ **DO EXERCISE 2.**

□ **Exercise 2**   Divide.

a. $\dfrac{x^2 + 5x + 6}{x + 2}$ ✓

$$
\begin{array}{r}
x + 3 \\
x + 2 \overline{)\, x^2 + 5x + 6} \\
\underline{x^2 + 2x} \\
3x + 6 \\
\underline{3x + 6} \\
0 \qquad 0
\end{array}
$$

b. $\dfrac{x^2 + 2x + 3}{x + 1}$

$$
\begin{array}{r}
x + 1 + \frac{2}{x+1} \\
x + 1 \overline{)\, x^2 + 2x + 3} \\
\underline{x^2 + x} \\
x + 3 \\
\underline{x \pm 1} \\
2
\end{array}
$$

c. $\dfrac{y^2 - 2y + 2}{y - 2}$

$$
\begin{array}{r}
y + \frac{2}{y-2} \\
y - 2 \overline{)\, y^2 - 2y + 2} \\
\underline{y^2 \mp 2y} \\
0 \qquad 2
\end{array}
$$

d. $\dfrac{x^3 + 1}{x + 1}$

$$
\begin{array}{r}
x^2 + x \pm 1 \\
x + 1 \overline{)\, x^3 + 0x^2 + 0x + 1} \\
\underline{x^3 \mp 1x^2} \\
x^2 + 0x \\
\underline{x^2 \mp x} \\
x + 1 \\
\underline{x - 1} \\
0
\end{array}
$$

□ **Exercise 3** Divide.

a. $\dfrac{2x^2 - 5x - 3}{2x + 1}$

$$
\begin{array}{r}
x - 3 \\
2x+1{\overline{\smash{\big)}\,2x^2-5x-3}} \\
\underline{2x^2 + x} \\
-6x - 3 \\
\underline{-\mp6x \mp 3} \\
0
\end{array}
$$

b. $\dfrac{9y^2 + 6y - 8}{3y - 2}$

$$
\begin{array}{r}
3y + 4 \\
3y-2{\overline{\smash{\big)}\,9y^2+6y-8}} \\
\underline{-9y^2 \mp 6y} \\
12y - 8 \\
\underline{12y - 8} \\
0
\end{array}
$$

EXAMPLE 3 Divide $\dfrac{12x^2 - 20x + 3}{2x - 3}$.

$$
\begin{array}{r}
6x - 1 \\
2x-3{\overline{\smash{\big)}\,12x^2-20x+3}} \\
\underline{12x^2 - 18x} \\
-2x + 3 \\
\underline{-2x + 3}
\end{array}
$$

$$\frac{12x^2 - 20x + 3}{2x - 3} = 6x - 1 \quad \blacksquare$$

□ **DO EXERCISE 3.**

Answers to Exercises

1. a. $x^3 - 2x$ b. $2x^3 - x + 3$ c. $x^2 + 2x - 1$
 d. $\dfrac{7}{2}x^3 - 4x^2 + \dfrac{3}{2}$ e. $3 + \dfrac{1}{x}$ f. $2x^2 + 4 + \dfrac{5}{2x}$

2. a. $x + 3$ b. $x + 1 + \dfrac{2}{x + 1}$ c. $y + \dfrac{2}{y - 2}$
 d. $x^2 - x + 1$

3. a. $x - 3$ b. $3y + 4$

PROBLEM SET 5.5

Divide.

1. $\dfrac{x^3 + 2x^2 + 3x}{x}$

2. $\dfrac{7y^3 + 3y^2 + y}{y}$

3. $\dfrac{8x^3 + 16x^2 - 4x}{4x}$

4. $\dfrac{9y^3 - 6y^2 + 6y}{3y}$

5. $\dfrac{5x^2 - 3x + 4}{5}$

6. $\dfrac{3x^2 - 4x + 8}{4}$

7. $\dfrac{8y^2 - 4y}{4y}$

8. $\dfrac{7x^3 + 14x + 3}{7x}$

9. $\dfrac{x^2 + 5x + 6}{x + 3}$

10. $\dfrac{x^2 + 6x + 8}{x + 4}$

11. $\dfrac{y^2 + 2y - 24}{y + 6}$

12. $\dfrac{y^2 + 11y + 24}{y + 8}$

13. $\dfrac{x^2 + x \quad 12}{x + 4}$

14. $\dfrac{x^2 - 7x - 18}{x + 2}$

15. $\dfrac{x^2 + 3x + 8}{x + 2}$

16. $\dfrac{y^2 + 7y - 1}{y + 2}$

17. $\dfrac{x^2 - 5x + 4}{x + 1}$

18. $\dfrac{y^2 - 2y + 4}{y + 1}$

19. $\dfrac{x^3 - 1}{x + 1}$

20. $\dfrac{x^3 + x - 1}{x - 1}$

21. $\dfrac{6y^2 + 5y - 6}{3y - 2}$

22. $\dfrac{20y^2 + 13y - 15}{5y - 3}$

23. $\dfrac{15x^2 + 19x + 10}{5x - 7}$

24. $\dfrac{15x^2 + 19x - 4}{3x + 8}$

25. To divide a polynomial by a monomial, divide each _____ of the polynomial by the monomial.

26. In the division $(x^2 + 3x - 2) \div (x + 4)$, the _____ is $x + 4$.

27. Long division may be used to divide a _____ by a binomial.

28. For the expression $(2x^2 - 4x + 5) \div (2x - 3)$ the _____ is $2x^2 - 4x + 5$.

Checkup

The following problems provide a review of some of Sections 4.5 and 5.3. Some of these problems will help you with the next section.

29. Mark and Mary left the hardware store at the same time going in opposite directions. If Mark walks 3 miles per hour and Mary walks 2 miles per hour, in how many hours will they be 2 miles apart?

30. Two cars leave Memphis at the same time traveling in the same direction. If one car averages 52 miles per hour and the other car travels 60 miles per hour, in how many hours will they be 90 miles apart?

31. A jet leaves Kelly Field flying due east at a speed of 850 kilometers per hour. It turns and flies due west at a speed of 700 kilometers per hour. If it remains in the air 5 hours, how long does it fly west?

Multiply.

32. $3(x + 2)$

33. $5(x - 3)$

34. $x^2(x + 1)$

35. $x^3(x - 2)$

36. $3x(2x^2 - 4x + 1)$

37. $5x^2(3x^3 + 2x - 4)$

Factor   to write an expression as the multiplication of expressions.

Greatest common factor of two or more expressions   the product of all *common factors* raised to the least power that they occur in any factorization.

5.6 FACTORING WHEN TERMS HAVE A COMMON FACTOR

1 The Greatest Common Factor

Factoring is used to solve equations. It reverses the process of multiplication. The distributive law may be used to multiply the following.

$$3(x - 2) = 3x - 6$$

Since any equation may be reversed, we also have

$$3x - 6 = 3(x - 2)$$

This is called factoring. When we **factor** we have written the expression as a multiplication of expressions.

> Factor out the largest number and the variable or variables with the highest exponent that will divide evenly into each term, without forming negative exponents. This is called factoring out the greatest common factor. This is often a first step in factoring. To find the greatest common factor:
>
> 1. Factor the coefficient of each term into prime factors. Do not factor the variables.
>
> 2. The **greatest common factor** is the product of all the *common factors* raised to the least power that they occur in any factorization.

EXAMPLE 1   Find the greatest common factor.

a. 14, 18, and 54

$$14 = \qquad\quad 2 \cdot 7$$
$$18 = 2 \cdot 3 \cdot 3 \quad = 2 \cdot 3^2$$
$$54 = 3 \cdot 3 \cdot 3 \cdot 2 = 3^3 \cdot 2$$

The greatest common factor is 2. The only common factor is 2, and the lowest power of 2 in any factorization is 1, so $2^1 = 2$ is the greatest common factor.

b. $16x^2$ and $84x^3$

$$16x^2 = 2 \cdot 2 \cdot 2 \cdot 2 \cdot x^2 = 2^4 x^2$$
$$84x^3 = 2 \cdot 2 \cdot 3 \cdot 7 = \qquad 2^2 \cdot 3 \cdot 7x^3$$

The greatest common factor is $2^2 x^2$ or $4x^2$. The common factors are 2 and x. The lowest power of 2 in any factorization is 2 and the lowest power of x is 2, so the greatest common factor is $2^2 x^2$ or $4x^2$.

c. $9(c + d)$, $y(c + d)$

$$9(c + d) = 3 \cdot 3(c + d)$$
$$y(c + d) = y(c + d)$$

The greatest common factor is $c + d$. The factor $(c + d)$ is the only factor that occurs in both factorizations, and it occurs to the first power. ■

☐ **DO EXERCISE 1.**

☐ **Exercise 1**   Find the greatest common factor.

a. 4, 12

$$4 = 2 \cdot 2$$
$$12 = 3 \cdot 2 \cdot 2$$
$$2^2 = 4$$

b. 36, 84, 24

$$36 = 2 \cdot 3 \cdot 2 \cdot 3 = 2^2 \cdot 3^2$$
$$84 = 3 \cdot 7 \cdot 2 \cdot 2 = 2^2 \cdot 3 \cdot 7$$
$$24 = 2 \cdot 2 \cdot 2 \cdot 3 = 2^3 \cdot 3$$
$$2^2 \cdot 3 = 12$$

c. z^3, z^2

$$z^2$$

d. $16yz^5$, $40y^2z^2$

$$2 \cdot 2 \cdot 2 \cdot 2 = 2^4 yz^5$$
$$2 \cdot 2 \cdot 2 \cdot 5 = 2^3 \cdot 5 y^2 z^2$$
$$2^3 yz^2$$

e. $4(a - b)$, $y(a - b)$

f. $x(a + b)$, $8(a + b)$

□ **Exercise 2**   Factor.

a. $5x - 10$

b. $x^4 + x^6$

c. $3x^3 - 6x^2 + 3x$

d. $7y - 14y^2 - 7y^3$

e. $4x^2y + 8xy$

f. $3x^2y^2 - 3xy$

EXAMPLE 2   Factor.

a. $6x + 12$

$$6 = \qquad 2 \cdot 3$$
$$12 = 2 \cdot 2 \cdot 3 = 2^2 \cdot 3$$

The greatest common factor is $2 \cdot 3 = 6$.
It is the largest number that will divide evenly into $6x$ and 12. Place it in front of a set of parentheses.

$$6(\qquad)$$

Now place in the parentheses the quotient of the given expression and the number that we factored out.

$$\frac{6x + 12}{6} = x + 2 \qquad \text{The quotient is } x + 2$$

Hence the result is

$$6x + 12 = 6(x + 2)$$

b. $x^3 + x^2$

The variable with the highest exponent that divides into each term is x^2.

$$x^2(\qquad) \qquad x^2 \text{ is the greatest common factor}$$

The quotient of the given expression and x^2 is

$$\frac{x^3 + x^2}{x^2} = x + 1 \qquad \text{Recall } x^3/x^2 = x \text{ and } x^2/x^2 = 1$$

Therefore, $x^3 + x^2 = x^2(x + 1)$.

In the following examples we factor out the greatest common factor. This is the largest number and the variable with the highest exponent that will divide evenly into each term.

c. $8x^5 - 16x^3 + 4x^2$

$$8x^5 = 2 \cdot 2 \cdot 2 \cdot x^5 = \qquad 2^3 x^5$$
$$16x^3 = 2 \cdot 2 \cdot 2 \cdot 2 \cdot x^3 = 2^4 x^3$$
$$4x^2 = 2 \cdot 2 \cdot x^2 \qquad = 2^2 x^2$$

The greatest common factor is $2^2 x^2 = 4x^2$.

$8x^5 - 16x^3 + 4x^2$
$= 4x^2(2x^3 - 4x + 1)$ Since $(8x^5 - 16x^3 + 4x^2)/4x^2 = 2x^3 - 4x + 1$

d. $15xy^3 - 5xy^2 = 5xy^2(3y - 1)$ ■

□ **DO EXERCISE 2.**

2 Using Factoring to Solve Formulas for a Variable

Recall that we solved some formulas in Section 3.6. It may be necessary to factor in order to solve formulas for a specific variable.

EXAMPLE 3 Solve for the given variable.

a. $A = p + prt$ for p

$\qquad A = p(1 + rt)$ Factoring out p

$\qquad \dfrac{A}{1 + rt} = p$ Dividing both sides of the equation by $1 + rt$

b. $ax = a - x$ for x

$\qquad ax + x = a$ All terms that contain the variable must be on one side of the equation

$\qquad x(a + 1) = a$ Factoring out x

$\qquad x = \dfrac{a}{a + 1}$ Dividing both sides of the equation by $a + 1$

Notice that the equation is not solved if both sides of the equation contain an x. ∎

☐ **DO EXERCISE 3.**

3 Factoring by Grouping

A polynomial with four terms can sometimes be factored by grouping the terms into groups of two which have common factors.

EXAMPLE 4 Factor.

a. $x^2 - 2x + 4x - 8$

$= (x^2 - 2x) + (4x - 8)$

$= x(x - 2) + 4(x - 2)$

Each term has a common factor of $x - 2$, so factor it out.

$\qquad x(x - 2) + 4(x - 2) = (x + 4)(x - 2)$

b. $bx - ax + by - ay$

$= (bx - ax) + (by - ay)$

$= x(b - a) + y(b - a)$

Now each term has a common factor of $b - a$, so factor it out.

$\qquad x(b - a) + y(b - a) = (x + y)(b - a)$

☐ **Exercise 3** Solve for the given variable.

a. $3 = x + xy$ for x

b. $y = b + by$ for y

c. $v = kt + gt$ for t

d. $S = 2r^2 + 2rh$ for h

□ **Exercise 4** Factor.

a. $x^2 - 3x + 2x - 6$

c. $ax + bx + a + b$

$= (ax + bx) + (a + b)$

$= x(a + b) + (a + b) = x(a + b) + 1(a + b)$

$= (x + 1)(a + b)$ Do not forget 1

d. $ay + by - a - b$

The factors in the parentheses must be the same so that we have a common factor. Hence we factor -1 out of the second group.

$$ay + by - a - b = y(a + b) - (a + b) = (y - 1)(a + b)$$ ■

□ **DO EXERCISE 4.**

b. $b + b^2 + a + ab$

DID YOU KNOW?

Our word *algebra* comes from the Arabic word *al-jabr*. *Al-jabr*, which can be translated as meaning "restoration," was used to indicate that terms subtracted from one side of an equation were restored on the other side of the equation. In Spain, *al-jabr* was translated to *algebrista* and came to mean "bonesetter," one who restored bones.

c. $1 - x - y + xy$

d. $4xz - 4yz - x + y$

Answers to Exercises

1. a. 4   **b.** 12   **c.** z^2 **d.** $8yz^2$ **e.** $a - b$ **f.** $a + b$

2. a. $5(x - 2)$ **b.** $x^4(1 + x^2)$ **c.** $3x(x^2 - 2x + 1)$

d. $7y(1 - 2y - y^2)$ **e.** $4xy(x + 2)$ **f.** $3xy(xy - 1)$

3. a. $x = \dfrac{3}{1 + y}$ **b.** $y = \dfrac{b}{1 - b}$ **c.** $t = \dfrac{v}{k + g}$ **d.** $h = \dfrac{S - 2r^2}{2r}$

4. a. $(x - 3)(x + 2)$ **b.** $(1 + b)(b + a)$ **c.** $(1 - x)(1 - y)$

d. $(x - y)(4z - 1)$

PROBLEM SET 5.6

A. *Factor.*

1. $4x + 8$

2. $3x - 9$

3. $x^3 + 4x^2$

4. $x^3 + 2x^5$

5. $6x^4 + 18x^3$

6. $7x^3 - 14x$

7. $4x^3 + 2x^2 + 2x$

8. $9x^4 - 3x^2 + 3x$

9. $10x^3 - 5x^2 + 5$

10. $8x^3 + 4x^2 + 5x$

11. $2x^2 - 4xy$

12. $5xy^2 + 10x^2y^2$

13. $8x^3y^3 + 12x^2y^2 + 4xy$

14. $3x^4 - 6x^3y$

15. $7x^2 - 7xy$

16. $2\pi rh + 2\pi r^2$

B. *Solve for the given variable.*

17. $x + xy = 7$ for x

18. $y = a - ay$ for y

19. $A = xt + gt$ for t

20. $9 = y - ay$ for y

21. $x = r + rx$ for x

22. $12 = 3y - by$ for y

C. *Factor.*

23. $y^2 + 5y + 2y + 10$

24. $x^2 - 4x + 2x - 8$

25. $x^2 + 5x - 2x - 10$

26. $2y^2 + 8y - 3y - 12$

27. $6x^2 - 9x - 2x + 3$

28. $5x^2 + 10x + 4x + 8$

29. $(a - b)x + (a - b)$

30. $a(x + y) + b(x + y)$

31. $y^2 + ay + xy + ax$

32. $xp + xq + yp + yq$

33. $3y + x - 6y^2 - 2xy$

34. $3xy - yz - 3vx + vz$

35. $x^3 + 2xy^2 - 2x^2y - 4y^3$

36. $2x^2y + 6x^2 - y - 3$

37. To factor out the greatest common factor, factor out the largest number and the variable with the largest _____ that will divide evenly into each term.

38. It may be necessary to factor in order to solve some formulas for a specific _____.

39. A polynomial with four terms can sometimes be factored by grouping the terms into groups of two which have _____ factors.

Checkup

The following problems provide a review of some of Section 5.3. These problems will help you with the next section. Multiply using FOIL.

40. $(x - 3)(x + 5)$

41. $(x + 2)(x + 4)$

42. $(x + 2)(2x + 3)$

43. $(2x - 1)(x + 1)$

44. $(3x - 2)(2x - 1)$

45. $(4x - 1)(2x - 1)$

46. $2(3x - 4)(x + 2)$

47. $3(x - 2)(2x + 3)$

48. $x(x - 1)(2x - 3)$

49. $x^2(3x - 1)(x + 5)$

50. $2x(x - 2)(x + 4)$

51. $3x^2(x - 3)(x - 1)$

5.7 FACTORING TRINOMIALS

OBJECTIVES

1 *Factor trinomials when the coefficient of x^2 is 1*

2 *Factor trinomials when the coefficient of x^2 is other than 1*

3 *Factor completely*

Many trinomials can be factored into the product of two binomials. If the trinomial has a common factor, it should be factored out first.

1 Factoring Trinomials When the Coefficient of x^2 Is 1

If we multiply $(x + 2)(x + 4)$, the result is $x^2 + 6x + 8$. We must learn to reverse the process and factor $x^2 + 6x + 8$ to get $(x + 2)(x + 4)$. Remember that to factor means to write the expression as a multiplication of expressions. Factoring is done by trial and error. It becomes easier with practice!

Recall that we may multiply two binomials as follows:

$$
\begin{array}{ccccc}
& \text{F} & \text{O} & \text{I} & \text{L} \\
(x + 2)(x + 4) & = x^2 & + 4x & + 2x & + 8 \\
& = x^2 + 6x + 8
\end{array}
$$

In general,

$$
\begin{array}{ccccc}
& \text{F} & \text{O} & \text{I} & \text{L} \\
(x + b)(x + a) & = x^2 & + ax & + bx & + ab \\
& = x^2 + (a + b)x + ab
\end{array}
$$

We factor by reversing the procedure.

$$x^2 + (a + b)x + ab = (x + b)(x + a)$$

Notice that we need to find numbers a and b whose sum is the coefficient of x and whose product is the last term.

EXAMPLE 1 Factor.

a. $x^2 + 6x + 8$

1. Factor the x^2 term. $x^2 = x \cdot x$. These factors will be the first terms in each binomial factor.

$$x^2 + 6x + 8 = (x \quad)(x \quad)$$

2. We need to find two numbers whose product is 8 (the last term) and whose sum is 6 (the coefficient of x). The two numbers whose product is 8 are factors of 8.

   Factor the constant term, $+8$. To do this, list all the possible pairs of factors of $+8$. We can do this in a systematic way by starting with 1 as the first number in our first pair of factors. So our first factors are 1 and 8. Since $(-1)(-8) = 8$, -1 and -8 are also factors. Then use as a first number in the next pair of factors the next number greater than 1 that divides 8 evenly. In this case the number is 2. So 2 and 4 are factors of $+8$ and since $(-2)(-4) = 8$, -2 and -4 are also factors. We stop with -2 and -4 since 4 and 2 is the same pair of factors as 2 and 4.

Factors of $+8$	Sum of Factors of $+8$
1, 8	9
$-1, -8$	-9
2, 4	6
$-2, -4$	-6

3. The numbers are 2 and 4 since their product is 8 and their sum is 6. Place them as the second terms in the binomial factors shown in step 1.

$$x^2 + 6x + 8 = (x + 2)(x + 4)$$

   This is the correct factorization. Notice that we may eliminate the negative factors in this case because the middle term, $6x$, is positive.

b. $x^2 - 4x - 21$.

1. $x^2 - 4x - 21 = (x \quad)(x \quad)$
2. We want two numbers whose product is -21 and whose sum is -4.

Factors of -21	Sum of Factors of -21
-1, 21	20
1, -21	-20
-3, 7	4
3, -7	-4

   Notice that 2 does not divide evenly into 21. The numbers are 3 and -7.

$$x^2 - 4x - 21 = (x + 3)(x - 7) \quad \blacksquare$$

□ **DO EXERCISE 1.**

2 Factoring Trinomials When the Coefficient of x^2 Is Other Than 1

The methods of Example 1 cannot be used for this type of factoring. The x^2 term must be factored and the last term must be factored. We use the factors that give the correct middle term.

EXAMPLE 2 Factor.

a. $2x^2 + 7x + 3$

    1. Factor $2x^2$.

Factors of $+2x^2$

$x, 2x$

Use only the positive factors of the first term since we usually want the first coefficients of the factors to be positive.

$$2x^2 + 7x + 3 = (x \quad)(2x \quad)$$

    2. Ignore the middle term, $7x$, until step 4.

    3. Factor $+3$.

Factors of $+3$

$$1, \quad 3$$
$$-1, -3$$

    4. Use the pair of factors that will give $7x$, which is the middle term of the original expression. Since this term is positive, the factors are 1 and 3.

$$2x^2 + 7x + 3 \overset{?}{=} (x + 1)(2x + 3)$$

The middle term of the expression, which is $7x$, is the sum of the products of the outer and inner terms of the binomial factors. We learned this when we used FOIL to multiply two binomials. Do these binomial factors work?

The sum of the products of the outer and inner terms is $5x$, not $7x$, which is the middle term of the original expression.

    In this type of problem, where the coefficient of x^2 in the first term is other than 1, it is sometimes necessary to reverse the second pair of factors.

$$2x^2 + 7x + 3 = (x + 3)(2x + 1)$$

This is the correct factorization since the sum of the products of the inner and outer terms is $7x$.

□ **Exercise 2**   Factor.

a. $3x^2 + 10x + 7$

b. $3x^2 + x - 10$

c. $15x^2 + x - 2$

d. $6x^2 + 7x + 2$

b. $8x^2 - 10x - 3$

  1. Factor $8x^2$.

Factors of $8x^2$

$$x, 8x$$
$$2x, 4x$$

We try $2x$ and $4x$.

$$8x^2 - 10x - 3 = (2x \quad)(4x \quad)$$

  2. Ignore $-10x$ until step 4.
  3. Factor -3.

Factors of -3

$$1, -3$$
$$-1, \quad 3$$

  4. We will try 1 and -3.

$$8x^2 - 10x - 3 \overset{?}{=} \overbrace{(2x + 1)(4x - 3)}^{O}_{I}$$

This gives us a middle term of $-2x$, which does not equal $-10x$, the middle term of the original expression. Reverse the second pair of factors.

$$8x^2 - 10x - 3 = (2x - 3)(4x + 1)$$

This is the correct factorization. If it had not been correct, we would have tried the factors -1 and 3. If those were not correct, we would try the other factors of $8x^2$, $8x$ and x, with the factors of -3. ■

An *alternative method* of factoring these trinomials is shown in *Appendix B*.

□ **DO EXERCISE 2.**

3 **Factoring Completely**

If the terms of the trinomial have a greatest common factor, it should be factored out first. This makes factoring easier and helps us to factor completely.

EXAMPLE 3 Factor.

a. $2x^2 + 4x - 6$

    1. Factor out the greatest common factor, 2.
$$2x^2 + 4x - 6 = 2(x^2 + 2x - 3)$$

    2. Now factor $x^2 + 2x - 3$ carrying along the greatest common factor, 2. Notice that the coefficient of the x^2 term is 1.
$$2x^2 + 4x - 6 = 2(x^2 + 2x - 3) = 2(x \quad)(x \quad)$$

Find two numbers whose product is -3 and whose sum is 2.

Factors of -3	Sum of Factors of -3
1, -3	-2
-1, 3	2

The numbers are -1 and 3.
$$x^2 + 2x - 3 = (x - 1)(x + 3)$$
$$2x^2 + 4x - 6 = 2(x - 1)(x + 3)$$

b. $2x^3 - 8x^2 - 10x$

    1. Factor out the greatest common factor, $2x$.
$$2x^3 - 8x^2 - 10x = 2x(x^2 - 4x - 5)$$

    2. Now factor $x^2 - 4x - 5$ carrying along the greatest common factor, $2x$.
$$2x(x^2 - 4x - 5) = 2x(x \quad)(x \quad)$$

Find two factors of -5 whose sum is -4.

Factors of -5	Sum of Factors of -5
1, -5	-4
-1, 5	4

The factors are 1 and -5.
$$2x^3 - 8x^2 - 10x = 2x(x^2 - 4x - 5) = 2x(x + 1)(x - 5)$$
■

We may be unable to factor some trinomials.

□ **DO EXERCISE 3.**

□ **Exercise 3** Factor.

a. $6x^2 - 14x + 4$

b. $2x^2 - 2x - 24$

c. $4x^2 + 2x - 12$

d. $3x^2 - 18x + 27$

e. $3x^3 - 3x^2 - 18x$

f. $9x^3y - 21x^2y + 6xy$

PROBLEM SET 5.7

Factor.

1. $x^2 + 6x + 5$ **2.** $x^2 - 8x + 7$ **3.** $x^2 + 9x + 20$ **4.** $x^2 + 7x + 12$

5. $y^2 - 14y + 24$ **6.** $y^2 - 7y - 18$ **7.** $y^2 - 4y - 12$ **8.** $y^2 - 6y + 8$

9. $x^2 + 3x - 28$ **10.** $x^2 + 8x + 12$ **11.** $x^2 + 2x + 1$ **12.** $x^2 - 2x + 1$

13. $x^4 - 4x^2 + 3$ **14.** $y^4 + 5y^2 + 6$ **15.** $y^2 + 10y + 9$ **16.** $y^2 - 10y - 11$

17. $x^2 - 8x + 15$ **18.** $x^2 + 3x - 40$ **19.** $y^4 - 5y^2 - 14$ **20.** $x^4 - 9x^2 + 18$

21. $y^2 + 13y + 42$ **22.** $x^2 - 17x + 72$ **23.** $6 + x - x^2$ **24.** $8 + 6x + x^2$

25. $y^2 - 7y + 10$ **26.** $y^2 + y - 56$ **27.** $x^2 - x + \dfrac{1}{4}$ **28.** $y^2 - \dfrac{8}{3}y + \dfrac{16}{9}$

29. $x^2 - 3x - 108$ **30.** $x^2 - 14x + 33$ **31.** $3x^2 + 18x + 15$ **32.** $2x^2 + 16x + 30$

33. $3y^2 + 21y - 24$ **34.** $2y^2 - 4y - 30$ **35.** $9x^2 + 18x + 8$ **36.** $15x^2 - 25x + 10$

37. $y^2 - 14y + 49$ **38.** $y^2 + y - 42$ **39.** $3y^2 + 9y + 6$ **40.** $2y^2 + 20y + 50$

41. $6x^2 + 8x + 2$

42. $6x^2 - 3x - 3$

43. $14y^2 + 35y + 14$

44. $35x^2 - 11x - 6$

45. $4x^2 + 20x - 24$

46. $2x^2 + 16x + 24$

47. $3x^2 + 27x + 60$

48. $4y^2 + 16y - 20$

49. $3x^3 + 12x^2 + 9x$

50. $p^3 + 7p^2 - 8p$

51. $3x^4 - 3x^3 - 90x^2$

52. $2y^3 - 8y^2 - 10y$

53. $14y^2 + 19y - 3$

54. $15y^2 + 19y - 10$

55. $18x^3 - 21x^2 - 9x$

56. $9x^5 + 18x^3 + 8x$

57. $18x^2 + x - 4$

58. $35x^2 - 11x - 6$

59. $6x^2 - 3x - 63$

60. $4x^2 - 30x + 56$

61. $2x^3 - 7x^2 - 15x$

62. $3x^4 - 5x^3 - 2x^2$

63. $2x^3 - 24x^2 + 64x$

64. $3x^4 + 15x^3 - 42x^2$

65. If a trinomial has a common factor, it is usually easier to factor it out _____.

66. To list all possible pairs of factors of a number, begin with _____ as the first number in the first pair.

Checkup

The following problems provide a review of some of Section 5.4. These problems will help you with the next section. Multiply. Do not use FOIL.

67. $(y - 3)^2$

68. $(x - 7)^2$

69. $(2x + 3)^2$

70. $(3y + 4)^2$

71. $(x + 5)(x - 5)$

72. $(x + 8)(x - 8)$

73. $(6x + 7)(6x - 7)$

74. $(5x + 2)(5x - 2)$

75. $(7x - 1)^2$

76. $(9x - 2)^2$

Perfect square trinomial the square of a binomial.

5.8 SPECIAL FACTORS OF POLYNOMIALS

OBJECTIVES

1 *Factor a perfect square trinomial*

2 *Factor the difference of two squares*

1 Factoring a Perfect Square Trinomial

A **perfect square trinomial** is the square of a binomial. Recall from Section 5.4 that there is a special method for squaring a binomial.

$$(A + B)^2 = A^2 + 2AB + B^2$$

and

$$(A - B)^2 = A^2 - 2AB + B^2$$

We want to reverse this procedure and factor as follows.

$$A^2 + 2AB + B^2 = (A + B)^2$$

and

$$A^2 - 2AB + B^2 = (A - B)^2$$

Notice that the first and last terms of the trinomial to be factored must be squares. Here is a list of some squares.

Number	Square	
1	1	
−1	1	
2	4	
−2	4	
±3	9	Both $(+3)^2$ and $(-3)^2$ equal 9
±4	16	
±5	25	
±6	36	
±7	49	
±8	64	
±9	81	
±10	100	

Variables may also be squares.

Variable	Square	
x	x^2	
x^2	x^4	Recall $(x^2)^2 = x^2 \cdot x^2 = x^4$
x^3	x^6	Recall $(x^3)^2 = x^3 \cdot x^3 = x^6$
x^4	x^8	
x^5	x^{10}	

We may continue the list indefinitely. Notice that a square is a number or variable multiplied times itself.

Since the first and last terms of the square of a binomial are squares, they must factor into identical factors.

EXAMPLE 1 Factor.

a. $x^2 + 8x + 16$

1. Factor the first term, x^2.

$$(x \quad)(x \quad) \qquad \text{Identical factors}$$

2. Ignore the middle term, $8x$, until later.
3. Factor the last term, 16. The factors must be identical. They are either $(+4)(+4)$ or $(-4)(-4)$. Since the middle term, $8x$, is positive, they must be $(+4)(+4)$.

$$(x + 4)(x + 4)$$

4. Check to see that the sum of the products of the outer terms and inner terms equals the middle term, $8x$.

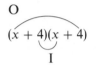

The sum does equal $8x$.

$$x^2 + 8x + 16 = (x + 4)(x + 4) = (x + 4)^2$$

b. $4x^2 - 12xy + 9y^2$

1. Factor $4x^2$.

$$(2x \quad)(2x \quad) \qquad \text{Identical factors}$$

2. Ignore the middle term until step 3.
3. Factor $9y^2$.

$$(2x - 3y)(2x - 3y) \qquad \text{Notice that we use } -3y \text{ since the middle term is } -12xy$$

4. Check to see that the sum of the products of the outer and inner terms is $-12xy$.

The sum is $-12xy$.

$$4x^2 - 12xy + 9y^2 = (2x - 3y)(2x - 3y)$$

Caution A trinomial may not be the square of a binomial even though the first and last terms of the trinomial are squares.

c. $x^2 + 2x + 4$

   1. Factor x^2.

$$(x \quad)(x \quad)$$

   2. Ignore $2x$.

   3. Factor 4.

$$(x + 2)(x + 2)$$

   4. Check the sum of the products of the outer and inner terms.

$$(x + 2)(x + 2)$$

$$\text{I}$$

   This sum is $4x$, which does not equal the middle term of the original expression, $2x$. This trinomial is not factorable. We must always check to see that the middle term is correct.

Some trinomials whose first and last terms are squares may be factorable by the general method.

d. $x^2 + 10x + 9$

   1. Factor x^2.

$$(x \quad)(x \quad)$$

   2. Find two factors of 9 whose sum is 10.

Factors of 9	Sum of Factors of 9
1, 9	10
−1, −9	−10
3, 3	6
−3, −3	−6

   The factors are 1 and 9.

$$x^2 + 10x + 9 = (x + 1)(x + 9)$$

   Notice that 3 and 3 are not the factors of 9 to use. ■

☐ **DO EXERCISE 1.**

☐ **Exercise 1**   Factor.

a. $x^2 + 10x + 25$

b. $x^2 - 6x + 9$

c. $4x^2 - 20xy + 25y^2$

d. $9x^2 + 12xy + 4y^2$

□ **Exercise 2** Factor.

a. $x^2 - 36$

b. $y^2 - 100$

c. $9x^2 - 49y^2$

d. $25x^2 - 81y^2$

2 **Factoring the Difference of Two Squares**

Recall from Section 5.4 that when we multiply two binomials that are identical except for the sign between them, the result is the difference of two squares.

$$(A - B)(A + B) = A^2 - B^2$$

We want to reverse this procedure and factor the difference of two squares as follows:

$$A^2 - B^2 = (A - B)(A + B)$$

Notice that if an expression is the difference of two squares, both terms must be squares and there must be a minus sign between the terms. *The sum of two squares, $A^2 + B^2$, does not factor.*

EXAMPLE 2 Factor.

a. $x^2 - 9$

    1. Factor x^2.

$$(x \quad)(x \quad)$$

    2. Factor 9. The factors must be identical.

$$(x \quad 3)(x \quad 3)$$

    3. Use a plus sign between one set of factors and a minus sign between the other set.

$$x^2 - 9 = (x + 3)(x - 3) \quad \text{or} \quad (x - 3)(x + 3)$$

b. $16x^4 - 25y^2$

    1. Factor $16x^4$.

$$(4x^2 \quad)(4x^2 \quad)$$

    2. Factor $25y^2$.

$$(4x^2 \quad 5y)(4x^2 \quad 5y)$$

    3. Insert a minus sign between one set of factors and a plus sign between the other set.

$$(4x^2 - 5y)(4x^2 + 5y)$$

$$16x^4 - 25y^2 = (4x^2 - 5y)(4x^2 + 5y) \quad ∎$$

□ **DO EXERCISE 2.**

Special Factors of Polynomials

$$A^2 + 2AB + B^2 = (A + B)(A + B) = (A + B)^2$$

$$A^2 - 2AB + B^2 = (A - B)(A - B) = (A - B)^2$$

$$A^2 - B^2 = (A + B)(A - B)$$

$A^2 + B^2$ does not factor

© 1989 by Prentice Hall

Prime polynomial cannot be factored using polynomials with integer coefficients.

3 A General Summary of Factoring

> There are certain steps that help you to factor quickly.
>
> 1. Factor out the greatest common factor first. If a factor in your answer still has a common factor, it is not factored completely.
> 2. If there are *two terms* to be factored: Try to factor the expression as the difference of squares. If there are *three terms*: Check to see if the expression is a perfect square trinomial. If it is not, use the general method for factoring a trinomial. If there are *four terms*: Try to separate the expression into groups of two terms that have a common factor and factor completely.

Some expressions are **prime** (cannot be factored using polynomials with integer coefficients).

EXAMPLE 3 Factor.

a. $5x^2 - 20$

    1. Factor out the greatest common factor, 5.

$$5(x^2 - 4)$$

    2. Factor the difference of two squares. Remember to keep the factor 5.

$$5(x + 2)(x - 2)$$
$$5x^2 - 20 = 5(x + 2)(x - 2)$$

b. $x^3 + x^2 - 6x$

    1. Factor out the greatest common factor, x.

$$x(x^2 + x - 6)$$

    2. Is the expression in parentheses the square of a binomial? No, 6 is not a square, so use the general method of factoring a trinomial. Do not forget to carry along the common factor.

    3. Factor x^2.

$$x(x \quad)(x \quad)$$

    4. Find two factors of -6 whose sum is 1 (the coefficient of x).

Factors of -6	Sum of Factors of -6
1, -6	-5
-1, 6	5
2, -3	-1
-2, 3	1

The correct factors are -2 and 3.

$$x(x - 2)(x + 3)$$
$$x^3 + x^2 - 6x = x(x - 2)(x + 3)$$

☐ **Exercise 3**   Factor.

a. $x^3 - 16x$

b. $xy + x + yz + z$

c. $4x^2 + 8x + 4$

d. $6x^2 - 3x - 9$

c. $ab + a + b + 1$

1. Is there a greatest common factor? No.
2. There are four terms, so group the expression into two binomials.

$$(ab + a) + (b + 1)$$

3. Factor out the greatest common factor.

$$a(b + 1) + (b + 1)$$

The greatest common factor of the second binomial is 1.

4. Factor the common factor from the two binomials.

$$(a + 1)(b + 1)$$
$$ab + a + b + 1 = (a + 1)(b + 1)$$

d. $3x^4 - 48$

1. Factor out the greatest common factor, 3.

$$3(x^4 - 16)$$

2. Factor x^4. Remember to keep 3 as a factor.

$$3(x^2 \quad)(x^2 \quad)$$

3. Factor -16. We are factoring the difference of two squares, so the factors must be $+4$ and -4.

$$3(x^2 + 4)(x^2 - 4)$$

4. This expression is not factored completely. The difference of two squares $(x^2 - 4)$ can be factored.

$$x^2 - 4 = (x + 2)(x - 2)$$

Hence

$$3(x^2 + 4)(x^2 - 4) = 3(x^2 + 4)(x + 2)(x - 2)$$

Notice that the sum of two squares cannot be factored.

$$3x^4 - 48 = 3(x^2 + 4)(x + 2)(x - 2) \quad \blacksquare$$

☐ **DO EXERCISE 3.**

Some equations can be solved by factoring. We will learn how to solve these equations in Chapter 11.

Answers to Exercises _____

1. **a.** $(x + 5)^2$ **b.** $(x - 3)^2$ **c.** $(2x - 5y)^2$ **d.** $(3x + 2y)^2$
2. **a.** $(x + 6)(x - 6)$ **b.** $(y - 10)(y + 10)$ **c.** $(3x + 7y)(3x - 7y)$
   d. $(5x - 9y)(5x + 9y)$
3. **a.** $x(x + 4)(x - 4)$ **b.** $(y + 1)(x + z)$ **c.** $4(x + 1)^2$
   d. $3(2x - 3)(x + 1)$

PROBLEM SET 5.8

A. *Factor.*

1. $x^2 + 4x + 4$

2. $x^2 - 14x + 49$

3. $x^2 - 16x + 64$

4. $x^2 + 18x + 81$

5. $4x^2 + 12x + 9$

6. $25x^2 - 10x + 1$

7. $16x^2 - 8x + 1$

8. $9x^2 + 12x + 4$

9. $x^2 - 4xy + 4y^2$

10. $x^2 + 6xy + 9y^2$

11. $9x^2 + 6xy + y^2$

12. $16x^2 - 8xy + y^2$

13. $x^4 - 4x^2 + 4$

14. $y^4 + 8y^2 + 16$

15. $y^6 + 10y^3 + 25$

16. $x^6 - 14x^3 + 49$

17. $36 - 12x + x^2$

18. $25 + 10y + y^2$

19. $x^2 - 25$

20. $y^2 - 16$

21. $x^2 - 1$

22. $y^2 - 64$

23. $4y^2 - 9$

24. $49x^2 - 4$

25. $36x^2 - 25$

26. $64x^2 - 49$

27. $9x^2 - y^2$

28. $25x^2 - y^2$

29. $36x^2 - 49y^2$

30. $16x^2 - 9y^2$

31. $x^4 - 4$

32. $y^4 - 25$

33. $x^6 - 25$

34. $y^8 - 49$

35. $9 - x^2$

36. $64 - y^2$

B. *Factor completely.*

37. $x^3 - 9x$

38. $x^4 - 16x^2$

39. $5x^2 + 10x + 5$

40. $3x^2 - 12x + 12$

41. $4x^2 + 4x - 8$ **42.** $3x^2 - 18x - 21$ **43.** $x^3 + 2x^2 - 8x$ **44.** $y^3 - 7y^2 - 8y$

45. $x^3 + 10x^2 + 25x$ **46.** $y^3 - 6y^2 + 9y$ **47.** $8x^4 - 72x^2$ **48.** $4y^5 - 4y^3$

49. $3x^2 + 5x - 2$ **50.** $21x^2 - 11x - 2$ **51.** $2bx + by + 2x + y$ **52.** $3ax + 3ay - x - y$

53. $x^4 - 1$ **54.** $y^4 - 81$ **55.** $4x^2 - 4xy + y^2$ **56.** $y^2 + 6xy - 9x^2$

57. $ax + 3ay + 4x + 12y$ **58.** $bx - 2by - 6x + 12y$ **59.** $2x^4 - 36x^3 + 162x^2$ **60.** $5x^3 - 40x^2 + 80x$

61. $10x - 10x^3$ **62.** $12y^2 - 3y^4$ **63.** $25 - 10x^2 + x^4$ **64.** $49 - 14y^3 + y^6$

65. A perfect square trinomial is the square of a _____.

66. A trinomial may not be the square of a binomial even though the first and last terms of the trinomial are _____.

67. The first step in factoring is to factor out the _____ _____ _____.

68. If there are two terms in the expression to be factored, after factoring out the greatest common factor, try to factor the expression as the _____ of squares.

Checkup

The following problems provide a review of some of Section 5.1. Are the following polynomials?

69. $x^2 + 3x - 2$ **70.** $x^{-1} + 2x - 3$ **71.** $\dfrac{9}{x} - 7$ **72.** $\dfrac{x}{3} + 8$

Find the degree of the polynomial.

73. $x^5 + x^2 - 2$ **74.** $x + x^4 + 3$ **75.** $6 - x^2$ **76.** $3x + x^3$

CHAPTER 5   ADDITIONAL EXERCISES (OPTIONAL)

Section 5.1

Are the following polynomials?

1. $x^2 + 3x$

2. $x^{-3} + 4$

3. $\dfrac{3}{x} - 8$

4. $\dfrac{x}{3} + 5$

Are the following trinomials, binomials, monomials, or none of these?

5. $x + 2$

6. 3

7. $6x^2 + 7x - 4$

8. $7x^3 + 4x^2 - 3x - 8$

Combine like terms and arrange in descending order.

9. $-5x + \dfrac{1}{3}x^2 + \dfrac{1}{4} - \dfrac{3}{5}x^2 - \dfrac{5}{16} - 8x$

10. $\dfrac{3}{8} - \dfrac{2}{5}x^2 + 7x^3 + \dfrac{5}{2} - \dfrac{9}{2}x^2 + 2x^3$

Find the degree of the polynomial.

11. $4x^2 - 6x + 8$

12. $7 - y$

Section 5.2

Add.

13. $(3x^3 - 5x^2 + 2) + (6x^3 + 9x^2 - 8)$

14. $(2x^3 - 7x - 3) + (-5x^3 - 2x + 1)$

15. $(7x^3 - 3.1x - 4.2) + (0.4x^3 - 7.8x^2 + 2.5)$

16. $(3.7x^2 - 0.6x - 8.1) + (4.2x^3 - 8.9x^2 + 5.7x)$

Subtract.

17. $(8x^4 + 3x^2 - 2) - (9x^4 - 6x^2 + 5)$

18. $(4x^2 - 7x + 5) - (5x^2 + 3x + 2)$

19. Subtract $3x^2 + 4x - 17$ from $-8x^3 + 9x^2 - 23$.

20. Subtract $-5x^3 + 3x^2 + 4x$ from $-9x^3 + 8x - 19$.

Section 5.3

Multiply. Assume that all variables in exponents represent positive integers.

21. $3x(x^2 - 2x + 4)$

22. $(y^2 - 5y + 7)(-2y)$

23. $-8x^{2n}y^{3n}(9x^{4n}y^{5n})$

24. $-9x^ny^{4n}(-7x^{3n}y^n)$

25. $4x^n(3x^{5n} - 8x^{2n})$

26. $-9x^{4n}(x^n + 6x^{3n})$

27. $(3x - 2y)(5x + 4y)$

28. $(7x - y)(6x + 2y)$

$12x^{6n} - 32x^{3n}$

29. $(8x^{2n} - 3y^{3n})(x^n - 2y^{2n})$

30. $(5x^{4n} - 2y^{7n})(3x^n - 2y^{4n})$

Section 5.4

Multiply. Assume that variables in exponents represent positive integers.

31. $(x + 8)(x - 8)$

32. $(7 - y)(7 + y)$

33. $\left(2x - \dfrac{3}{4}\right)\left(2x + \dfrac{3}{4}\right)$

34. $\left(5x + \dfrac{8}{9}\right)\left(5x - \dfrac{8}{9}\right)$

35. $(2x + 5)^2$

36. $(3x - 7)^2$

37. $(x^{2n} - 9y^{3n})^2$

38. $(x^{4n} + 7y^{6n})^2$

Section 5.5

Divide.

39. $\dfrac{8x^2 + 12x - 3}{4}$

40. $\dfrac{14x^3 - 9x^2 + 21}{7x}$

41. $\dfrac{x^2 + x - 15}{x + 5}$

42. $\dfrac{x^2 - 9x + 9}{x - 2}$

43. $\dfrac{12a^4 + 13a^2 - 16}{3a^2 - 2}$

44. $\dfrac{21x^3 + 28x^2 - 6x - 8}{7x^2 - 2}$

Section 5.6

Find the greatest common factor.

45. $27x^5y^3$, $81x^2y^4$, $54xy^5$

46. $24x^2y^8$, $48x^3y^4$, $96xy^3$

Factor. Assume that variables in exponents represent positive integers.

47. $10x^{2n} - 25x^{4n}$

48. $9x^{3n} + 6x^n$

Solve for the given variable.

49. $A = \dfrac{bh}{2} + \dfrac{ch}{2}$ for h

50. $\dfrac{ax}{3} = b - \dfrac{cx}{3}$ for x

Factor.

51. $4xy - 3yz - 4wx + 3wz$

52. $5y + x - 10y^2 - 2xy$

53. $3x^2y + 6x^2 - y - 2$

54. $x^3 + 4xy^2 - x^2 - 4y^2$

Section 5.7

Factor. Assume that variables in exponents represent positive integers.

55. $x^2 - 7x + 6$

56. $x^2 + x - 12$

57. $6x^2 - x - 15$

58. $8x^2 + 22x + 5$

59. $4x^3 - 22x^2 + 30x$

60. $6x^4 - 21x^3 + 15x^2$

61. $6x^{4n} + 5x^{3n} - 4x^{2n}$

62. $5x^{4n} + 10x^{2n} - 40$

Section 5.8

Factor. Assume that variables in exponents represent positive integers.

63. $9x^2 - 24x + 16$

64. $x^{4n} + 4x^{2n} + 4$

65. $y^{6n} - 10y^{3n} + 25$

66. $x^{2n} - y^{8n}$

67. $25x^2 - 4$

68. $16x^{10n} - 81$

69. $96x^2 + 4x - 15$

70. $105y^2 - 3y - 36$

71. $x^6 - 81x^2$

72. $x^5 - 625x$

CHAPTER 5   PRACTICE TEST

1. Find the degree of the polynomial: $x^2 - 6x + 4$.

1. _____

2. Combine the like terms and arrange in descending order: $-8y^2 + 6y^4 - 3y^2 + 4y - 7y^4$.

2. _____

3. Add: $(5x^2 + 3xy - y^2) + (7x^2 - 6xy - 8y^2)$.

3. _____

Subtract.

4. $\begin{array}{r} 8x^2 - 6x \\ -\ 8x^2 + 4x \\ \hline \end{array}$

4. _____

5. $(-3x^3 - 6x^2 + x) - (-2x^3 + 4x^2 - x)$

5. _____

Multiply.

6. $-2x^2(3x^2 - 4x + 8)$

6. _____

7. $(3y - 2)(4y + 5)$

7. _____

8. $(x - 3)(2x^2 - 3x + 4)$

8. _____

9. $(2x - 4)(2x + 4)$

9. _____

10. _____

10. $(2x - 4)^2$

11. _____

11. Divide:

$$\frac{x^2 - 3x + 5}{x - 2}.$$

12. _____

12. Solve for x: $x = b - bx$.

Factor completely.

13. _____

13. $6x^2 - 3xy + 9xy^2$

14. _____

14. $py + qy + px + qx$

15. _____

15. $x^2 + x - 20$

16. _____

16. $18x^2 + 24x + 6$

17. _____

17. $3x^3 - 12x^2y + 12xy^2$

18. _____

18. $x^4 - 16$

Graphing Straight Lines

6.1 THE RECTANGULAR COORDINATE SYSTEM

It is helpful to be able to draw the graph of an equation. It is usually easier to understand a relationship on a graph than it is to see it as an equation. To draw the graph of an equation, we must learn to plot points on a rectangular coordinate system.

The Rectangular Coordinate System

The **rectangular coordinate system** consists of two number lines at right angles to each other. These number lines are called **axes**.

The following terms are used when we graph a point in this system.

1. An **ordered pair** of numbers is a pair of real numbers given in a definite order. This ordered pair corresponds to a point that is plotted on the rectangular coordinate system. The numbers are called ordered pairs because it makes a difference which number comes first. The point corresponding to the ordered pair (5, 2) is

X-axis   the horizontal number line of a rectangular coordinate system.
First coordinate, X-coordinate, Abscissa   The first number in an ordered pair.
Y-axis   the vertical number line of a rectangular coordinate system.
Second coordinate, Y-coordinate, Ordinate   the second number in an ordered pair.
Origin   the point, (0, 0), where the number lines cross.

placed in a different position than the point corresponding to the ordered pair (2, 5). The phrase "the point corresponding to the ordered pair (*a, b*)" is usually abbreviated "the point (*a, b*)."

2. The horizontal number line is called the **x-axis**. It is labeled *x*. The first number in the ordered pair may be called the **first coordinate**, the **x-coordinate**, or the **abscissa**.

3. The vertical number line is called the **y-axis**. It is labeled *y*. The second number in the ordered pair may be called the **second coordinate**, the **y-coordinate**, or the **ordinate**.

4. The point where the number lines cross is called the **origin**. It has coordinates (0, 0).

▪ Plotting Points

To plot the point (*x, y*), we start at the origin and move *x* units to the right or left (right if *x* is positive and left if *x* is negative). Then from that position we move *y* units up or down (up if *y* is positive and down if *y* is negative). We plot a point at this location.

EXAMPLE 1   Plot the points.

a. (5, 2)

Notice that the first coordinate is the *x*-coordinate and the second coordinate is the *y*-coordinate.

$$(x, y)$$

$$(5, 2)$$

1. Start at the origin.
2. Move 5 units from the origin in a positive direction on the *x*-axis since the *x* coordinate is +5. Do *not* plot a point here.
3. From our position on the *x*-axis we move 2 units in a positive direction parallel to the *y*-axis since the *y*-coordinate is +2.
4. Plot the point.

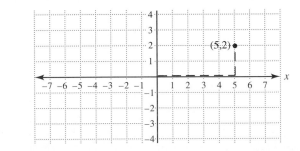

b. (−3, −2)

1. Start at the origin.
2. Move 3 units in a negative direction on the *x*-axis.
3. From this position move 2 units in a negative direction parallel to the *y*-axis.
4. Plot the point.

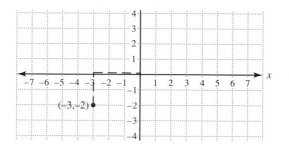

c. (0, 4)

    1. Start at the origin.

    2. Do not move any units along the *x*-axis.

    3. Move 4 units in a positive direction along the *y*-axis.

    4. Plot the point.

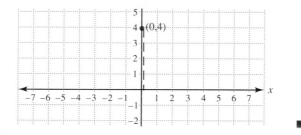

☐ **DO EXERCISE 1.**

2 **Locating Coordinates**

Sometimes we want to know the coordinates of a point that has been plotted on a rectangular coordinate system.

EXAMPLE 2   Find the coordinates of point *P*.

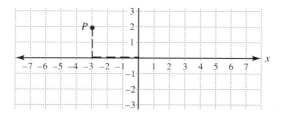

    1. Start at the origin.

    2. Count the units that we move along the *x*-axis until we are directly above or below the point. We moved 3 units in a negative direction along the *x*-axis. The *x*-coordinate is −3.

    3. Count the units that we must move from the *x*-axis to reach the point. We must move 2 units in a positive direction. The *y*-coordinate is 2.

    4. The coordinates of the point *P* are (−3, 2).

Notice that in locating coordinates we move only in horizontal or vertical directions. No diagonal movements are allowed. ■

☐ **DO EXERCISE 2.**

☐ **Exercise 1** Plot the following points.

a. (3, 1)   **b.** (1, 3)

c. (−1, 2)   **d.** (−2, −2)

e. (0, −3)   **f.** (−4, 0)

g. (−2, −5)   **h.** (2, −1)

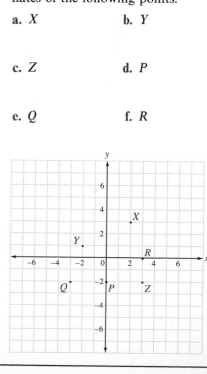

☐ **Exercise 2** Find the coordinates of the following points.

a. *X*   **b.** *Y*

c. *Z*   **d.** *P*

e. *Q*   **f.** *R*

257

3 Quadrants

The regions bordered by the coordinate axes are designated the first, second, third, and fourth **quadrants**, as shown here.

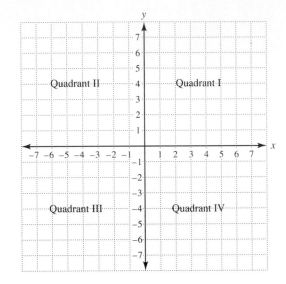

In quadrant I, the signs on both coordinates of any point are positive. In quadrant II, the *x*-coordinate of a point is negative and the *y*-co-ordinate is positive. The coordinates of a point are both negative in quadrant III. The *x*-coordinate is positive and the *y*-coordinate of a point is negative in quadrant IV.

EXAMPLE 3   In which quadrant is the point $(-2, -4)$?

Since both the *x* and *y* coordinates are negative, the point is in quadrant III.   ■

□ **DO EXERCISE 3.**

A point on an axis is not in any quadrant.

Answers to Exercises

1.

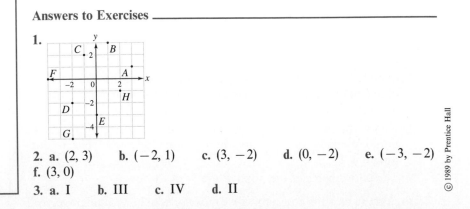

2. **a.** $(2, 3)$ **b.** $(-2, 1)$ **c.** $(3, -2)$ **d.** $(0, -2)$ **e.** $(-3, -2)$
f. $(3, 0)$
3. **a.** I   **b.** III   **c.** IV   **d.** II

□ **Exercise 3**   In which quadrant are the following points located?

a. $(9, 5)$

b. $(-10, -7)$

c. $(1.5, -11)$

d. $(-4, 8.7)$

PROBLEM SET 6.1

A. *Plot the following points.*

1. (4, 5)
2. (−4, 1)
3. (−1, −2)
4. (1, −5)
5. (0, 4)
6. (5, 0)
7. (−2, 0)
8. (0, −5)

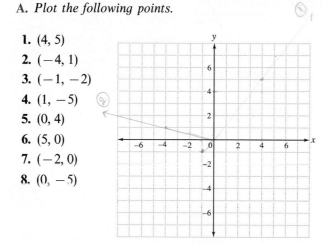

B. *Plot the following points.*

9. (−2, −2)
10. (−3, 1)
11. (3, 3)
12. (4, −2)
13. (0, −2)
14. (−1, 0)
15. (3, 0)
16. (0, 5)

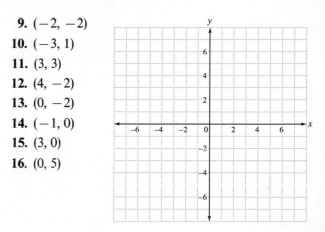

C. *Find the coordinates of these points.*

17. P
18. Q
19. R
20. T
21. U
22. V
23. W
24. X

D. *Find the coordinates of these points.*

25. Y
26. Z
27. A
28. B
29. C
30. D
31. E
32. F

E. *Without plotting, find in which quadrant the following points are located.*

33. (−2, −4)
    III

34. (12, 1)
    I

35. (−5, 10)
    II

36. (50, −7)
    IV

37. (−3, 10)
    II

38. (17, 6)

39. (−8, −8)
    III

40. (9, −2)

41. The horizontal number line in the rectangular co-ordinate system is the _____.

42. The first coordinate in an ordered pair may also be called the *x*-coordinate or the _____.

43. The second coordinate or *y*-coordinate of an ordered pair is also called the _____.

44. The point in the rectangular coordinate system where the number lines cross is called the _____.

45. In quadrant II, the *x*-coordinate of a point is negative and the *y*-coordinate is _____.

46. A point not in any quadrant is on an _____.

47. In quadrant IV, the *x*-coordinate of a point is _____ and the *y*-coordinate is negative.

Checkup

The following problems provide a review of some of Section 5.2.
Add or subtract as indicated.

48. $(x^3 - 3x) + (2x^2 + 5x)$

49. $(2x^2 - 4x) - (7x^2 + 3)$

50. $\left(\dfrac{2}{5}y^2 + 3y - \dfrac{3}{4}\right) - \left(\dfrac{1}{3}y^2 + 2y - \dfrac{5}{2}\right)$

51. $\left(\dfrac{3}{8}x^2 - 2x + \dfrac{2}{3}\right) + \left(\dfrac{1}{5}x^2 - 4x - \dfrac{1}{6}\right)$

52. $(3x^3 + 2x) - (5x^2 + 4x) + (6x^3 - 7x^2)$

53. $(7y^4 - 8y^2) - (3y^4 + y^3) - (6y^2 + 4)$

54. $(3x^2 - 2xy + 4y^2) + (7x^2 - 5xy - 9y^2)$

55. $(2a^2 + 3ab - 7b^2) - (4a^2 - 6ab - 3b^2)$

Constant a letter that stands for a specific number.

Linear equations in two variables Equations that can be put in the form $ax + by = c$ where a, b, and c are constants, a and b are not both zero and x and y are variables.

6.2   GRAPHING LINEAR EQUATIONS

In Section 3.1 we said that a linear equation in one variable must have an exponent of one on the variable. In this chapter we study **linear equations in two variables**. In order to describe these equations we need to define the word "constant."

A **constant** is a letter that stands for a specific number. Letters at the beginning of the alphabet are often used as constants, and letters at the end of the alphabet are frequently used as variables. Linear equations in two variables can be put in the form

$$ax + by = c$$

where a, b, and c are constants, a and b are not both zero, and x and y are variables.

For an equation in two variables to be linear, the following must be true:

1. There must be an exponent of 1 on each variable.
2. We may not have a term that contains the product or quotient of two variables. The equations $xy = 7$ and $y/x + x = 4$ are *not* linear.

■ Solutions of Equations in Two Variables

> The solutions of equations in two variables are ordered pairs of numbers.

For example, the ordered pair $(1, 5)$ is a solution to the equation $x + y = 6$ since $1 + 5 = 6$. The ordered pair $(2, 4)$ is also a solution. Notice that we may substitute a number without using parentheses if the substitution does not result in two successive operation symbols. If we substitute $(-3, -5)$ in the equation $x - y = 2$, we must use parentheses around -5.

$$x - y = 2 \quad \text{or} \quad -3 - (-5) = 2$$

EXAMPLE 1   Find three solutions to the equation $x + 2y = 4$.

1. Choose any number for x, say -4.
2. Substitute this value for x in the equation to find the corresponding y value.

$$x + 2y = 4$$
$$-4 + 2y = 4$$
$$2y = 8$$
$$y = 4$$

---

1 *Identify solutions to linear equations in two variables*

2 *Graph linear equations by plotting points*

3 *Graph equations with one variable*

4 *Graph linear equations using intercepts*

□ **Exercise 1** Are the ordered pairs (3, −1) and (5, 3) solutions to the equations?

a. $x + 3y = 0$

b. $x - y = 2$

□ **Exercise 2** Graph.

a. $y = 2x$

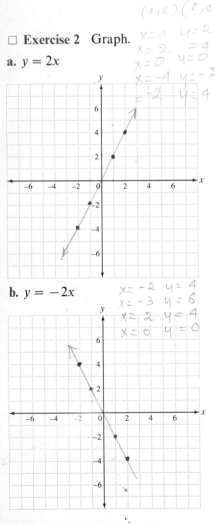

(1,2)(2,4)
x=1 y=2
x=2 =4
x=0 y=0
x=−1 y=−2
−2 y=4

b. $y = -2x$

x=−2 y=4
x=−3 y=6
x=2 y=4
x=0 y=0

Exercise 2 continues on pages 263 and 264.

262

3. Make a table and put in it the number we have chosen for x and the corresponding value of y. We have found an ordered pair that is a solution to the equation.

x	y	Ordered Pair
−4	4	(−4, 4)

Repeat steps 1 through 3 until you have found three ordered pairs. We chose the additional x values of 0 and 2. We could have chosen other x values. There are an unlimited number of solutions.

x	y	Ordered Pairs
−4	4	(−4, 4)
0	2	(0, 2)
2	1	(2, 1)

We say that the number of solutions of an equation in two variables is infinite. ■

□ **DO EXERCISE 1.**

2 Graphing Equations

Remember that to graph an equation means to draw a picture of it. We graph an equation in two variables by making a drawing of all its solutions. We need to find ordered pairs that will make the equation true. To do this, we choose a value for one variable, usually x, and find the corresponding value of the other variable. The technique we will describe here may also be used for equations that are not linear.

> The graph of a linear equation $ax + by = c$ is a straight line.

EXAMPLE 2 Graph.

a. $y = 3x$

1. Choose values for x and find the corresponding y values. Make a table of these values and the ordered pairs.

x	y	Ordered Pairs
−2	−6	(−2, −6)
−1	−3	(−1, −3)
0	0	(0, 0)
1	3	(1, 3)
2	6	(2, 6)

2. Plot these ordered pairs. Remember the letters in the ordered pair are written in alphabetical order with the x-coordinate first.

3. Since the number of solutions is infinite, we could plot enough
points to fill in a straight line. As we show in Example **b** below,
five points are more than sufficient to sketch the line shown.

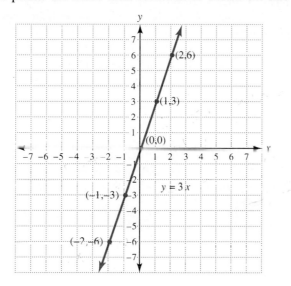

b. $y = -3x$

    1. Choose values for x and find the corresponding y values. Make
a table. We only need to find two ordered pairs since two points
determine a line, but you should find a third point as a check.

x	y	Ordered Pairs
2	6	$(-2, 6)$
0	0	$(0, 0)$
2	-6	$(2, -6)$

    2. Plot these points.
    3. Sketch the line.

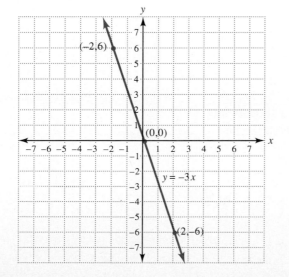

c. $y = x$

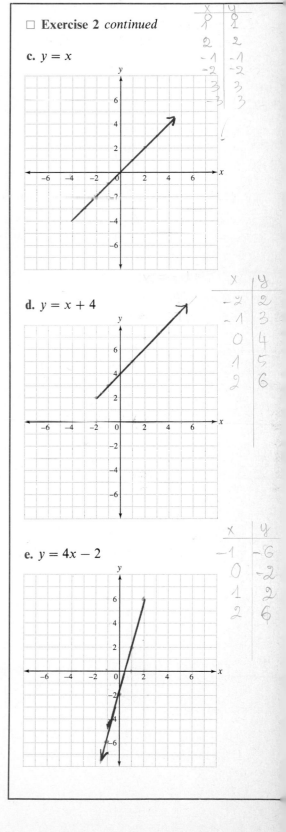

d. $y = x + 4$

e. $y = 4x - 2$

□ **Exercise 2** *continued*

f. $y = 3x - 1$

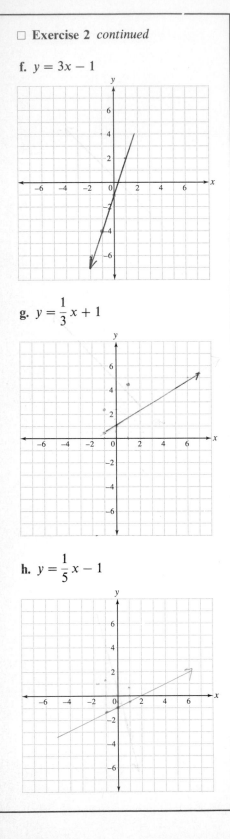

g. $y = \dfrac{1}{3}x + 1$

h. $y = \dfrac{1}{5}x - 1$

c. $y = 2x + 3$

   1. Choose values for *x*. Make a table. Notice that we choose some negative values for *x*. If the ordered pairs that we get are outside the rectangular coordinate system that we have drawn we disregard them.

x	y	Ordered Pairs
−3	−3	(−3, −3)
0	3	(0, 3)
1	5	(1, 5)

   2. Plot these points.
   3. Sketch the line.

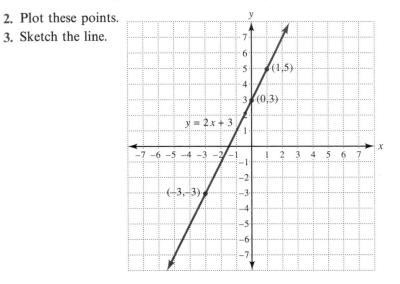

d. $y = \dfrac{1}{2}x - 3$

   1. Choose values for *x*. We chose $x = -2$, 0, and 4 since they are divisible by 2 and it is easy to find the corresponding *y* values. Make a table.

x	y
−2	−4
0	−3
4	−1

   2. Plot these points.
   3. Sketch the line.

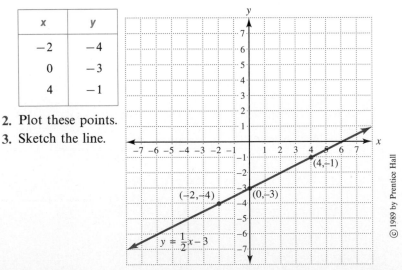

264

□ **DO EXERCISE 2.** Exercise 2 begins on page 262.

3 Equations Containing Only One Variable

These equations are of the type $x = a$ or $y = b$. They may also be graphed using a table.

EXAMPLE 3 Graph.

a. $x = 3$

This is an equation of the form $x = a$.

1. Notice that the equation says that x is always 3 for any value of y. We make a table using this fact. Use any values of y within the rectangular coordinate system that we have drawn.

x	y
3	−2
3	0
3	3

2. Plot these points, reading the ordered pairs from the table.
3. Sketch the line.

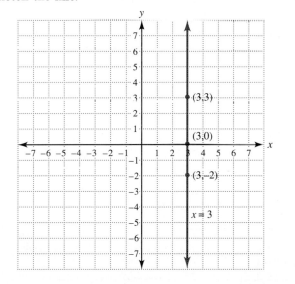

Notice that this is a line parallel to the y axis through $(3, 0)$. We may use this fact to graph it without using the table.

All equations of the form $x = a$ are vertical lines passing through the point $(a, 0)$.

X-intercept of a line   the point at which a line crosses the x-axis.
Y-intercept of a line   the point at which a line crosses the y-axis.

□ **Exercise 3**   Graph.

a. $x = -2$

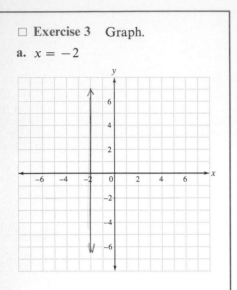

b. $y = 4$

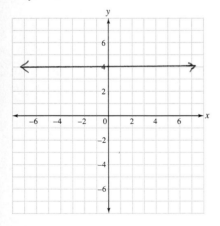

b. $y = -4$

This is an equation of the form $y = b$.

1. Make a table using the fact that $y = -4$ for any x values.

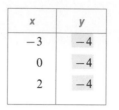

x	y
-3	-4
0	-4
2	-4

2. Plot these points.
3. Sketch the line.

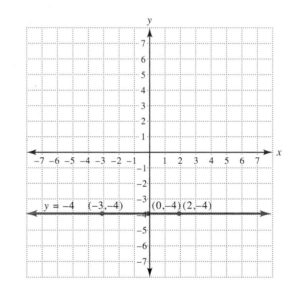

We may also draw the graph by observing that this is a line through $(0, -4)$ parallel to the x-axis.

> All equations of the form $y = b$ are horizontal lines passing through the point $(0, b)$.

■

□ **DO EXERCISE 3.**

4 **Graphing First-Degree Equations Using Intercepts**

Many linear equations can be graphed more quickly using intercepts. The **x-intercept** of a line is a point where it intercepts or crosses the x-axis. The y value at this point is 0. Hence the coordinates of the x intercept are $(a, 0)$. Similarly, the **y-intercept** is the point where the line crosses the y-axis. The coordinates of the y-intercept are $(0, b)$.

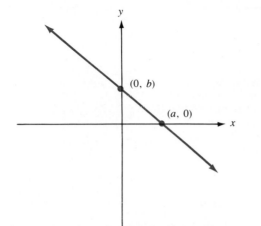

Since we know that the coordinates of the x-intercept are $(a, 0)$, we may let $y = 0$ in the equation and find a. Since the coordinates of the y-intercept are $(0, b)$, we may let $x = 0$ in the equation and find b. Choose an additional value of x as a check point. This method is faster than the methods of Example 2 because we are using zeros.

EXAMPLE 4 Graph $2x + 3y = 6$.

1. Make a table using $x = 0$ and $y = 0$. Use a third point as a check point.

x	y
0	2
3	0
−3	4

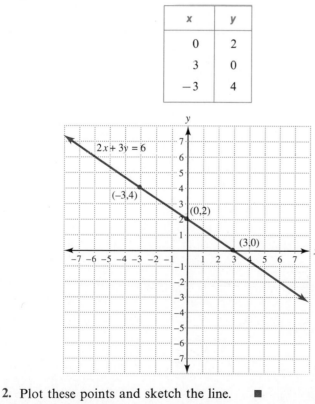

2. Plot these points and sketch the line. ■

☐ **DO EXERCISE 4.**

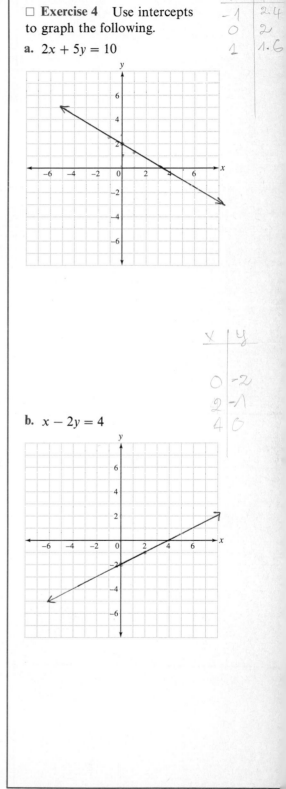

☐ **Exercise 4** Use intercepts to graph the following.

a. $2x + 5y = 10$

x	y
−1	2.4
0	2
1	1.6

b. $x − 2y = 4$

x	y
0	−2
2	−1
4	0

Answers to Exercises

1. a. $(3, -1)$ is a solution   **b.** $(5, 3)$ is a solution

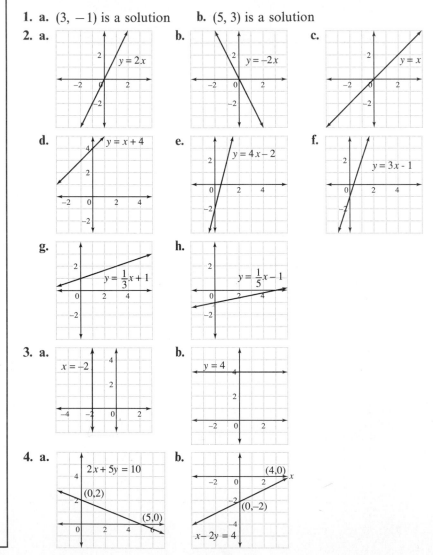

2. a. $y = 2x$ **b.** $y = -2x$ **c.** $y = x$

d. $y = x + 4$ **e.** $y = 4x - 2$ **f.** $y = 3x - 1$

g. $y = \frac{1}{3}x + 1$ **h.** $y = \frac{1}{5}x - 1$

3. a. $x = -2$ **b.** $y = 4$

4. a. $2x + 5y = 10$, $(0,2)$, $(5,0)$ **b.** $(4,0)$, $(0,-2)$, $x - 2y = 4$

PROBLEM SET 6.2

A. *Is the given ordered pair a solution of the equation?*

1. $(1, 4)$ $x + 2y = 9$ Yes

2. $(3, 2)$ $2x + y = 8$ Yes

3. $(-1, 5)$ $y = -4x + 7$ No

4. $(3, -2)$ $y = x - 5$ Yes

5. $(-1, -3)$ $y = x - 2$ No

6. $(-1, -8)$ $2x + y = 3$ NO

B. *Graph using three different x values.*

7. $y = 4x$

8. $y = -4x$

9. $y = x$

10. $y = -x$

11. $y = x + 2$

12. $y = -x + 2$

13. $y = 3x + 1$

14. $y = -2x + 2$

15. $y = 5x$

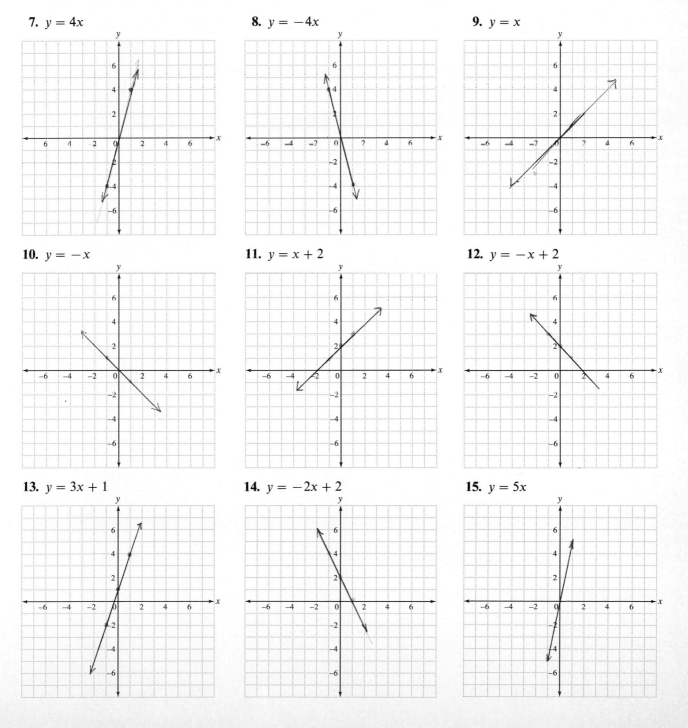

16. $y = -5x$

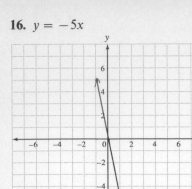

17. $2y + x = 4$

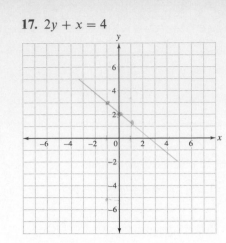

18. $2y + 3x = 6$

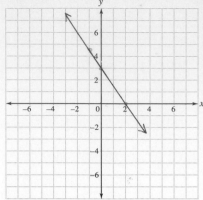

C. *Graph.*

19. $x = 2$

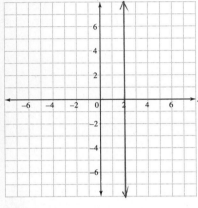

20. $x = -1$

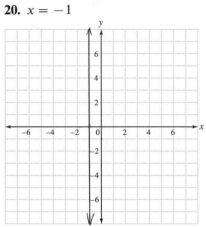

21. $y = -2$

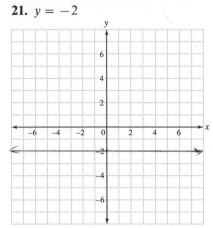

22. $y = 2$

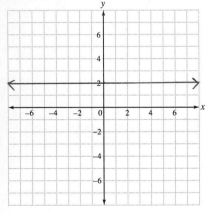

23. $x = 0$

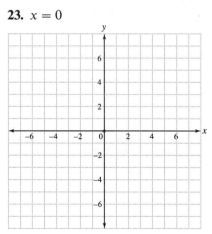

24. $y = 0$

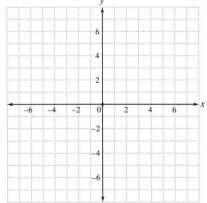

25. $x = -\dfrac{3}{2}$

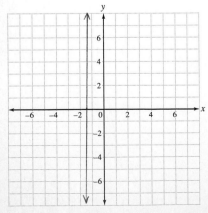

26. $y = \dfrac{7}{2}$

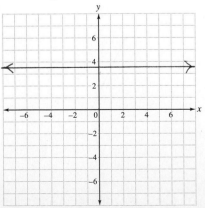

D. *Graph using intercepts.*

27. $x + y = 4$

28. $x - y = 3$

29. $2x + y = 6$

30. $x - 2y = 4$

31. $3x - 2y = 6$

32. $4x + 3y = 12$

33. $x - 5y = 5$

34. $x + 3y = 3$

35. $y = 2x + 2$

36. $y = 4x - 4$

37. $x = 3y - 6$

38. $x = 2y + 4$

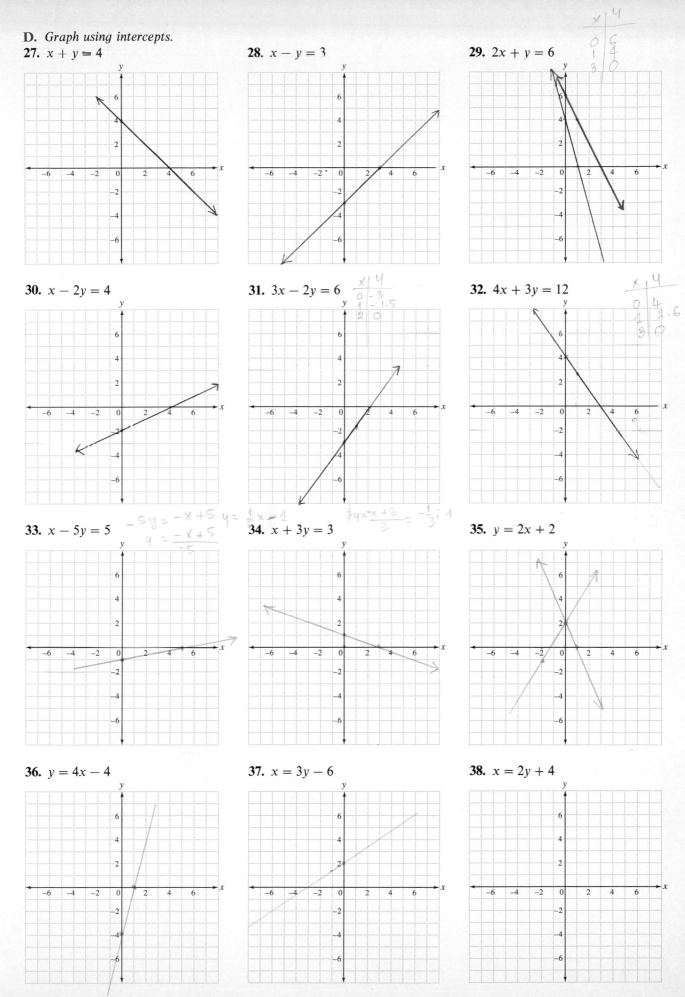

39. A letter that stands for a specific number is a _____.

40. Equations that can be put in the form $ax + by = c$ are _____ equations in two variables.

41. The solutions of equations in two variables are _____ _____ of numbers.

42. The number of solutions of one equation in two variables is _____.

43. The graph of a linear equation, $ax + by = c$, is a _____ _____.

44. Equations of the form $x = a$ are _____ lines passing through the point $(a, 0)$.

45. The y-intercept of a line is the point where the line crosses the _____.

Checkup

The following problems provide a review of some of Section 5.5.
Divide.

46. $\dfrac{6x^3 + 5x^2 - 2x}{2x}$

47. $\dfrac{9y^2 + 3y + 4}{3y}$

48. $\dfrac{x^2 - 2x - 7}{x + 2}$

49. $\dfrac{x^2 - 5x - 21}{x + 3}$

50. $\dfrac{2x^2 + x - 1}{2x + 3}$

51. $\dfrac{3x^2 + 10x - 5}{3x - 2}$

52. $\dfrac{x^3 - x + 1}{x - 1}$

53. $\dfrac{x^2 + 1}{x + 1}$

Slope The vertical change (difference of y-coordinates) divided by the horizontal change (difference of x-coordinates) of two points on a line.

Subscripts numbers written below the variable.

6.3 SLOPES OF STRAIGHT LINES

OBJECTIVES

1 *Use the geometric interpretation of slope to find the slope of a line*

2 *Use the algebraic definition of slope to find the slope of a line through two points*

1 Geometric Interpretation of Slope

It is useful in mathematics and the applications of it to know the slant of a line. A number called the **slope**, designated m, tells us how the line slants. Geometrically, the *slope of a line* is the vertical change (difference of y-coordinates) divided by the horizontal change (difference of x-coordinates) between two points P_1 and P_2 on the line:

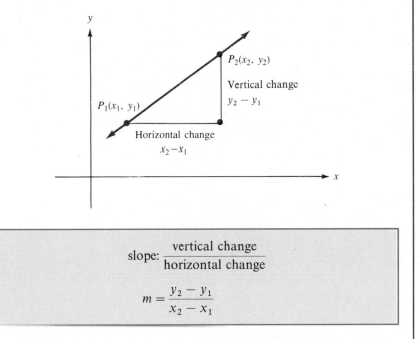

$$\text{slope:} \quad \frac{\text{vertical change}}{\text{horizontal change}}$$

$$m = \frac{y_2 - y_1}{x_2 - x_1}$$

The numbers written below the variable are called **subscripts**. They indicate specific points on the line. We may choose any two points on the line to find its slope.

EXAMPLE 1

a. Find the slope of the line shown in the graph.

    1. Choose two points on the line where it crosses a corner of the unit squares. We will choose $P_1(0, 2)$ and $P_2(2, 5)$.

    2. The vertical change between the two points is $5 - 2$ or 3.

    3. The horizontal change between the two points is $2 - 0$ or 2.

    4. Slope: $\dfrac{\text{vertical change}}{\text{horizontal change}}$

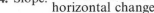

$$m = \frac{3}{2}$$

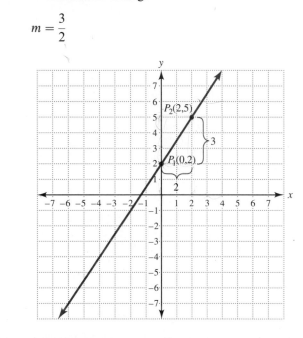

Lines that slope *upward* to the right have *positive* slope.

b. Find the slope of the line shown in the graph.

    1. We choose points on the line $P_1(0, 2)$ and $P_2(5, -2)$.

    2. The vertical change between the two points is $-2 - 2$ or -4.

    3. The horizontal change between the two points is $5 - 0$ or 5.

    4. Slope: $m = \dfrac{-4}{5}$

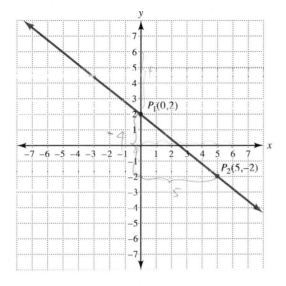

If we subtract the coordinates in the opposite order, we get the same result, since the slope from P_2 to P_1 is the same as the slope from P_1 to P_2.

The change in vertical direction is $2 - (-2)$ or 4.

The change in horizontal direction is $0 - 5$ or -5.

$$m = \frac{4}{-5} = \frac{-4}{5}$$

> Lines that slope *downward* to the right have *negative* slope.

☐ **DO EXERCISE 1.**

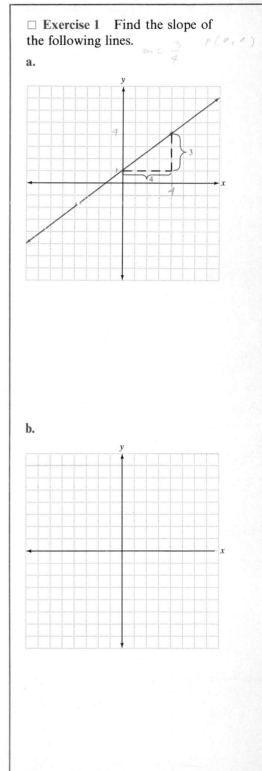

☐ **Exercise 1** Find the slope of the following lines.

a.

b.

The absolute value of a slope tells us how steeply the line is rising to the right or left. The greater the absolute value, the steeper the line is rising. Examples follow.

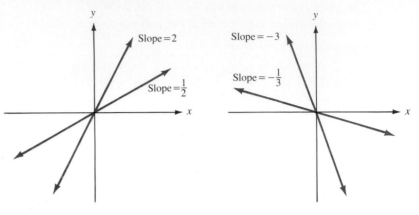

The absolute value of 2 is greater than the absolute value of $\frac{1}{2}$ since $|2| = 2$ and $\left|\frac{1}{2}\right| = \frac{1}{2}$. The line with slope 2 is steeper than the line with slope $\frac{1}{2}$.

The absolute value of -3 is greater than the absolute value of $\frac{1}{3}$ since $|-3| = 3$ and $\left|-\frac{1}{3}\right| = \frac{1}{3}$. The line with slope -3 is steeper than the line with slope $-\frac{1}{3}$.

2 Algebraic Definition of Slope

If we know two points on a line it is easier to find the slope using the algebraic definition:

$$\text{slope:} \quad m = \frac{y_2 - y_1}{x_2 - x_1}$$

EXAMPLE 2   Find the slope of the lines through the following points.

a. (1, 5), (4, 7)

   1. Label the points.
      Be sure to use the same subscripts on each ordered pair.

$$(x_1, y_1), \quad (x_2, y_2)$$
$$(1, 5), \quad (4, 7)$$

   2. $m = \dfrac{y_2 - y_1}{x_2 - x_1} = \dfrac{7 - 5}{4 - 1} = \dfrac{2}{3}$

The slope is positive, so the line slopes upward to the right.

b. (3, 7), (5, 2)

    1. Label the points.

$$(x_2, y_2), \quad (x_1, y_1)$$
$$(3, 7), \quad (5, 2)$$

    2. $m = \dfrac{y_2 - y_1}{x_2 - x_1} = \dfrac{7 - 2}{3 - 5} = \dfrac{5}{-2} = -\dfrac{5}{2}$

The slope is negative, so the line slopes downward to the right.   ■

☐ **DO EXERCISE 2.**

Slopes of Horizontal and Vertical Lines

All horizontal lines have slope zero.

EXAMPLE 3   Find the slope of the line $y = 4$.

    1. Choose two points on the line, say (0, 4) and (4, 4).
    2. Label the points.

$$(x_1, y_1), \quad (x_2, y_2)$$
$$(0, 4), \quad (4, 4)$$

    3. $m = \dfrac{y_2 - y_1}{x_2 - x_1} = \dfrac{4 - 4}{4 - 0} = \dfrac{0}{4} = 0$

■

☐ **DO EXERCISE 3.**

☐ **Exercise 2**   Find the slope of the line through the pairs of points.

a. (−4, 1), (2, 8)

b. (7, 9), (2, 10)

c. (3, 7), (5, 2)

d. (4, 6), (−2, −3)

☐ **Exercise 3**   Find the slope of the following lines.

a. $y = 6$

b. $y = -2$

□ **Exercise 4** Find the slope of the following lines.

a. $x = 3$

b. $x = -5$

All vertical lines have no slope.

EXAMPLE 4 Find the slope of the line $x = -1$.

1. Choose two points on the line, say $(-1, 3)$ and $(-1, 0)$.
2. Label the points.

$$(x_2, y_2), \quad (x_1, y_1)$$
$$(-1, 3), \quad (-1, 0)$$

3. $m = \dfrac{y_2 - y_1}{x_2 - x_1} = \dfrac{3 - 0}{-1 - (-1)} = \dfrac{3}{0}$

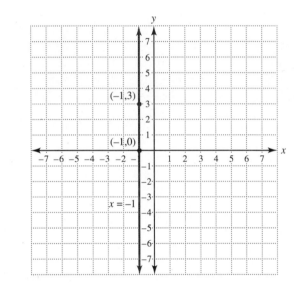

Since division by zero is undefined, we say that the line has no slope or the slope is undefined. ■

□ **DO EXERCISE 4.**

Answers to Exercises

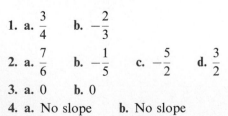

1. **a.** $\dfrac{3}{4}$ **b.** $-\dfrac{2}{3}$

2. **a.** $\dfrac{7}{6}$ **b.** $-\dfrac{1}{5}$ **c.** $-\dfrac{5}{2}$ **d.** $\dfrac{3}{2}$

3. **a.** 0   **b.** 0

4. **a.** No slope   **b.** No slope

PROBLEM SET 6.3

A. *Find the slope of the following lines.*

1. L_1 $(0,2)$ $\frac{2}{3}$

2. L_2 -2

3. L_3 $(0,4)$ 0

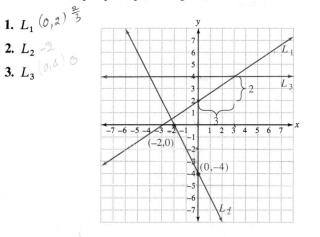

4. L_4

5. L_5

6. L_6

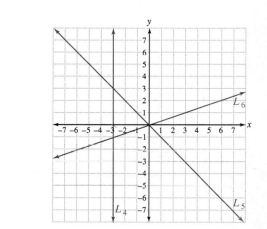

B. *Find the slope of the line through the pairs of points.*

7. $(9, 5)$, $(3, 4)$

$$m = \frac{y_2 - y_1}{x_2 - x_1} = \frac{4-5}{3-9} = \frac{-1}{-6} = \frac{1}{6}$$

8. $(7, 2)$, $(4, 1)$

$x_1\ y_1\ \ x_2\ y_2$

$$m = \frac{1-2}{4-7} = \frac{-1}{-3} = \frac{1}{3}$$

9. $(-4, 5)$, $(7, 8)$

$x_1\ y_1\ \ x_2\ y_2$

$$m = \frac{8-5}{7-(-1)} = \frac{3}{11} = \frac{3}{11}$$

10. $(-3, 1)$, $(-5, 2)$

$$m = \frac{2-1}{(-5)-(-3)} = -\frac{1}{2}$$

11. $(0, 4)$, $(3, 4)$

$$m = \frac{y_2 - y_1}{x_2 - x_1} = \frac{4-4}{3-0} = \frac{0}{3} = 0$$

12. $(-5, 0)$, $(-5, 4)$

$$m = \frac{4-0}{(-5)-(-5)} = \frac{4}{0} =$$

13. $(-4, -2)$, $(3, 5)$

$$m = \frac{5-(-2)}{3-(-4)} = \frac{7}{7} = 1$$

14. $(2, 7)$, $(-1, -8)$

$$m = \frac{(-8)-7}{(-1)-2} = \frac{-1}{-1} = 1$$

15. $(-7, 4)$, $(6, -3)$

$$m = \frac{(-3)-(4)}{(6)-(-7)} = \frac{-7}{13}$$

16. $(8, 0)$, $(0, 5)$

$$m = \frac{5-0}{0-8} = \frac{5}{0} =$$

17. $(0, 3)$, $(7, -5)$

$$m = \frac{(-5)-(3)}{(7)-0} = \frac{-8}{7}$$

18. $(4, 0)$, $(-7, -6)$

$$m = \frac{(-6)-(4)}{(-7)-4} = \frac{-10}{-11} = \frac{10}{11}$$

19. $(0, -3)$, $(5, -3)$

$$m = \frac{(-3)-(-3)}{5-0} = \frac{0}{5} = 0$$

20. $(-6, 0)$, $(-6, -2)$

$$m = \frac{(-2)-0}{-6-(-6)} = \frac{-2}{0} = 0$$

C. *Find the slope of the following lines.*

21. $y = 7$

22. $x = -4$

23. $x = 0$

24. $y = 0$

25. $y + 8 = 0$

26. $x - 2 = 0$

27. The slope of a line is the _____ change divided by the _____ change between two points on the line.

28. Lines that slope upward to the right have _____ slope.

29. All _____ lines have slope zero.

30. The slope of a _____ line is undefined.

Checkup

The following problems provide a review of some of Section 5.6.
Factor.

31. $7x^2 + 14x^3$

32. $16x^4 - 8x^2$

33. $12x^2 + 24x - 18$

34. $15x^2 - 10x + 35$

35. $4x^2y^2 + 16xy + 8x$

36. $9x^3y^3 - 3xy + 12x^2y^2$

37. $(a + b)x - (a + b)$

38. $(x + y) + a(x + y)$

39. $y^2 - y - xy + x$

40. $px + p - qx - q$

41. $2x^2 + 8x - 3x - 12$

42. $5x^2 - 10x + x - 2$

6.4 GRAPHING USING SLOPE

OBJECTIVES

1 *Graph a line with a given slope and y-intercept*

2 *Find the slope of a line from the equation of the line*

3 *Graph a linear equation using slope*

1 **Using the Slope and y-Intercept**

If we know the slope of a line and the *y*-intercept, we can graph the line quickly.

EXAMPLE 1

a. Graph the line with slope $\frac{2}{3}$ that passes through the point (0, 1).

1. Plot the point (0, 1).

2. Slope: $\dfrac{\text{vertical change}}{\text{horizontal change}} = \dfrac{2}{3}$

   Notice that both the vertical change and the horizontal change are positive. So from the point you have plotted, move 2 units in the positive vertical direction and then 3 units in the positive horizontal direction. Plot another point. This is another point on the line.

3. Draw the line.

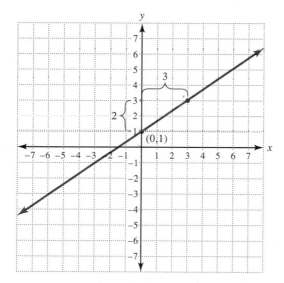

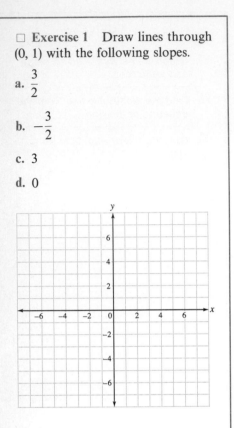

□ **Exercise 1** Draw lines through (0, 1) with the following slopes.

a. $\dfrac{3}{2}$

b. $-\dfrac{3}{2}$

c. 3

d. 0

b. Graph the line with slope $-\frac{2}{3}$ that passes through the point (0, 1).

1. Plot the point (0, 1).
2. Recall that $-\frac{2}{3} = \frac{-2}{3} = \frac{2}{-3}$.

$$\text{Slope: } \frac{\text{vertical change}}{\text{horizontal change}} = \frac{-2}{3} = \frac{2}{-3}$$

From the point (0, 1) that we have plotted, we may choose to move either in a negative vertical direction or in a negative horizontal direction.

Let us move 2 units in a negative vertical direction and then 3 units in a positive horizontal direction. Plot a point.

3. Draw the line.

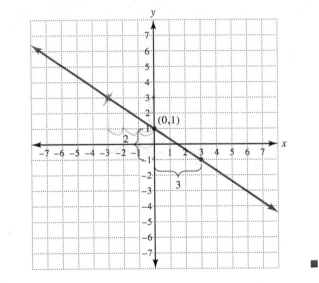

□ **DO EXERCISE 1.**

2 Finding the Slope of a Line from Its Equation

If a linear equation has intercepts that are fractions, often the easiest way to draw its graph is to use the slope of the line and a point on the line. We can find the slope from the equation.

Consider the linear equation $y = 2x + 5$. We will find two points on the line. Let $x = 1$, then $y = 7$. Let $x = 0$, then $y = 5$. Two points on the line are (1, 7) and (0, 5). Then the slope of the line is

$$m = \frac{y_2 - y_1}{x_2 - x_1} = \frac{7 - 5}{1 - 0} = \frac{2}{1} = 2$$

Notice that in the given equation, $y = 2x + 5$, the coefficient of x is also 2.

Slope-intercept form of the equation of a line $y = mx + b$, where m is the slope and (x_1, y_1) is a point on the line.

If a linear equation is solved for y, the coefficient of the x term is the slope.

$$y = mx + b$$

is the **slope-intercept form of the equation of a line** where m is the slope and $(0, b)$ is the y-intercept.

Notice that the y-intercept is $(0, b)$ because if we let $x = 0$ in the equation $y = mx + b$, then $y = b$.

EXAMPLE 2   Find the slope and the y-intercept of the lines.

a. $y = \dfrac{3}{4}x + 6$

  1. The slope is the coefficient of x. It is $\frac{3}{4}$.
  2. The y-intercept is $(0, 6)$.

b. $y = 3x - 4$

  1. The slope is 3.
  2. The y-intercept is $(0, -4)$.

c. $3y + 2x = 6$

  1. Put the equation in slope-intercept form (solved for y) in order to find the slope and y-intercept.

$$3y + 2x = 6$$
$$\underline{\quad -2x = -2x} \quad \text{Adding } -2x \text{ to each side of the equation}$$
$$3y \quad\quad = -2x + 6$$

$$\dfrac{3y}{3} = \dfrac{-2x + 6}{3} \quad \text{Dividing both sides by 3 to solve for } y$$

$$y = -\dfrac{2}{3}x + \dfrac{6}{3}$$

$$y = -\dfrac{2}{3}x + 2$$

  2. The slope is $-\frac{2}{3}$ and the y-intercept is $(0, 2)$. ∎

☐ **DO EXERCISE 2.**

☐ **Exercise 2**   Find the slope and y-intercept of the following lines.

a. $y = \dfrac{3}{2}x - 2$

b. $y = -5x + 8$

c. $4y + 3x = 8$

d. $2y - 5x = 4$

□ **Exercise 3** Graph.

a. $y = 3x - 1$

b. $y = -3x - 1$

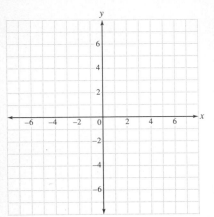

c. $2x + 3y = 6$

d. $y - \dfrac{3}{2}x = -2$

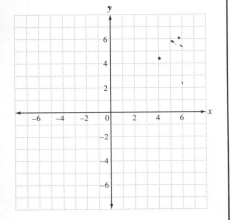

3 **Graphing a Linear Equation Using Slope**

We know how to put an equation in slope-intercept form. This gives us the slope and the y-intercept, which we can use to graph the line.

EXAMPLE 3 Graph $y = 2x - 1$.

1. The y-intercept is $(0, -1)$. Plot this point.
2. Slope: vertical change/horizontal change $= 2 = \dfrac{2}{1}$. From the point $(0, -1)$, move 2 units in the positive vertical direction and 1 unit in the positive horizontal direction. Plot a point.
3. Draw the line.

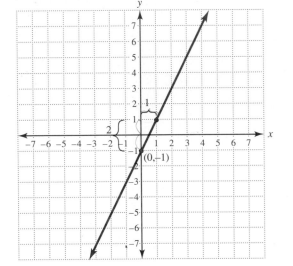

■

□ **DO EXERCISE 3.**

> **Suggestions for Graphing Linear Equations**
>
> 1. Is the equation of the form $y = b$ or $x = a$? The graphs of each of these equations is a line parallel to an axis.
> 2. Determine by inspection if one or both of the intercepts will be fractions. If they are not, graph the line using intercepts unless the equation is of the form $y = kx$, where k is a constant. (If $y = kx$, the line passes through the origin.) If the points are too close together, choose another point farther from the intercepts.
> 3. If one or both of the intercepts are fractions or if the equation is of the form $y = kx$, it is often easier to graph the line using the slope and y-intercept.

EXAMPLE 4 Graph.

a. $y = 3x$

1. This equation is of the form $y = kx$. If we try to graph it using intercepts, we get only one point, $(0, 0)$, and we have to choose two more values for x. Graph using the slope and the y-intercept.

2. The slope is 3 and the y-intercept is $(0, 0)$. Notice that $b = 0$.

3. The graph is as shown.

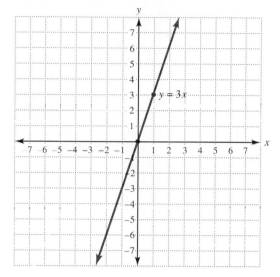

b. $4x + 3y = 6$

1. If we let $y = 0$, we get $x = \frac{3}{2}$. The x-intercept is a fraction. Graph using the slope and y-intercept.

2. Put the equation into slope-intercept form.

$$4x + 3y = 6$$

$$3y = -4x + 6$$

$$y = -\frac{4}{3}x + 2$$

3. The slope is $-\frac{4}{3}$ and the y-intercept is $(0, 2)$.

4. The graph is as shown.

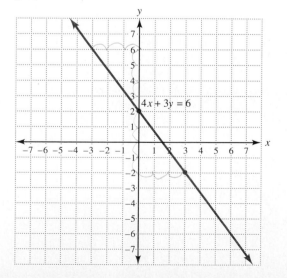

□ **Exercise 4** Graph.

a. $3x + 4y = 8$

b. $2x + y = 4$

c. $y = x$

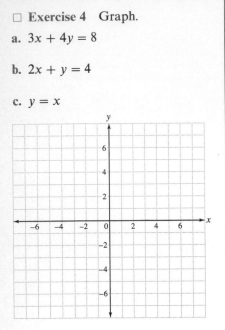

d. $x = 4$

e. $3x - 2y = 8$

f. $y = 0$

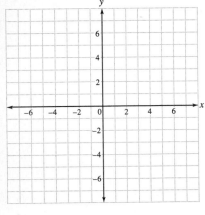

c. $x = -2$

This is an equation of the type $x = a$. It is a line through $(-2, 0)$ parallel to the y-axis.

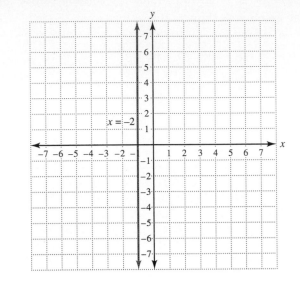

□ **DO EXERCISE 4.**

Answers to Exercises _____

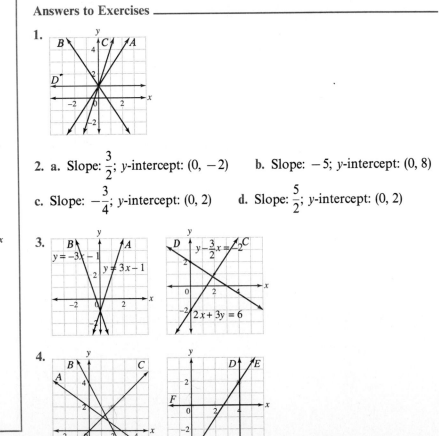

1.

2. a. Slope: $\dfrac{3}{2}$; y-intercept: $(0, -2)$ **b.** Slope: -5; y-intercept: $(0, 8)$

c. Slope: $-\dfrac{3}{4}$; y-intercept: $(0, 2)$ **d.** Slope: $\dfrac{5}{2}$; y-intercept: $(0, 2)$

3.

4.

PROBLEM SET 6.4

A. *Draw lines through* $(0, -1)$ *with the following slopes.*

1. $\dfrac{3}{4}$

2. $\dfrac{2}{3}$

3. -3

4. -2

5. No slope

6. 0

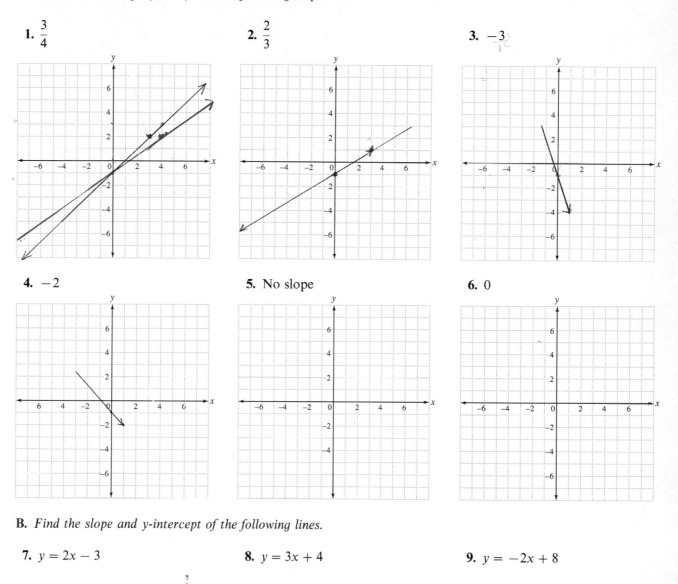

B. *Find the slope and y-intercept of the following lines.*

7. $y = 2x - 3$

8. $y = 3x + 4$

9. $y = -2x + 8$

10. $y = -5x - 7$

11. $3x + 2y = 8$

12. $4x - 5y = 10$

13. $6x - 9y = 15$

14. $8x + 4y = 26$

C. *Use the slope and y-intercept to graph.*

15. $y = 3x + 4$

16. $y = -2x - 1$

17. $y = -2x + 3$

18. $y = 4x + 2$

19. $2x - 5y = 10$

20. $3x + 4y = 4$

21. $y = \dfrac{3}{2}x - 1$

22. $y = \dfrac{1}{3}x + 2$

23. $5x - 6y = 12$

24. $4x - 5y = 5$

25. $3x + 4y = 8$

26. $2x + 5y = 5$

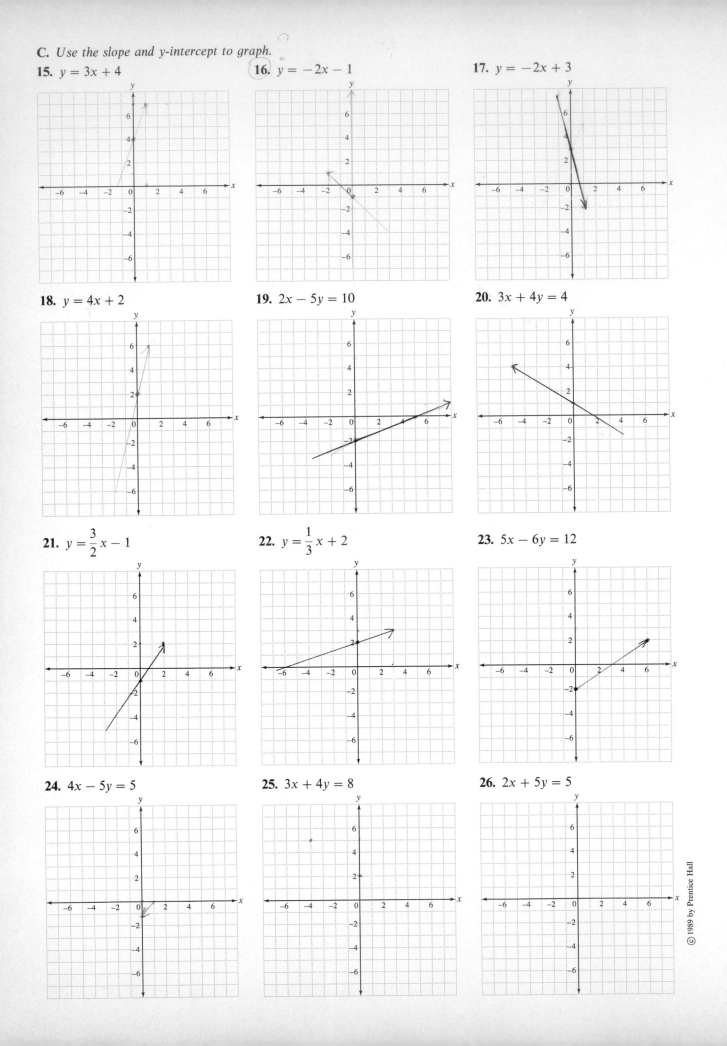

D. *Graph.*

27. $x = 3$

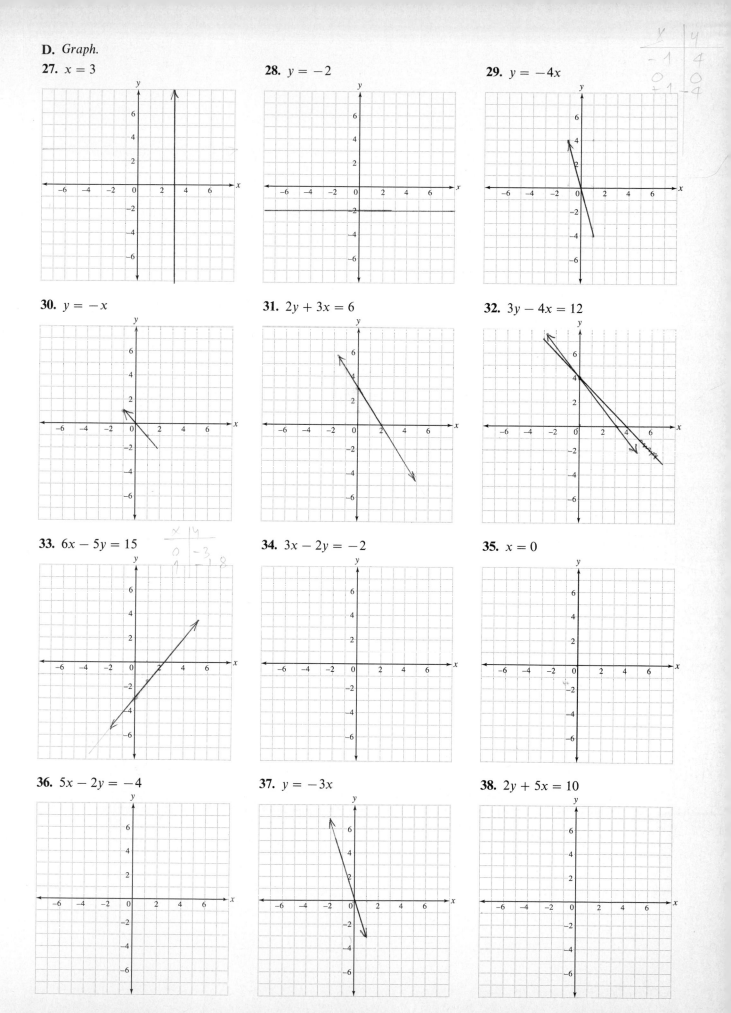

28. $y = -2$

29. $y = -4x$

30. $y = -x$

31. $2y + 3x = 6$

32. $3y - 4x = 12$

33. $6x - 5y = 15$

34. $3x - 2y = -2$

35. $x = 0$

36. $5x - 2y = -4$

37. $y = -3x$

38. $2y + 5x = 10$

39. For the graph of the equation $y = mx + b$, m is the _____ and b is the _____.

40. The graph of the line $x = a$ is a line parallel to the _____.

Checkup

The following problems provide a review of some of Sections 2.5 and 5.7. Some of these problems will help you with the next section.
Find the reciprocal.

41. $\dfrac{3}{2}$

42. $\dfrac{-5}{2}$

43. -3

44. 7

45. $\dfrac{-1}{8}$

46. $\dfrac{1}{6}$

Factor.

47. $x^2 + 3x - 10$

48. $x^2 + 10x + 21$

49. $2x^2 - x - 6$

50. $8x^2 - 6x + 1$

51. $3x^2 - 27x + 24$

52. $5x^2 - 25x + 30$

53. $4x^3 + 18x^2 - 10x$

54. $12x^3 + 28x^2 - 24x$

6.5 EQUATIONS OF STRAIGHT LINES

We may be given information about a straight line and asked to write an equation of it. The method that we use depends on the information that we are given.

1 Using the Slope and *y*-intercept to Write an Equation for a Line

Recall that the slope-intercept form of an equation is

$$y = mx + b$$

where m is the slope and b is the y-intercept. If we know the slope and y-intercept, it is easy to write an equation.

EXAMPLE 1

a. Find an equation for the straight line in slope-intercept form with y-intercept (0, 2) and slope -3.

$$m = -3 \quad \text{and} \quad b = 2$$

$$y = mx + b \qquad \text{Slope intercept form}$$

$$y = -3x + 2$$

An equation of the line is $y = -3x + 2$.

b. Find an equation for the straight line with y-intercept (0, -3) and slope 0.

$b = -3$ and lines with slope 0 are of the form $y = b$; hence

$$y = -3$$

An equation of the line is $y = -3$.   ∎

□ **DO EXERCISE 1.**

We may be asked to write an equation of a line when we are given two points on the line.

2 Point-Slope Form of an Equation of a Line

Consider a line with slope m and passing through a given point (x_1, y_1). If we give the coordinates (x, y) to any other point on the line, the slope of the line is

$$\frac{y - y_1}{x - x_1} = m \qquad (x \neq x_1)$$

Multiplying both sides of the equation by $x - x_1$ gives us

$$(x - x_1) \cdot \frac{y - y_1}{x - x_1} = m(x - x_1)$$

or

$$y - y_1 = m(x - x_1)$$

1 *Write the equation of a line given the slope and y-intercept*

2 *Write the equation of a line given one point and the slope or two points on the line*

3 *Identify two parallel lines from their slope*

4 *Identify two perpendicular lines from their slope*

□ **Exercise 1**   Find an equation for the line with the following y-intercept and slope.

a. (0, 1), $m = 5$

b. (0, -2), $m = -4$

c. (0, 4), $m = 0$

d. (0, -3), $m = 1$

Point-slope form of an equation of a line $y - y_1 = m(x - x_1)$, where m is the slope and (x_1, y_1) is a point on the line.

> The **point-slope form for a linear equation** is
>
> $$y - y_1 = m(x - x_1)$$

We can use it to write an equation of a line if we are given the slope and one point on the line or two points on the line.

EXAMPLE 2

a. Find an equation for the line through the point $(3, -2)$ with slope 4.

    1. $m = 4$ and $(x_1, y_1) = (3, -2)$

    2. Use the point-slope form.

$$y - y_1 = m(x - x_1)$$

    3. Substitute for m and (x_1, y_1) in the point-slope form.

$$y - (-2) = 4(x - 3)$$

$$y + 2 = 4(x - 3)$$

$$y + 2 = 4x - 12$$

$$y = 4x - 12 - 2 \qquad \text{Adding } -2 \text{ to both sides of the equation}$$

$$y = 4x - 14 \qquad \text{(In slope-intercept form)}$$

An equation of the line is $y = 4x - 14$.

b. Find an equation for the line through the points $(4, -3)$ and $(-2, -1)$.

    1. Find the slope.

$$m = \frac{y_2 - y_1}{x_2 - x_1} = \frac{-3 - (-1)}{4 - (-2)} = -\frac{2}{6} = -\frac{1}{3}$$

    2. Use either point as (x_1, y_1). We will use $(4, -3)$.

    3. Substitute for m and (x_1, y_1) in the point-slope form.

$$y - y_1 = m(x - x_1)$$

$$y - (-3) = -\frac{1}{3}(x - 4)$$

$$y + 3 = -\frac{1}{3}(x - 4)$$

$$y + 3 = -\frac{1}{3}x + \frac{4}{3} \qquad \text{Using a distributive law}$$

$$y = -\frac{1}{3}x + \frac{4}{3} - 3 \qquad \text{Adding } -3 \text{ to both sides of the equation}$$

$$y = -\frac{1}{3}x + \frac{4}{3} - \frac{9}{3} \qquad \text{Since } 3 = \frac{9}{3}$$

$$y = -\frac{1}{3}x - \frac{5}{3} \qquad \text{(In slope-intercept form)}$$

An equation of the line is $y = -\frac{1}{3}x - \frac{5}{3}$.

Parallel lines distinct lines with the same slope (or all vertical lines because they have no slope).

c. Find an equation for the line through the points $(4, -1)$ and $(3, -1)$.

    1. Find the slope.

$$m = \frac{y_2 - y_1}{x_2 - x_1} = \frac{-1 - (-1)}{4 - 3} = \frac{0}{1} = 0$$

    2. We will choose $(3, -1)$ for (x_1, y_1).

    3. Substitute in the point-slope formula.

$$y - y_1 = m(x - x_1)$$
$$y - (-1) = 0(x - 3)$$
$$y + 1 = 0 \qquad \text{Notice that } 0(x - 3) = 0$$
$$y = -1$$

An equation of the line is $y = -1$. Recall that a line with zero slope is of the form $y = b$. ∎

□ **DO EXERCISE 2.**

3 **Parallel Lines**

> Two lines with the same slope are **parallel**.

EXAMPLE 3 Find the slope of each line for the pair of lines $4x + 3y = 4$ and $8x + 6y = 0$ and decide if they are parallel.

    1. Find the slope of $4x + 3y = 4$. Put the equation into slope-intercept form.

$$4x + 3y = 4$$
$$3y = -4x + 4$$
$$y = -\frac{4}{3}x + \frac{4}{3}$$

The slope of the line is $-\frac{4}{3}$.

□ **Exercise 2** Find equations for the following lines.

a. $m = 3$, through $(-1, 2)$

b. $m = -\dfrac{8}{5}$, through $(5, 3)$

c. Through $(-1, 3), (2, -2)$

d. Through $(4, 2), (5, 3)$

e. Through $(-2, 1), (3, 4)$

f. Through $(4, 2), (-2, 2)$

g. Through $(3, -3), (3, 5)$

h. Through $(-8, -2), (-1, -7)$

Perpendicular lines   the slope of one line is the negative reciprocal of the slope of the line perpendicular to it. A line with no slope is perpendicular to a line with zero slope.

□ **Exercise 3**   Find the slope of each line in the pair of lines. Are the lines parallel?

a. $x + y = 2$
$\quad x + y = 5$

b. $y - x = 4$
$\quad y + x = 1$

c. $3x - 2y = 7$
$\quad 2x + 3y = 1$

d. $2x - 5y = 3$
$\quad 4x - 10y = 1$

2. Find the slope of $8x + 6y = 0$.

$$8x + 6y = 0$$

$$6y = -8x$$

$$y = -\frac{4}{3}x$$

The slope of the line is $-\frac{4}{3}$.

The lines are parallel since the slopes are the same. The graph of the lines is shown below.

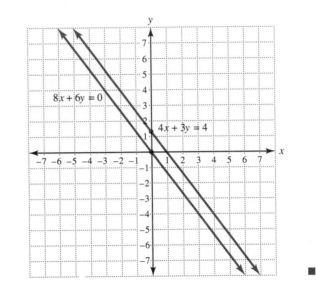

□ **DO EXERCISE 3.**

4 **Perpendicular Lines**

> If the slope of one line is the negative reciprocal of another line, the lines are **perpendicular**. A line with no slope is perpendicular to a line with zero slope.

Slope of Line	*Slope of Perpendicular Line*	
$\dfrac{2}{3}$	$-\dfrac{3}{2}$	
$-\dfrac{5}{4}$	$\dfrac{4}{5}$	
3	$-\dfrac{1}{3}$	
$-\dfrac{1}{5}$	$\dfrac{5}{1}$ or 5	
0	No slope	Since 0 does not have a reciprocal

EXAMPLE 4

a. Find the slopes of the lines $3x - 2y = 6$ and $2x + 3y = 9$ and decide if the lines are perpendicular.

1. Find the slope of $3x - 2y = 6$.

$$3x - 2y = 6$$

$$-2y = -3x + 6$$

$$y = \frac{3}{2}x - 3$$

The slope of the line is $\frac{3}{2}$.

2. Find the slope of $2x + 3y = 9$.

$$2x + 3y = 9$$

$$3y = -2x + 9$$

$$y = -\frac{2}{3}x + 3$$

The slope of the line is $-\frac{2}{3}$. The lines are perpendicular since $-\frac{2}{3}$ is the negative reciprocal of $\frac{3}{2}$.

The graph of the lines is as shown.

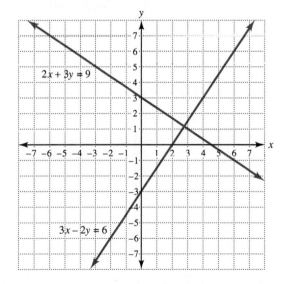

□ **Exercise 4** Find the slope of each line in the pair of lines and decide if they are perpendicular.

a. $3x - 5y = 4$
$5x + 3y = -1$

b. $2x + y = 0$
$x - 2y = 5$

c. $x - 4y = 1$
$2x + 4y = 0$

d. $y - x = 3$
$y - x = 5$

e. $y - x = 4$
$y + x = 3$

f. $8x - 9y = 3$
$3x + 6y = 4$

b. Find the slopes of the lines $5x + 2y = 3$ and $2x - 5y = 8$ and decide if the lines are perpendicular.

   1. Find the slope of $5x + 2y = 3$.
$$5x + 2y = 3$$
$$2y = -5x + 3$$
$$y = -\frac{5}{2}x + \frac{3}{2}$$

   The slope of the line is $-\frac{5}{2}$.

   2. Find the slope of $2x - 5y = 8$.
$$-5y = -2x + 8$$
$$y = \frac{2}{5}x - \frac{8}{5}$$

   The slope of the line is $\frac{2}{5}$.

   The lines are perpendicular since $\frac{2}{5}$ is the negative reciprocal of $-\frac{5}{2}$. ■

□ **DO EXERCISE 4.**

DID YOU KNOW?

Pierre de Fermat discovered analytic geometry at about the same time as Descartes. Fermat worked as a lawyer and studied mathematics as a hobby. He is known as the prince of amateur mathematics because of his many discoveries.

PRINCE OF AMATEUR MATHEMATICS

Answers to Exercises _____

1. a. $y = 5x + 1$ **b.** $y = -4x - 2$ **c.** $y = 4$ **d.** $y = x - 3$

2. a. $y = 3x + 5$ **b.** $y = -\frac{8}{5}x + 11$ **c.** $y = -\frac{5}{3}x + \frac{4}{3}$

d. $y = x - 2$ **e.** $y = \frac{3}{5}x + \frac{11}{5}$ **f.** $y = 2$ **g.** $x = 3$

h. $y = -\frac{5}{7}x - \frac{54}{7}$

3. a. $-1, -1$; yes **b.** $1, -1$; no **c.** $\frac{3}{2}, -\frac{2}{3}$; no **d.** $\frac{2}{5}, \frac{2}{5}$; yes

4. a. $\frac{3}{5}, -\frac{5}{3}$; yes **b.** $-2, \frac{1}{2}$; yes **c.** $\frac{1}{4}, -\frac{1}{2}$; no **d.** $1, 1$; no

e. $1, -1$; yes **f.** $\frac{8}{9}, -\frac{1}{2}$; no

PROBLEM SET 6.5

A. *Find an equation for the line with the following y-intercept and slope.*

1. $(0, -1)$, $m = 2$ **2.** $(0, 4)$, $m = -3$ **3.** $(0, 0)$, $m = -4$ **4.** $(0, 0)$, $m = 6$

5. $(0, -8)$, $m = 0$ **6.** $(0, 7)$, $m = 0$ **7.** $(0, 5)$, $m = \dfrac{1}{2}$ **8.** $(0, -2)$, $m = \dfrac{5}{3}$

B. *Find an equation for the line with the given slope through the given point. Use the point-slope form.*

9. $m = 2$, $(-1, 4)$ **10.** $m = 5$, $(1, 3)$ **11.** $m = -3$, $(2, -5)$ **12.** $m = -4$, $(-2, -3)$

13. $m = 5$, $(0, -4)$ **14.** $m = -1$, $(3, 0)$ **15.** $m = \dfrac{3}{2}$, $(4, -1)$ **16.** $m = -\dfrac{3}{5}$, $(-1, -2)$

17. $m = 0$, $(8, -2)$ **18.** $m = 0$, $(7, 6)$

C. *Find an equation for the line through the given two points.*

19. $(7, 4)$, $(8, 5)$ **20.** $(3, -4)$, $(-2, -1)$ **21.** $(-7, -1)$, $(-9, -2)$ **22.** $(-2, 5)$, $(3, 4)$

23. $(0, 1)$, $(4, 0)$ **24.** $(3, 0)$, $(0, -1)$ **25.** $(8, 4)$, $(6, 4)$ **26.** $(7, -1)$, $(4, -1)$

27. $(-1, -4)$, $(-2, 0)$ **28.** $(-5, 2)$, $(0, 3)$

D. *Find the slope of each line in the pair of lines to decide if the lines are parallel, perpendicular, or neither.*

29. $x + 2y = 3$
$\quad 2x + 4y = 7$

30. $3x - y = 4$
$\quad 2x + 3y = 5$

31. $5x - 2y = 3$
$\quad 2x + 5y = 1$

32. $3x + y = 8$
$\quad 4x + 5y = 1$

33. $8x - y = 0$
$\quad x - 8y = 3$

34. $x - y = 3$
$\quad x + y = 2$

35. $y - x = 4$
$\quad y - x = -1$

36. $4x - 2y = 9$
$\quad 3x + 2y = 5$

37. $7x - y = 4$
$\quad x + 7y = 3$

38. $2x - 4y = 1$
$\quad 4x - 8y = 3$

39. The form $y = mx + b$ is the _Slope –_ _intercept_ form of the equation of a line.

40. The equation $y - y_1 = m(x - x_1)$ is the _point_-_slope_ form of the equation of a line.

41. Two lines with the same slope are _parralel_.

42. If the slope of one line is the negative reciprocal of the slope of another line, the lines are _perpendicular_.

Checkup

The following problems provide a review of some of Section 5.8.
Factor.

43. $9x^2 - 12x + 4$

44. $25x^2 + 10x + 1$

45. $x^2 - 49$

46. $y^2 - 81$

47. $6x^2 - 17x + 12$

48. $8x^2 - 2x - 3$

49. $x^4 - 9x^2$

50. $3x^2 - 48x + 192$

51. $bx + 4by + 3x + 12y$

52. $6ax + 6ay - x - y$

Linear relationship when the relationship between two variables can be
represented by a linear equation.

6.6 APPLIED PROBLEMS

OBJECTIVE

1 Linear equations in two variables are used in business, the physical
sciences, and other areas. If the relationship between two variables can
be represented by a linear equation, we say that this is a **linear relationship**.

1 *Solve applied problems using linear equations in two variables*

EXAMPLE 1 The relationship between Celsius and Fahrenheit tem-
peratures is a linear relationship. The freezing point of water is 0° Celsius
or 32° Fahrenheit. The boiling point of water is 100° Celsius or 212°
Fahrenheit. Find an equation relating Fahrenheit degrees to Celsius
degrees. Find the Fahrenheit temperature when the Celsius temperature
is 20°.

1. Let x = number of Celsius degrees
      y = number of Fahrenheit degrees
2. We are given two points that are solutions to our linear equation.
    When the Celsius temperature is 0°, the Fahrenheit temperature
    is 32°. This gives us the point (0, 32). We are also given the point
    (100, 212). Since we are given two points on the line, we use the
    point-slope form:

$$y - y_1 = m(x - x_1)$$

3. We can find m using the two points (0, 32) and (100, 212).

$$m = \frac{212 - 32}{100 - 0} = \frac{180}{100} = \frac{9}{5}$$

4. We may use either point for (x_1, y_1). We chose (0, 32) because our
    calculations are easier when we work with zero.

$$y - y_1 = m(x - x_1)$$

$$y - 32 = \frac{9}{5}(x - 0)$$

$$y - 32 = \frac{9}{5}x$$

$$y = \frac{9}{5}x + 32$$

The equation is $y = \frac{9}{5}x + 32$.

5. When the Celsius temperature, x, is 20°,

$$y = \frac{9}{5}(20) + 32$$

$$y = 36 + 32 = 68$$

The Fahrenheit temperature, y, is 68° when the Celsius temperature
is 20°.

The positive slope in the equation $y = \frac{9}{5}x + 32$ means that as the Celsius temperature increases, the Fahrenheit temperature increases. We can graph the relationship as shown. Notice that the *y*-intercept is 32 and the slope is $\frac{9}{5}$.

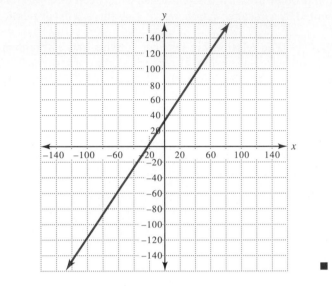

□ **DO EXERCISE 1.**

EXAMPLE 2   For income tax purposes, Mrs. Green computes the depreciation of her rental house using straight-line depreciation. If the house cost $80,000 and is depreciated at a rate of $4000 per year, find the equation involving the book value *V* and the time *t* that it has depreciated. What is the book value of the house after 7 years?

1. Use the point-slope form. Since we want an equation that will give us book value *V*, we write the variable *V* first. We say that *V* is the dependent variable since *V* is dependent on *t* for its value.

$$V - V_1 = m(t - t_1)$$

2. Slope is the rate of change of the book value with respect to time. Hence $m = -4000$ dollars per year.

3. Since the house cost $80,000 at time $t_1 = 0$ when it has not depreciated, $V_1 = 80,000$, so we are given the point $(0, 80,000)$.

4. $V - V_1 = m(t - t_1)$

   $V - 80,000 = -4000(t - 0)$

   $V = -4000t + 80,000$

   The equation is $V = -4000t + 80,000$.

5. After 7 years, $t = 7$ and

$$V = -4000(7) + 80,000$$

$$V = 52,000$$

The book value after 7 years is $52,000.   ■

300

□ **DO EXERCISE 2.**

Notice that in Example 2 the slope is negative, showing that as the years increase, the book value decreases.

EXAMPLE 3 For each foot of depth, the pressure in ocean water increases by $\frac{5}{11}$ pounds per square inch (psi). If the pressure at 33 feet is 30 psi, find an equation for the linear relationship between pressure p and depth d below the surface of the ocean. What is the pressure on a diver 44 feet below the surface?

1. If p is the pressure and d is the distance below the surface, the equation is

$$(p - p_1) = m(d - d_1)$$

2. The slope is $\frac{5}{11}$ since the pressure increases by $\frac{5}{11}$ square foot for each foot of depth.

3. Since the pressure at 33 feet is 30 pounds per square foot, we are given the point (33, 30).

4. Using the equation from step 1 and substituting yields

$$p - 30 = \frac{5}{11}(d - 33)$$

$$p - 30 = \frac{5}{11}d - 15$$

$$p = \frac{5}{11}d + 15$$

The equation is $p = \frac{5}{11}d + 15$.

5. The pressure on a diver 44 feet below the surface is

$$p = \frac{5}{11}(44) + 15$$

$$p = 35$$

The pressure 44 feet below the surface is 35 pounds per square inch.

∎

□ **DO EXERCISE 3.**

□ **Exercise 3** A steel bridge cable increases 0.024 foot in length for each degree Celsius that the temperature rises. If the length of the cable is 2000 feet at 0° Celsius, write an equation to express the relationship between the length L and the temperature t. What is the length of the cable at 30° Celsius?

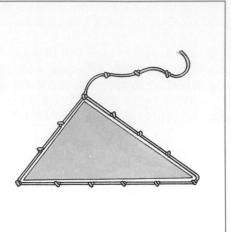
Answers to Exercises

1. $y = -\dfrac{1}{100}x + 80$; $20°$

2. $V = -2000t + 12{,}000$; $4000

3. $L = 0.024t + 2000$; 2000.72 ft

© 1989 by Prentice Hall

1. If the Fahrenheit temperature is 50° when the Celsius temperature is 10° and the Fahrenheit temperature is 95° when the Celsius temperature is 35°, find the linear equation expressing Celsius degrees y in terms of Fahrenheit degrees x. What is the temperature in degrees Celsius when the Fahrenheit temperature is 77°?

2. For the Rich Company the relationship between the number of units sold and the profit is linear. If the profit is $1500 when 500 units are sold and $5000 when 1200 units are sold, write the equation relating profit y to units sold x. What is the profit when 725 units are sold?

3. A house costing $60,000 is depreciated using straight-line depreciation at a rate of $3000 per year. Find the equation involving book value V and time t that it has depreciated. What is the book value after 11 years?

4. A football player signs a contract that pays him $1000,000 plus $5000 for each game the team wins. Find the linear equation that gives his total wages w in terms of the number of games g the team won. How much will he make if the team wins 9 games?

5. Jane wishes to build a doghouse. She will need lumber and hardware. If she uses 44 square feet of plywood, the total cost is $20. If she uses 68 square feet of plywood, the total cost is $26. Write the linear equation that expresses the total cost c in terms of the number of square feet of plywood L used. What is the cost of using 88 square feet of plywood?

6. Assume that the temperature f feet above the surface of the earth is $t°$ Celsius and the relationship between f and t is linear. Find the number of feet above the earth that the temperature is 70° given that at 7000 feet the temperature is 10° Celsius and at 4000 feet the temperature is 40° Celsius.

7. An airline company computes the cost of a ticket on a 125-seat charter flight as $120 plus $7 for each empty seat. Write an equation to express the cost per ticket c in terms of the number of empty seats e. What is the cost of a ticket if there are 116 passengers?

8. Kevin decides that he will save $\frac{2}{5}$ of the amount of his income tax refund plus $500. Write a linear equation to express the amount s that he will save in terms of the amount of his refund r. How much will he save if his refund is $425.25?

9. If the depth in ocean water increases $\frac{11}{5}$ feet for each psi of pressure and the depth is 88 feet when the pressure on a diver is 55 psi, find the depth of a diver when the pressure on her is 75 psi.

10. As an appliance salesman receives $1250 per month plus $50 for each appliance he sells. How many appliances did he sell if he earns $2200?

11. Find the linear equation that relates the book value V to the number of years of depreciation t of a machine which cost $21,000 and is being depreciated at $3000 per year. What is the book value after $3\frac{1}{2}$ years?

12. Joan receives a salary of $1200 per month plus a commission of 4% on sales. What was her sales total if her wages were $1400?

Checkup

These problems provide a review of some of Sections 6.2 and 6.4 and will help you with the next section.
Graph using intercepts.

13. $x - y = 4$

14. $x + y = -3$

15. $2x - 3y = 6$

16. $4x - 3y = 12$

17. $x + 4y = 4$

18. $2x + y = 4$

Graph using the slope and y-intercept.

19. $y = 2x$

20. $y = -3x$

21. $y = 3x - 2$

22. $y = 4x + 1$

23. $y = -2x + 1$

24. $y = -5x + 4$

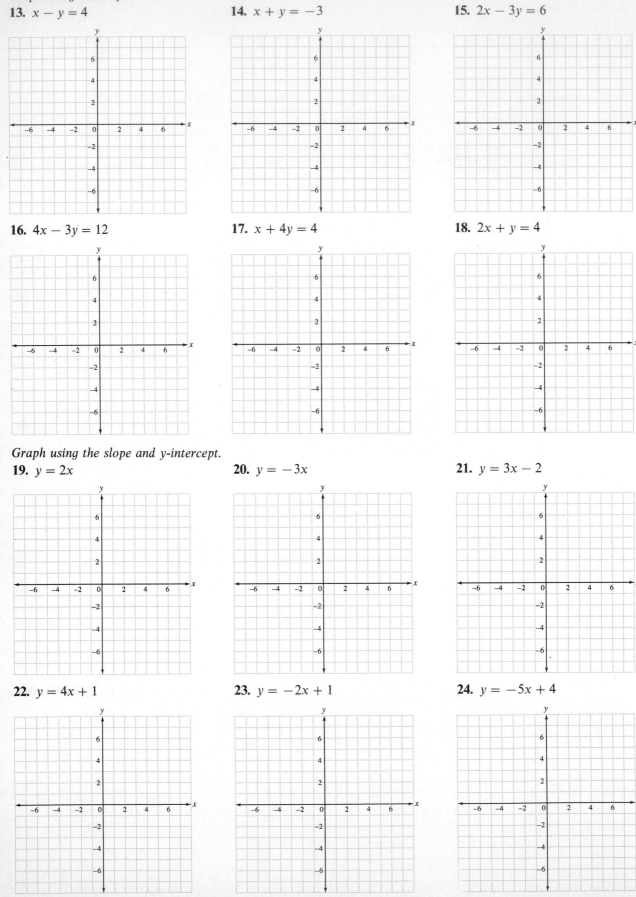

CHAPTER 6 ADDITIONAL EXERCISES (OPTIONAL)

Section 6.1

Plot the following points.

1. $(4, -2)$
2. $(5, 3)$
3. $(-3, -4)$
4. $(-2, 2)$

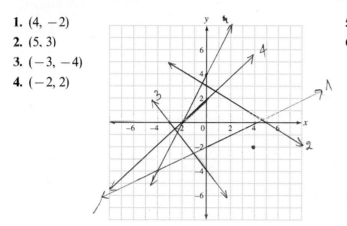

Find the coordinates of these points.

5. A $(4, 2)$
6. B $(-4, 1)$

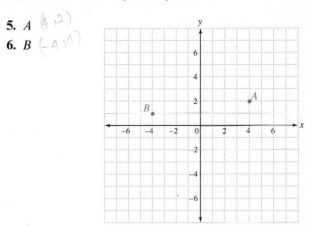

In which quadrant are the following points located?

7. $(3, -7)$ $\left(\text{IV}\right)$

8. $(-4, -4)$ $\left(\text{III}\right)$

9. $(-8, 5)$ $\left(\text{II}\right)$

10. $(9, 7)$ I

Section 6.2

Is the given ordered pair a solution of the equation?

11. $4x - 5y = 7$ $(-2, -3)$ Yes

12. $3x + 2y = 6$ $(-1, 4)$ No

13. $y = 3x + 6$ $(3, -5)$ No

14. $x = 4y - 8$ $(0, 2)$ Yes.

Graph using three different x values.

15. $y = 3x - 1$

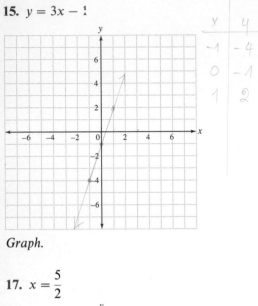

16. $y = -2x + 4$

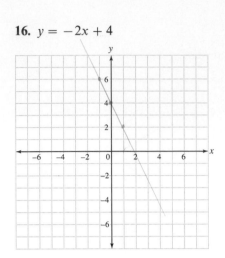

Graph.

17. $x = \dfrac{5}{2}$

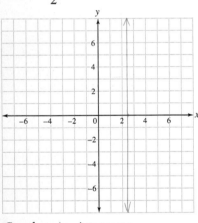

18. $y = -\dfrac{7}{2}$

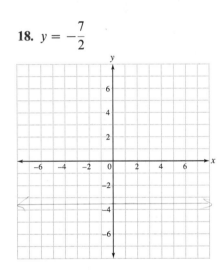

Graph, using intercepts.

19. $x = y + 3$

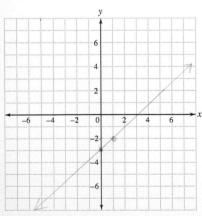

20. $y = -x + 2$

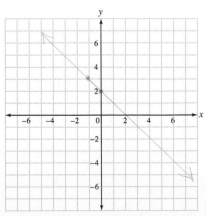

21. $3x + 5y = 15$

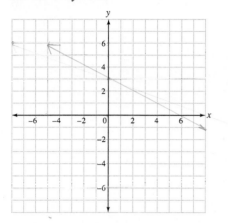

22. $2x - 4y = 8$

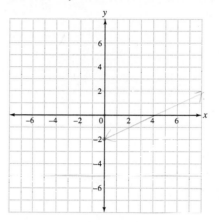

Section 6.3

Find the slope of the line through the pairs of points.

23. $(6, -2), \quad (4, -1)$

$$m = \frac{y_2 - y_1}{x_2 - x_1} = \frac{(-1) - (-2)}{4 - 6} = \frac{1}{2}$$

24. $(-3, 5), \quad (-7, -2)$

25. $\left(-\frac{5}{8}, \frac{3}{4}\right), \quad \left(\frac{1}{3}, -2\right)$

$$m = \frac{(-2) - \left(\frac{3}{4}\right)}{\left(\frac{1}{3}\right) - -\frac{5}{8}} = \frac{\frac{11}{4}}{\frac{23}{24}} = \frac{11}{4} \cdot \frac{24}{23} = = \frac{11 \cdot 6}{23} = \frac{66}{23}.$$

26. $\left(\frac{9}{5}, -3\right), \quad \left(-\frac{7}{10}, \frac{5}{6}\right)$

27. $(0.9, -3.4), \quad (2.1, -1)$

28. $(4.8, -3.7), \quad (2.2, 4.1)$

Find the slope of the following lines.

29. $3x - 6 = 0$

30. $4y + 8 = 0$

Section 6.4

Find the slope and y-intercept of the following lines.

31. $y = 3x - 7$

32. $y = -2x + 5$

33. $7x - 2y = 14$

34. $8y - 9x = 27$

Graph. Use the slope and y-intercept.

35. $5x + 3y = 6$

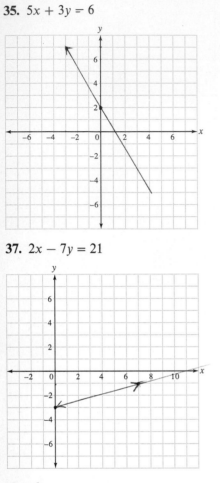

36. $-4x + 5y = 5$

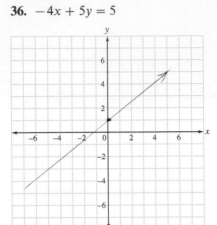

37. $2x - 7y = 21$

38. $6x - y = 4$

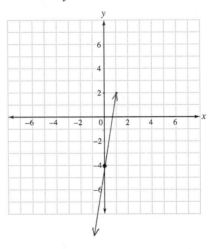

Graph.

39. $x + y = 4$

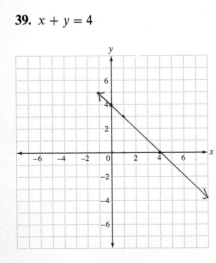

40. $x = -\dfrac{3}{2}$

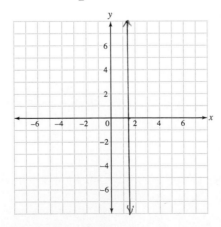

41. $3x - 5y = 20$

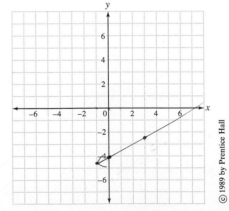

42. $2x - 3y = 6$

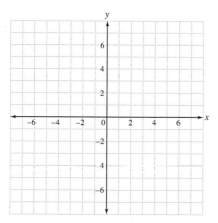

43. $y = -\dfrac{3}{2}x$

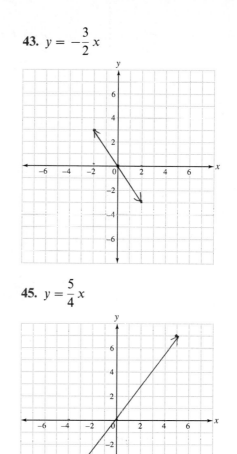

44. $2x + 3y = 9$

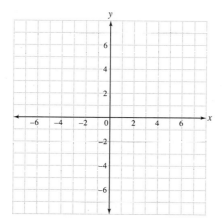

45. $y = \dfrac{5}{4}x$

Section 6.5

Find an equation for the line with the following y-intercept and slope.

46. $(0, -3), \quad m = 4$

47. $(0, 6), \quad m = -2$

48. $\left(0, \dfrac{7}{8}\right), \quad m = \dfrac{1}{3}$

49. $\left(0, -\dfrac{9}{5}\right), \quad m = \dfrac{7}{4}$

Find an equation for the line with the given slope through the given point.

50. $m = -3$, $(4, -2)$

51. $m = 4$, $\left(\dfrac{3}{4}, -\dfrac{1}{5}\right)$

52. $m = -\dfrac{1}{3}$, $(-4, 5)$

Find an equation for the line through the given two points.

53. $(3, 5)$, $(6, 8)$

54. $(3, -1)$, $(-3, 4)$

55. $(6.7, -2.5)$, $(7.8, 3)$

56. $(-2.5, -0.9)$, $(-3.2, 4)$

Decide if the following pairs of lines are parallel, perpendicular, or neither.

57. $3x + 5y = 7$
$6y - 15x = 8$

58. $7x + 2y = 4$
$4y + 14x = 3$

Section 6.6

59. For the James Company the relationship between the number of units sold and the profit is linear. If the profit is $2000 when 500 units are sold and $3500 when 800 units are sold, write the equation relating profit y to units sold x. What is the profit when 625 units are sold?

60. Susan wants to make drapes. She will need fabric and supplies. If she uses 8 yards of fabric, the total cost is $75. If she uses 12 yards of fabric, the total cost is $115. Write the linear equation that expresses the total cost c in terms of the yards of fabric f used. What is the cost of using 9 yards of fabric?

61. Find the linear equation that relates book value, V, to the number of years of depreciation, t, of a new car that cost $14,000 and is being depreciated at $1500 per year. What is its book value after 7 years?

62. Karen receives a salary of $900 per month plus a commission of 5% on sales. How much did her sales total if she earned $2400 in one month?

CHAPTER 6   PRACTICE TEST

1. Find the coordinates of the points P and Q.

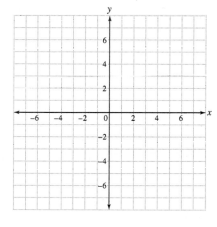

1. _____

2. In question 1, in which quadrant is the point P located?

2. _____

3. Graph $2x + 3y = 6$.

3. _____

4. Find the slope of the line through the points $(7, 5)$ and $(-2, 6)$.

4. _____

5. Find the slope of the line $3x - 4y = 2$.

5. _____

6. _____

6. Graph $y = \dfrac{2}{3}x + 2$.

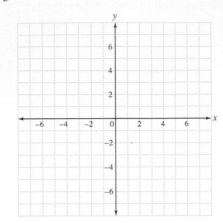

7. _____

7. Graph $x + 4 = 0$.

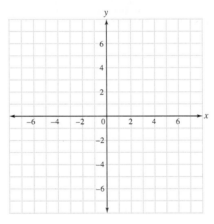

8. _____

8. Find an equation for the line with slope 3 and y-intercept $(0, -4)$.

9. _____

9. Find an equation for the line through the points $(-4, -1)$ and $(0, -3)$.

10. _____

10. Mary sells rings. Her profit is $6 per ring but her annual overhead is $1500. Write an equation to express her profit p in terms of the number r of rings sold.

© 1989 by Prentice Hall

CUMULATIVE REVIEW   CHAPTERS 5 AND 6

1. Which of the following are polynomials?
   (a) $x^3 + 3x^2 + 4$
   (b) y^{-1}
   (c) 7
   (d) $\dfrac{x}{5} - 2$

2. Classify the following as a monomial, binomial, or trinomial.
   (a) $x + 5$
   (b) $x^3 + 4x + 8$
   (c) $7a^2$

3. Find the degree of the polynomials.
   (a) $7x^2 + 2x - 5$
   (b) $3x^2 + 7x^3 - 8$

Add or subtract as indicated.

4. $\left(\dfrac{1}{5} x^2 - \dfrac{7}{15} x + 2\right) + \left(\dfrac{3}{8} x^2 - \dfrac{3}{5} x - 7\right)$

5. $(3.9x^2 + 2.5x - 3) - (6.4x^2 - 9.3x - 0.7)$

6. $(6.7x^2 + 2.4x + 7.3) + (5.2x^2 - 7.8x - 4.5)$

7. $\left(\dfrac{11}{4} y^2 - \dfrac{3}{4} y - 2\right) - \left(\dfrac{7}{2} y^2 + \dfrac{9}{5} y - 8\right)$

Multiply.

8. $7x^4(3x^3 - 4x^2 + 8x)$

9. $(3y + 4)(7y - 8)$

10. $(8y - 7)(9y - 8)$

11. $(x - 3)(x^3 - 4x^2 + 2)$

12. $\left(\dfrac{3}{5} x - 8\right)\left(\dfrac{3}{5} x + 8\right)$

13. $(0.3y - 5)^2$

14. $(x^2 - 3)^2$

Divide.

15. $\dfrac{x^2 + 5x - 24}{x + 8}$

16. $\dfrac{2x^2 + 7x - 8}{2x - 3}$

17. Solve for y: $3 + ny = y$.

Factor completely.

18. $5x^3y + 3x^2y^4 - 7xy^3$

19. $a^2 - ay - ax + xy$

20. $x^2 - 2x - 63$

21. $6x^2 - 27x - 105$

22. $6y^4 - 2y^3 - 8y^2$

23. $9x^2 - 12x + 4$

24. $\dfrac{16}{81} y^4 - 1$

25. Without plotting the points, determine in which quadrant they are located.
    (a) $(-2, -3)$
    (b) $(7, -6)$

Graph.

26. $y = -3$

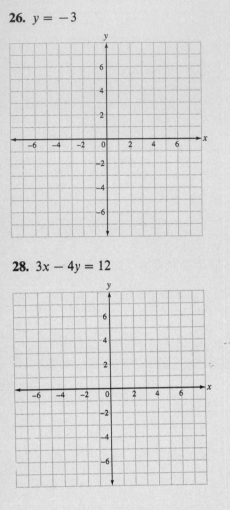

27. $2x - 3y = 9$

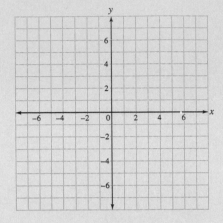

28. $3x - 4y = 12$

29. $y = 4x$

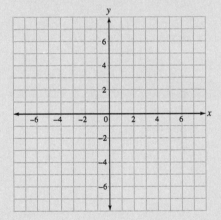

Find the slope of the following lines.

30. $3x - 2y = 8$

31. $x = 6$

32. Through the points $(-4, -8)$ and $(3, -6)$

33. Find an equation for the line through the points $(8, 5)$ and $(2, -4)$.

Find the slope of each line in the pair of lines. Decide if the lines are parallel, perpendicular, or neither.

34. $4x + 2y = 9$
$2x + \ y = 8$

35. $3x - 5y = 4$
$10x + 6y = 7$

36. Find the linear equation that relates the book value V to the number of years of depreciation t of a motorcycle that cost \$6000 and is being depreciated at \$800 per year. What is the book value after $3\frac{3}{4}$ years?

Graphing and Linear Systems of Equations

7.1 SOLUTION BY GRAPHING

In this chapter we study methods of solution of pairs of linear equations in two variables. An example of a pair of linear equations in two variables is

$$x + 2y = 4$$
$$3x - y = 7$$

Sets of linear equations in more than two variables are left for more advanced courses.

In Section 7.4 we see that it is often easier to solve word problems using pairs of linear equations in two variables. Before we discuss solution by graphing we want to see what the solution looks like on the graph.

1 Solutions of Linear Equations in Two Variables

Consider the equations

$$x - y = 3$$
$$2x + 3y = 6$$

We will graph the first equation $x - y = 3$. Using the intercept method, we find the following ordered pairs.

x	y	Ordered Pairs
0	-3	$(0, -3)$
3	0	$(3, 0)$
-2	-5	$(-2, -5)$

Plot the points and graph the equation. Notice that each *ordered pair* is a solution to the equation. Recall that any point on the line is a solution to the equation. For example, (5, 2) is a point on the line, so it is a solution. We can check it in the equation.

$$x - y = 3$$

Let $x = 5$ and $y = 2$.

$$5 - 2 \stackrel{?}{=} 3$$

$$3 = 3$$

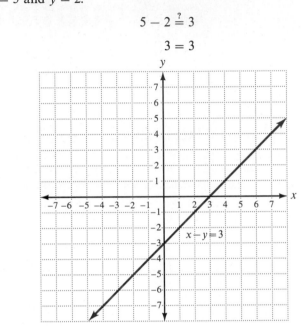

Graph the second equation, $2x + 3y = 6$, using the intercept method. The graph is as shown. All points on the line are solutions to the equation.

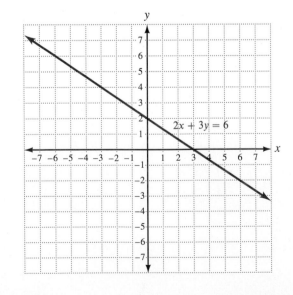

The solutions to pairs of linear equations are points where the graphs of the equations intersect.

The solution to two equations in two variables is usually an ordered pair that satisfies both equations and corresponds to a point that lies on both lines. We will show in Example 3 that the equations may also have many solutions. If we graph both equations on the same set of axes, it appears that the solution is (3, 0). We can check this solution in the given pair of equations $x - y = 3$ and $2x + 3y = 6$.

$$x - y = 3$$

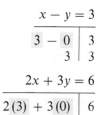

$$2x + 3y = 6$$

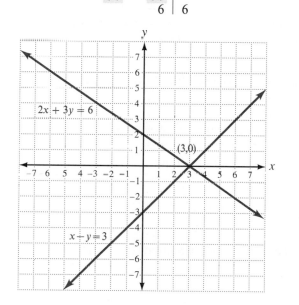

The ordered pair (3, 0) checks, so it is a solution to the system of equations.

□ **Exercise 1** Decide if the given ordered pair is a solution of the system of equations.

a. $(-1, 6)$; $2x + y = 4$
$\qquad\qquad\ 3x + 2y = 9$

b. $(2, 0)$; $2x + 4y = 5$
$\qquad\qquad 4x + 2y = 7$

EXAMPLE 1

a. Decide if $(2, -5)$ is a solution to the system

$$3x + \ y = \quad 1$$
$$2x + 3y = -11$$

1. Check the first equation.

$$
\begin{array}{c|c}
\multicolumn{2}{c}{3x + y = 1} \\
\hline
3(2) - 5 & 1 \\
6 - 5 & 1 \\
1 & 1
\end{array}
$$

2. Check the second equation.

$$
\begin{array}{c|c}
\multicolumn{2}{c}{2x + 3y = -11} \\
\hline
2(2) + 3(-5) & -11 \\
4 + (-15) & -11 \\
-11 & -11
\end{array}
$$

$(2, -5)$ is a solution to the system of equations.

b. Decide if $(4, -1)$ is a solution to the system

$$x + \ y = 3$$
$$2x + 5y = 2$$

1. Check the first equation.

$$
\begin{array}{c|c}
\multicolumn{2}{c}{x + y = 3} \\
\hline
4 - 1 & 3 \\
3 & 3
\end{array}
$$

2. Check the second equation.

$$
\begin{array}{c|c}
\multicolumn{2}{c}{2x + 5y = 2} \\
\hline
2(4) + 5(-1) & 2 \\
8 + (-5) & 2 \\
3 & 2
\end{array}
$$

The ordered pair $(4, -1)$ is not a solution to the second equation. Therefore, it is not a solution to the system of equations. ■

□ **DO EXERCISE 1.**

Consistent equations systems of equations that have at least one solution.

▨ Solution by Graphing

Most linear equations can be solved by graphing both equations. The coordinates of the points where they intersect is the solution.

EXAMPLE 2 Solve by graphing.

$$y = x$$
$$2x - y = 6$$

1. Graph $y = x$.
   This line has slope 1 and y-intercept $(0, 0)$. Graph using the slope and y-intercept.
2. Graph $2x - y = 6$ using intercepts.
3. From the graph the point of intersection appears to be $(6, 6)$.
4. Check:

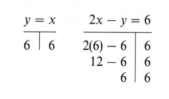

$$
\begin{array}{c|c}
y = x & 2x - y = 6 \\
\hline
6 \mid 6 & \begin{array}{c|c} 2(6) - 6 & 6 \\ 12 - 6 & 6 \\ 6 & 6 \end{array}
\end{array}
$$

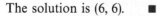

The solution is $(6, 6)$. ■

> Two linear equations in two variables that have at least one solution are called **consistent equations**.

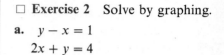

□ **Exercise 2** Solve by graphing.

a. $y - x = 1$
   $2x + y = 4$

b. $x + y = 4$
   $y - x = 2$

Inconsistent equations   systems of equations that have no solution.

EXAMPLE 3   Solve by graphing.

a. $y = 2x - 1$
    $y = 2x + 1$

1. Graph the first equation, $y = 2x - 1$. Use the slope and the y-intercept.
2. Graph the second equation using the slope and the y-intercept.
3. Notice that the lines do not cross. They are parallel. There is no solution.

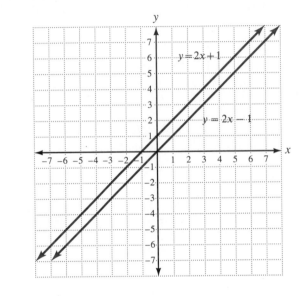

> If the graphs of a pair of linear equations are parallel lines, there is no solution to the equations and the equations are said to be **inconsistent equations**.

b. $3x + y = 3$
    $6x + 2y = 6$

1. Graph the first equation, $3x + y = 3$. Use the intercept method.
2. Graph the second equation using intercepts.
3. Notice that the graphs of the two equations are the same line. All the solutions of one equation are solutions of the second equation. We say that there are an infinite number of solutions.

Dependent equations   For a pair of linear equations in two variables, the graphs are the same line and there are an infinite number of solutions.

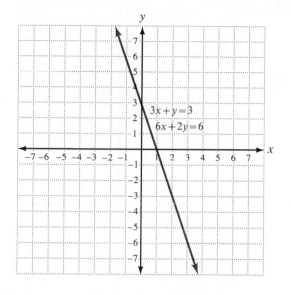

If the graphs of a pair of linear equations are the same line, there are an infinite number of solutions (all points on the line satisfy the equation). The two equations are **dependent**.   ∎

□ **DO EXERCISE 3.**

Solution of Pairs of Linear Equations in Two Variables by Graphing

If the graphs of the equations are two different intersecting lines, the point of intersection is the solution. If the graphs of the equations are parallel lines, there is no solution. The equations are inconsistent. If the graphs of the equations are the same line, there are infinite solutions. The equations are dependent.

□ **Exercise 3**   Solve by graphing.

a. $y = -2x + 2$
    $y = -2x + 4$

b. $x - y = 2$
    $2x - 2y = 4$

Answers to Exercises _____

1. a. Yes   **b.** No

2. a. (1, 2)   **b.** (1, 3)

3. a. No solution   **b.** Infinite solutions.

© 1989 by Prentice Hall

PROBLEM SET 7.1

A. *Decide if the given ordered pair is a solution of the system of equations.*

1. $(4, 6)$; $x + y = 10$ *Yes*
                $y - x = 2$

2. $(5, 1)$; $x + 2y = 7$ *Yes*
                $x - y = 4$

3. $(1, 0)$; $3x + 4y = 6$ *No*
                $4x + 2y = 7$

4. $(0, -4)$; $2x - 5y = 20$
                  $3x + 4y = -15$

5. $(-5, 2)$; $3x - 5y = -20$
                  $2x + 3y = -4$

6. $(3, -2)$; $4x - 5y = 0$
                  $3x + 2y = 5$

7. $(3, -2)$; $x - 3 = 0$
                  $y + 2 = 0$

8. $(-7, 1)$; $x - 7 = 0$
                  $y - 1 = 0$

B. *Solve by graphing.*

9. $x + y = 2$ $(0,2)(2,0)$
      $y - x = 4$ $(0,4)(-4,0)$

10. $x + y = 6$ $(0,6)(0,6)$
       $x - y = 2$ $(0,-2)(2,0)$

11. $y = x$ $(0,0)(0,0)$
       $2x + y = 6$ $(0,6)(3,0)$

12. $x = 2y$ $(0,0)(0,0)$
       $x + 2y = 4$ $(0,2)(4,0)$

13. $y = -3x + 5$ $(0,5)$
       $y = -3x + 3$ $(0,3)($

14. $y = -x + 1$
       $y = -x - 1$

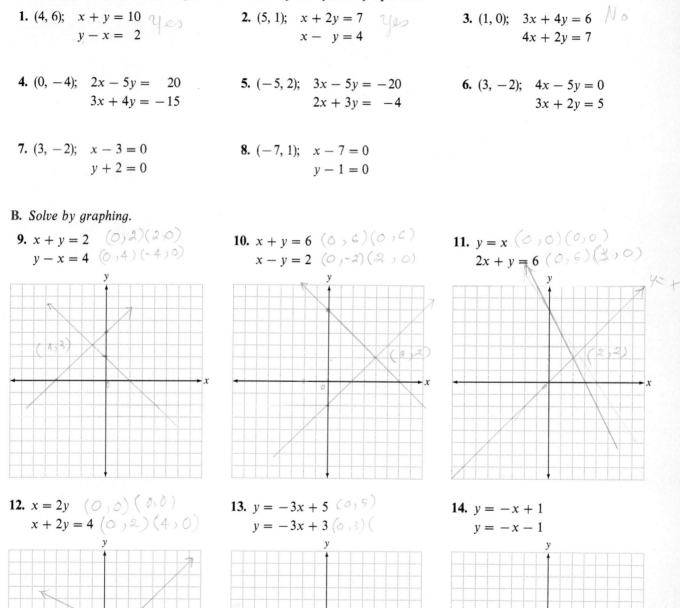

15. $x = 1$
$\quad y = 2x - 1$

16. $x + 3y = 3$
$\quad y = 1$

17. $x + y = 1$
$\quad 4x = 4 - 4y$

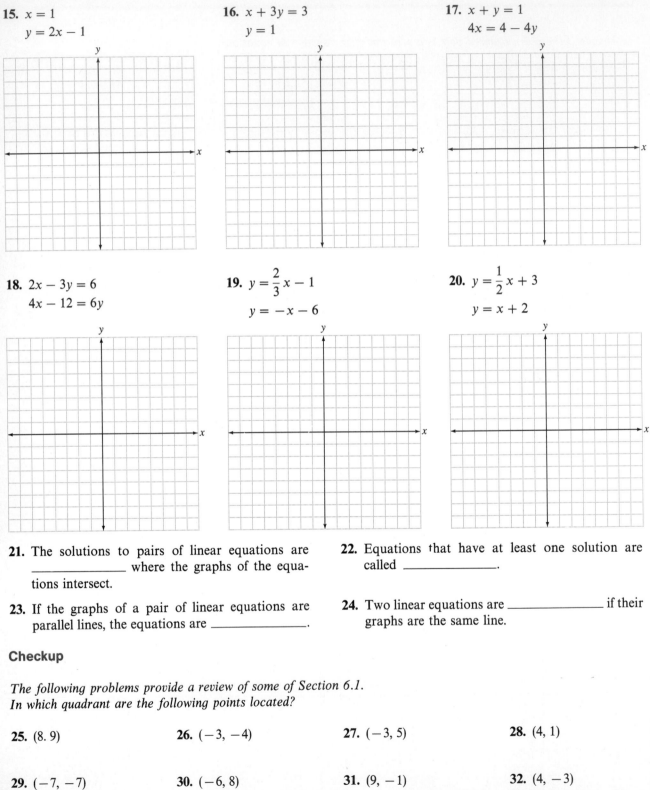

18. $2x - 3y = 6$
$\quad 4x - 12 = 6y$

19. $y = \dfrac{2}{3}x - 1$
$\quad y = -x - 6$

20. $y = \dfrac{1}{2}x + 3$
$\quad y = x + 2$

21. The solutions to pairs of linear equations are _____ where the graphs of the equations intersect.

22. Equations that have at least one solution are called _____.

23. If the graphs of a pair of linear equations are parallel lines, the equations are _____.

24. Two linear equations are _____ if their graphs are the same line.

Checkup

The following problems provide a review of some of Section 6.1.
In which quadrant are the following points located?

25. (8. 9)

26. (−3, −4)

27. (−3, 5)

28. (4, 1)

29. (−7, −7)

30. (−6, 8)

31. (9, −1)

32. (4, −3)

7.2 THE ADDITION METHOD

1 Solution by graphing allows us to see the geometric solution of a pair of equations. However, this method is tedious and slow. If the x-coordinate or y-coordinate of the point of intersection contains fractions, it is difficult to read them from the graph. The method most often used to solve pairs of equations is called the *addition method*.

We want to add the two equations so that one variable disappears and we have an equation in one variable that we can solve. Remember that an equation is a balance, so we may add the left sides and the right sides of the equations and get a new equation.

□ **Exercise 1**   Solve.

a. $x + y = 8$
    $x - y = 2$

EXAMPLE 1   Solve.

$$x - y = 3$$
$$x + y = 7$$

1. Add the two equations.

$$\begin{array}{r} x - y = 3 \\ x + y = 7 \\ \hline 2x = 10 \end{array}$$

The variable y has disappeared. We have an equation in one variable.

2. Solve for x.

$$x = 5$$

3. Remember that the solution must be an ordered pair. We must solve for y. Use either of the original equations. Choose the equation that makes the arithmetic easier. We found $x = 5$.

$$x - y = 3$$
$$(5) - y = 3$$
$$-y = -2$$
$$y = 2$$

b. $y - x = -5$
    $x + y = 1$

4. Check in both equations.

Let $x = 5$ and $y = 2$.

$$\begin{array}{c|c} x - y = 3 & x + y = 7 \\ \hline (5) - (2) \mid 3 & (5) + (2) \mid 7 \\ 3 \mid 3 & 7 \mid 7 \end{array}$$

5. The solution is (5, 2). ∎

□ **DO EXERCISE 1.**

Removing a Variable

When we add two equations, in order for one variable to drop out we must have the same coefficient and different signs on the variable in each equation. To do this, we multiply each side of one of the equations by a number. Often the number that we multiply by is -1.

EXAMPLE 2 Solve.

a. $x + 2y = 8$
$3x + 2y = 12$

1. We will eliminate the variable y. To do this we multiply both sides of the second equation by -1.

$$\begin{array}{r} x + 2y = 8 \\ -3x - 2y = -12 \\ \hline -2x \quad = -4 \qquad \text{Adding} \\ x \quad = 2 \end{array}$$

2. Solve for y by substituting $x = 2$ in one of the original equations.

$$x + 2y = 8$$
$$(2) + 2y = 8$$
$$2y = 6$$
$$y = 3$$

3. Check $(2, 3)$ in both equations.

$$\begin{array}{c|c} x + 2y = 8 \\ \hline (2) + 2(3) & 8 \\ 8 & 8 \end{array} \qquad \begin{array}{c|c} 3x + 2y = 12 \\ \hline 3(2) + 2(3) & 12 \\ 12 & 12 \end{array}$$

4. The solution is $(2, 3)$.

The number that we multiply by may be other than -1.

b. $x + y = 7$
$-3x + 3y = -9$

1. We may eliminate the variable x by multiplying both sides of the first equation by 3.

$$\begin{array}{r} 3x + 3y = 21 \\ -3x + 3y = -9 \\ \hline 6y = 12 \\ y = 2 \end{array}$$

2. Solve for x in one of the original equations.

$$x + y = 7$$
$$x + (2) = 7$$
$$x = 5$$

3. Check $(5, 2)$ in both equations.

$$
\begin{array}{c|c}
\multicolumn{2}{c}{x + y = 7} \\
\hline
(5) + (2) & 7 \\
7 & 7
\end{array}
\qquad
\begin{array}{c|c}
\multicolumn{2}{c}{-3x + 3y = -9} \\
\hline
-3(5) + 3(2) & -9 \\
-9 & -9
\end{array}
$$

4. The solution is $(5, 2)$. ∎

☐ **DO EXERCISE 2.**

Sometimes we multiply both sides of one equation by a number and both sides of the other equation by a different number to eliminate a variable.

EXAMPLE 3 Solve.

$$3x + 5y = -7$$

$$5x + 4y = 10$$

1. Multiply the first equation by 5 and the second equation by -3 to eliminate the x variable. Notice that 5, the coefficient of x in the second equation, is multiplied by the first equation, and the opposite of the coefficient of x in the first equation is multiplied by the second equation. We used the opposite of 3 in order to change the sign.

$$5(3x + 5y) = 5(-7)$$

$$-3(5x + 4y) = -3(10)$$

$$
\begin{array}{ll}
\;\;\;15x + \;\;\;\;25y = -35 & \text{First equation multiplied by 5} \\
-15x + (-12)y = -30 & \text{Second equation multiplied by } -3 \\
\hline
\qquad\qquad 13y = -65 \\
\qquad\qquad\;\; y = \;-5
\end{array}
$$

We could have eliminated y and solved for x in step 1.

2. Substitute for y in either of the original equations. We use $5x + 4y = 10$.

$$5x + 4y = 10$$

$$5x + 4(-5) = 10$$

$$5x - 20 = 10$$

$$5x = 30$$

$$x = 6$$

3. Check $(6, -5)$ in both equations.

$$
\begin{array}{c|c}
\multicolumn{2}{c}{3x + 5y = -7} \\
\hline
3(6) + 5(-5) & -7 \\
18 - 25 & -7 \\
-7 & -7
\end{array}
\qquad
\begin{array}{c|c}
\multicolumn{2}{c}{5x + 4y = 10} \\
\hline
5(6) + 4(-5) & 10 \\
30 - 20 & 10 \\
10 & 10
\end{array}
$$

4. The solution is $(6, -5)$. ∎

☐ **DO EXERCISE 3.**

☐ **Exercise 2** Solve.

a. $x - y = 1$
$\;\; 2x - y = 5$

b. $2x - y = 7$
$\;\; 3x + 2y = 0$

☐ **Exercise 3** Solve.

a. $2x + 3y = -1$
$\;\; 3x + 5y = -2$

b. $4x + 3y = 13$
$\;\; 3x - 7y = -18$

Some systems have no solution and some systems have an infinite number of solutions.

EXAMPLE 4 Solve.

a. $2x + y = 5$
$\underline{2x + y = -3}$

1. Multiply the second equation by -1 to eliminate the variable y.

$$\begin{array}{r} 2x + y = 5 \\ -2x - y = 3 \\ \hline 0 \neq 8 \end{array}$$

Both variables disappear and we get a false equation. The graphs of the equations are parallel lines.

2. There is no solution when we get a false equation. Therefore, the system is inconsistent.

b. $2x - y = 4$
$8x - 4y = 16$

1. Multiply the first equation by -4 to eliminate the variable y.

$$\begin{array}{r} -8x + 4y = -16 \\ 8x - 4y = 16 \\ \hline 0 = 0 \end{array}$$

2. Both variables are eliminated and we obtain a true equation. The graphs of the equations are the same line.

3. There are an infinite number of solutions when we obtain a true equation. The system is dependent. ■

□ **DO EXERCISE 4.**

DID YOU KNOW?

The Chinese solved pairs of linear equations by about 300 B.C. The Chinese people liked to study magic squares. According to legend, Emperor Yu was shown the first magic square on the back of a turtle around 2200 B.C. A magic square adds up to the same number when added along any row, column, or main diagonal.

16	3	2	13
5	10	11	8
9	6	7	12
4	15	14	1

Answers to Exercises

1. a. $(5, 3)$ b. $(3, -2)$
2. a. $(4, 3)$ b. $(2, -3)$
3. a. $(1, -1)$ b. $(1, 3)$
4. a. No solution   b. Infinite solutions

PROBLEM SET 7.2

Solve using the addition method.

1. $x - y = 7$
$x + y = 11$

2. $x + y = 4$
$x - y = 8$

3. $x + y = 2$
$2x - y = 4$

4. $6x - y = 1$
$-6x + 5y = 1$

5. $2x - 5y = 17$
$3x + y = 0$

6. $3x + 4y = 7$
$x - 4y = 5$

7. $x + y = 12$
$y - x = 4$

8. $y - x = 2$
$x + y = 8$

9. $2x + 3y = 6$
$2x + 3y = -3$

10. $2x + 3y = 8$
$x + 3y = 7$

11. $x - y = 7$
$x - y = 4$

12. $x - y = -4$
$x + 2y = -5$

13. $x + 4y = -2$
$2x + 8y = -4$

14. $x - 3y = 1$
$2x - 6y = 2$

15. $3x - 5y = 4$
$-2x + 2y = 1$

16. $3x - 4y = 1$
$-6x + 8y = -1$

17. $3x + 2y = 12$
$5x - 3y = 1$

18. $5x - 4y = -1$
$-7x + 5y = 8$

19. $3x + 5y = -7$
$5x + 4y = 10$

20. $5x - 2y = 0$
$2x - 3y = -11$

21. $3x + 5y = 6$
$5x + 3y = 4$

22. $7x - 5y = 4$
$-3x - 2y = -10$

23. $2x + 3y = -15$
$5x + 2y = 1$

24. $2x - 5y = 4$
$-3x + 7y = -2$

25. To solve two equations in two variables by the addition method, first we want to eliminate one _____.

26. To eliminate one variable from two equations, we must have the same coefficient and _____ signs on the variable in each equation.

27. Sometimes each equation must be _____ by a different number to eliminate a variable.

Checkup

The following problems provide a review of some of Section 6.3.
Find the slope of the line through the pairs of points.

28. (3, 7), (2, 5)

29. (4, 8), (6, 3)

30. (−3, 2), (4, −7)

31. (1, −4), (9, −5)

32. (−2, −4), (−5, −6)

33. (−1, 0), (−7, −8)

34. (6, 2), (6, 7)

35. (5, 4), (−8, 4)

Find the slope of the following lines.

36. $x = 7$

37. $y = 3$

38. $y = -4$

39. $x = -1$

7.3 THE SUBSTITUTION METHOD

OBJECTIVE

1 Most systems of linear equations are solved by the addition method, but there are certain cases where the substitution method is easier. If one of the equations is already solved for one variable in terms of the other variable, it is easier to use the substitution method.

1 Solve two linear equations in two variables by the substitution method

EXAMPLE 1 Solve.

$$x + y = 10 \qquad \text{First equation}$$

$$y = x + 8 \qquad \text{Second equation}$$

1. Notice that the second equation is solved for y in terms of x. We can substitute this expression for y in the first equation. When we substitute for y we remove y and replace it with $x + 8$.

$$x + y = 10 \qquad \text{First equation}$$

$$x + (x + 8) = 10 \qquad y \text{ in the first equation is replaced by } x + 8$$

2. Solve the equation for x.

$$2x + 8 = 10$$

$$2x = 2$$

$$x = 1$$

3. Use the original equation that is solved for y in terms of x to find y.

$$y = x + 8$$

We found $x = 1$.

$$y = 1 + 8$$

$$y = 9$$

4. Check the point $(1, 9)$ in both of the original equations.

$x + y = 10$		$y = x + 8$	
$(1) + (9)$	10	(9)	$(1) + 8$
10	10	9	9

5. The solution is $(1, 9)$. ■

□ **DO EXERCISE 1.**

There are some equations that are solved by substitution even though one of the original equations is not solved for one variable in terms of the other variable. Therefore, we will study this method, although the following problems would usually be done by the addition method.

□ **Exercise 1** Solve.

a. $x + y = -2$
   $y = x - 6$

b. $y = 2x - 5$
   $3y - x = 5$

c. $x + y = 6$
   $y = 2x$

d. $x + 5y = 3$
   $x = 2y + 10$

□ **Exercise 2** Solve by substitution.

a.
$$x + y = 7$$
$$-3x + 3y = -9$$

EXAMPLE 2 Solve.

$$2x + y = 5 \quad \text{First equation}$$
$$5x + 3y = 11 \quad \text{Second equation}$$

1. We must solve one of the equations for one variable in terms of the other variable. If one of the equations has a variable with a coefficient of 1, it is easier to solve for this variable. The coefficient of the y variable in the first equation is 1, so solve this equation for y.

$$2x + y = 5$$
$$y = -2x + 5$$

2. Substitute this value for y in the second original equation. Be sure to use parentheses around the value that is substituted for y.

$$5x + 3y = 11 \quad \text{Second equation}$$
$$5x + 3(-2x + 5) = 11$$
$$5x - 6x + 15 = 11$$
$$-x + 15 = 11$$
$$-x = -4$$
$$x = 4$$

b. $4x - 3y = 8$
$2x + y = 14$

3. Substitute $x = 4$ in the equation that was solved for y in terms of x in step 1.

$$y = -2x + 5$$
$$y = -2(4) + 5$$
$$y = -3$$

4. Check the point $(4, -3)$ in the original equations.

$2x + y = 5$		$5x + 3y = 11$	
$2(4) + (-3)$	5	$5(4) + 3(-3)$	11
5	5	11	11

5. The solution is $(4, -3)$. ∎

□ **DO EXERCISE 2.**

It is important to notice whether the addition or the substitution method is easier. If one of the original equations is solved for one variable in terms of the other variable, the substitution method is faster. Otherwise, use the addition method.

EXAMPLE 3 Solve.

a. $2x + y = 2$
$-x - y = 1$

    1. The addition method is easier. If we add the two equations, the y variable disappears.

$$\begin{array}{r} 2x + y = 2 \\ -x - y = 1 \\ \hline x \quad\quad = 3 \end{array}$$

    2. Use one of the original equations to solve for y.

$$2x + y = 2$$

We found $x = 3$.

$$2(3) + y = 2$$
$$y = -4$$

    3. Check the point $(3, -4)$ in both original equations.

$$\begin{array}{c|c} 2x + y = 2 \\ \hline 2(3) + (-4) & 2 \\ 2 & 2 \end{array} \qquad \begin{array}{c|c} -x - y = 1 \\ \hline -(3) - (-4) & 1 \\ 1 & 1 \end{array}$$

    4. The solution is $(3, -4)$.

b. $3x + 4y = -10$
$x = -2$

    1. This system of equations is easy to solve by substitution. Since the second equation is solved for x, substitute $x = -2$ in the first equation.

$$3x + 4y = -10$$
$$3(-2) + 4y = -10$$
$$-6 + 4y = -10$$
$$4y = -4$$
$$y = -1$$

    2. We are given that $x = -2$ in the original equations.
    3. Check the point $(-2, -1)$.

$$\begin{array}{c|c} 3x + 4y = -10 \\ \hline 3(-2) + 4(-1) & -10 \\ -10 & -10 \end{array} \qquad \begin{array}{c|c} x = -2 \\ \hline -2 & -2 \end{array}$$

    4. The solution is $(-2, -1)$. ■

☐ **DO EXERCISE 3.**

☐ **Exercise 3** Solve using either the addition or the substitution method. Indicate which method is easier.

a. $x - y = 6$
    $y = -x - 2$

b. $x - 2y = 0$
    $4x - 3y = 15$

c. $x + 2y = 7$
    $x = y + 5$

d. $8x - 5y = -9$
    $3x + 5y = -2$

© 1989 by Prentice Hall

□ **Exercise 4** Solve by either the addition method or the substitution method. First remove fractions.

a. $3x + \dfrac{y}{4} = 2$

$\dfrac{x}{2} + \dfrac{3y}{4} = -\dfrac{5}{2}$

b. $x - \dfrac{2y}{3} = -2$

$-\dfrac{3x}{4} - 2y = -6$

If there are fractions in the equation, multiply both sides of the equation by the lowest common denominator of each term to remove the fractions.

EXAMPLE 4 Solve.

$$\frac{x}{3} + \frac{y}{6} = \frac{5}{6}$$

$$\frac{5x}{12} + \frac{y}{4} = \frac{11}{12}$$

The lowest common denominator of the terms of the first equation is 6. Multiply both sides of the equation by 6.

$$6\left(\frac{x}{3} + \frac{y}{6}\right) = 6\left(\frac{5}{6}\right)$$

$$6\left(\frac{x}{3}\right) + 6\left(\frac{y}{6}\right) = 6\left(\frac{5}{6}\right)$$

$$2x + y = 5$$

Multiply both sides of the second equation by the lowest common denominator of its terms, 12.

$$12\left(\frac{5x}{12} + \frac{y}{4}\right) = 12\left(\frac{11}{12}\right)$$

$$12\left(\frac{5x}{12}\right) + 12\left(\frac{y}{4}\right) = 12\left(\frac{11}{12}\right)$$

$$5x + 3y = 11$$

We now have the system

$$2x + y = 5$$

$$5x + 3y = 11$$

This system is solved in Example 2. ■

□ **DO EXERCISE 4.**

Answers to Exercises

1. a. $(2, -4)$ b. $(4,$ c. $(2, 4)$ d. $(8, -1)$
2. a. $(5, 2)$ b. $(5, 4)$
3. a. $(2, -4)$; substitution   b. $(6, 3)$; addition
   c. $\left(\dfrac{17}{3}, \dfrac{2}{3}\right)$; substitution   d. $\left(-1, \dfrac{1}{5}\right)$; addition
4. a. $(1, -4)$ b. $(0, 3)$

PROBLEM SET 7.3

A. *Solve by substitution.*

1. $3x + 4y = 2$
  $y = x - 3$

2. $2x + y = 6$
  $x = 2y - 2$

3. $x = 2y$
  $3x - 2y = 4$

4. $y = x - 5$
  $2x + 3y = 15$

5. $x + y = -4$
  $y = x + 2$

6. $3x + 5y = -11$
  $x = 2y + 11$

7. $x = 2$
  $5x - 2y = 6$

8. $y = 3$
  $3x - 4y = 0$

9. $2x - y = 4$
  $x = y - 3$

10. $2x - 3y = 1$
  $x = y + 2$

11. $x - y = 1$
  $y = 2x - 3$

12. $2x - y = 3$
  $y = 2x + 1$

13. $2x - y = -4$
  $x + y = -2$

14. $x + 4y = 4$
  $x - 2y = -2$

15. $2x + y = 0$
  $x - 3y = 0$

B. *Solve by either the substitution or the addition method.*

16. $2x + y = 6$
  $x - y = 0$

17. $x - 3y = 6$
  $4x + 3y = 9$

18. $x = -2$
  $2x - 7y = 3$

19. $5x + 6y = 0$
  $y = -5$

20. $x + 4y = 4$
  $x = 2y - 2$

21. $\dfrac{2}{3}x + \dfrac{1}{2}y = \dfrac{1}{3}$
  $\dfrac{3}{5}x + y = -\dfrac{4}{5}$

22. $\dfrac{3}{4}x + \dfrac{1}{3}y = 8$
  $\dfrac{1}{2}x - \dfrac{5}{6}y = -1$

23. $\dfrac{x}{6} + \dfrac{y}{6} = 1$
  $-\dfrac{x}{2} - \dfrac{y}{3} = -5$

24. $\dfrac{x}{2} + \dfrac{3y}{10} = \dfrac{1}{5}$
  $\dfrac{3x}{10} + \dfrac{y}{2} = -\dfrac{1}{5}$

25. It is usually easier to solve two linear equations in two variables by the _____ method if one of the equations is already solved for one variable in terms of the other variable.

26. If there are fractions in an equation, first multiply both sides of the equation by the _____ _____ _____ of the fractions to remove the fractions.

Checkup

The following problems provide a review of some of Section 6.4.
Find the slope and y-intercept.

27. $y = 3x - 4$ **28.** $y = \dfrac{1}{2}x + 7$ **29.** $4x - 3y = 9$ **30.** $5x + 2y = 10$

Use the slope and y-intercept to graph.

31. $y = \dfrac{5}{2}x - 4$ **32.** $y = -\dfrac{3}{5}x + 5$ **33.** $y = -4x$

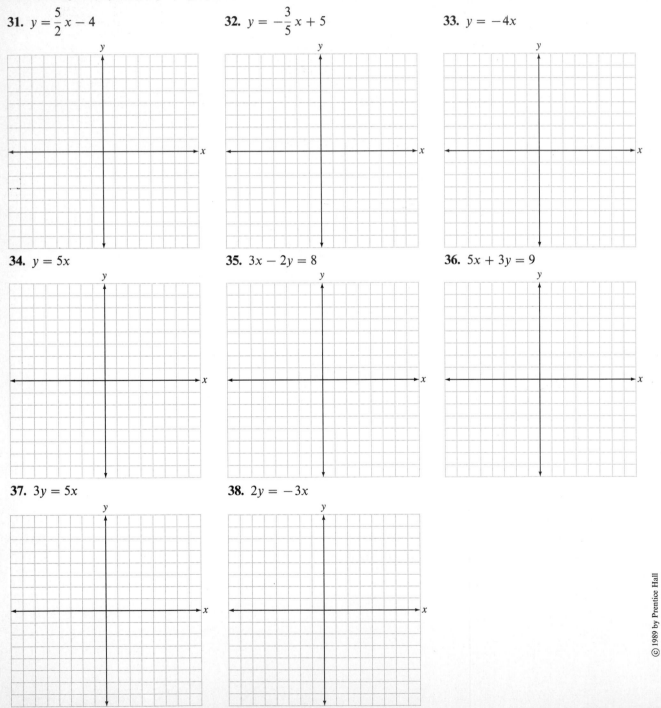

34. $y = 5x$ **35.** $3x - 2y = 8$ **36.** $5x + 3y = 9$

37. $3y = 5x$ **38.** $2y = -3x$

7.4 APPLIED PROBLEMS

OBJECTIVE

 We practiced solving applied problems in Chapter 4 using one variable because this work is helpful in more advanced mathematics. However, it is often easier to solve word problems using two variables.

■ *Solve geometry, percent, interest, and mixture problems using two linear equations in two variables*

> The steps in solving word problems with two variables are similar to the steps used in solving applied problems with one variable.
>
> 1. Read the problem carefully.
> 2. Make a drawing, if applicable.
> 3. Write variables for the unknowns.
> 4. Separate sentences into phrases to write two equations, since we are using two variables.
> 5. Solve the equations.
> 6. Check your answers in the word problem.

Geometry Problems

EXAMPLE 1   The length of a rectangle is three times the width. If the perimeter is 64 centimeters, find the dimensions.

Drawing

Variables Let x = width of the rectangle
y = length of the rectangle

Equations Use the formula for the perimeter of a rectangle

$$P = 2L + 2W$$

$$64 = 2y + 2x \qquad \text{First equation}$$

Since the length is three times the width,

$$y = 3x \qquad \text{Second equation}$$

We have two equations that we can solve.

$$64 = 2y + 2x \qquad \text{First equation}$$

$$y = 3x \qquad \text{Second equation}$$

Substitute for y in the first equation.

Solve $64 = 2(3x) + 2x$

$$64 = 6x + 2x$$

$$64 = 8x$$

$$8 = x$$

$$y = 3x = 3(8) = 24 \qquad \text{Using the second equation}$$

The dimensions of the rectangle are 8 centimeters by 24 centimeters.

Check The perimeter of the fence is $2L + 2W$ or $2(24) + 2(8) = 64$, so
the answer checks. ■

□ **DO EXERCISE 1.**

Percent Problems

EXAMPLE 2 The price of a truck was reduced by 10% and the truck
was sold for \$7200. What was the price before the sale?

Variables Let $x =$ price before the sale
$\qquad\qquad y =$ amount of the reduction

Equation The price before the sale minus the reduction equals the sale
price.

$$x - y = 7200 \qquad \text{First equation}$$

The reduction is 10% of the price before the sale.

$$y = (10\%) \cdot x$$

$$y = 0.10x \qquad \text{Second equation}$$

The two equations are

$$x - y = 7200 \qquad \text{First equation}$$

$$y = 0.10x \qquad \text{Second equation}$$

Substitute for y in the first equation.

Solve $\qquad 1x - 0.10x = 7200 \qquad \text{Remember that } x = 1x$

$$0.9x = 7200$$

$$x = 8000$$

The price before the sale was \$8000.

Check Ten percent of \$8000 is \$800. When we subtract \$800 from \$8000
we get \$7200, which is the sale price of the truck. ■

□ **DO EXERCISE 2.**

© 1989 by Prentice Hall

Interest Problems

EXAMPLE 3   If a sum of money is invested at 8% per year and $3000 less is invested at 9% per year, how much is invested at each rate if the total annual interest is $1430?

Variables Let x = amount invested at 8%

y = amount invested at 9%

Principal ·	rate	· time =	interest	
x	· (0.08) ·	(1)	$= x(0.08)(1)$	Use these facts in the first equation
y	· (0.09) ·	(1)	$= y(0.09)(1)$	

Equation Interest at 8% + interest at 9% = total interest

$$x(0.08)(1) \quad + \quad y(0.09)(1) \quad = 1430 \qquad \text{First equation}$$

Since $3000 less is invested at 9% than at 8%,

$$y = x - 3000 \qquad \text{Second equation}$$

The two equations are

$$x(0.08)(1) + y(0.09)(1) = 1430 \qquad \text{First equation}$$

$$y = x - 3000 \qquad\qquad\qquad \text{Second equation}$$

Substitute the value for y in the first equation.

Solve

$$x(0.08)(1) + (x - 3000)(0.09)(1) = 1430$$

$$0.08x \quad + \quad 0.09x - 270 \quad = 1430$$

$$0.17x - 270 \quad = 1430$$

$$0.17x = 1700$$

$$x = 10,000$$

$$y = x - 3000$$

$$y = 10,000 - 3000 = 7000$$

The amount invested at 8% is $10,000 and the amount invested at 9% is $7000.

Check The interest for 1 year on the $10,000 invested at 8% is 0.08($10,000) = $800. The interest for 1 year at 9% is 0.09($7000) = $630. The total interest is $800 + $630 = $1430, which checks. ■

☐ **DO EXERCISE 3.**

☐ **Exercise 3**   Linda earns a total annual interest of $387 on two sums of money. How much money does she have invested at each rate if four times as much money is invested at 9% per year as is invested at 7% per year?

Mixture Problems

EXAMPLE 4 Jon wants to make 60 liters of 50% sodium chloride solution by mixing a 40% solution with a 70% solution. How many liters of the 40% and 70% solutions should he use?

Variables Let x = liters of 40% solution
y = liters of 70% solution

Constant 60 = liters of final 50% solution

$\begin{pmatrix} \text{Percent of acid} \\ \text{in solution} \end{pmatrix}$	$\begin{pmatrix} \text{liters of} \\ \text{solution} \end{pmatrix}$	$\begin{pmatrix} \text{liters of} \\ \text{pure acid} \end{pmatrix}$	
40	x	0.40(x)	Use these
70	y	0.70(y)	facts in the
50	60	0.50(60)	first equation

The liters of pure acid in the solutions that are mixed must equal the liters of pure acid in the final solution.

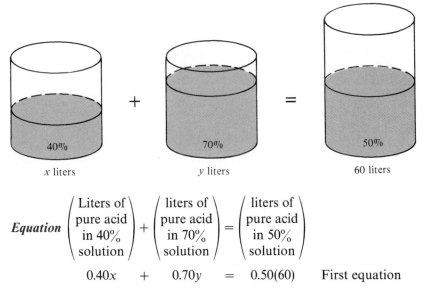

Equation $\begin{pmatrix} \text{Liters of} \\ \text{pure acid} \\ \text{in 40\%} \\ \text{solution} \end{pmatrix} + \begin{pmatrix} \text{liters of} \\ \text{pure acid} \\ \text{in 70\%} \\ \text{solution} \end{pmatrix} = \begin{pmatrix} \text{liters of} \\ \text{pure acid} \\ \text{in 50\%} \\ \text{solution} \end{pmatrix}$

$$0.40x \quad + \quad 0.70y \quad = \quad 0.50(60) \qquad \text{First equation}$$

Remember that we are combining x liters with y liters to get 60 liters. This gives us the equation

$$x + y = 60 \qquad \text{Second equation}$$

We have two equations in two variables that we can solve.

$$0.40x + 0.70y = 0.50(60) \qquad \text{First equation}$$

$$x + \quad y = 60 \qquad \text{Second equation}$$

Solve Multiply both sides of the first equation by 10 to remove the decimal points. The result is

$$4x + 7y = 5(60)$$

$$4x + 7y = 300$$

We have the system of linear equations

$$4x + 7y = 300$$

$$x + y = 60 \qquad \text{Second equation}$$

Use the addition method. We may multiply the second equation by -4. Then add.

$$
\begin{array}{r}
4x + 7y = 300 \\
-4x - 4y = -240 \\
\hline
3y = 60 \\
y = 20
\end{array}
$$

Substitute for y in the original second equation.

$$x + y = 60$$

$$x + (20) = 60$$

$$x = 40$$

Jon should use 40 liters of 40% solution and 20 liters of 70% solution.

Check The liters of pure sodium chloride in the 40 liters of 40% solution is $0.40(40) = 16$. The liters of pure sodium chloride in the 20 liters of 70% solution is $0.70(20) = 14$. The total liters of pure sodium chloride mixed together is $16 + 14 = 30$. This is the same as the number of liters of pure sodium chloride in the final 60 liters of 50% solution, which is $0.50(60) = 30$. ∎

☐ **DO EXERCISE 4.**

☐ **Exercise 4** A laboratory assistant has a 50% alcohol solution and an 80% alcohol solution. She wants to make 100 grams of a solution that is 68% alcohol. How much of each solution should she use?

1. Let $y = $ length
   $x = $ width
   $P = 2L + 2W$
   $144 = 2y + 2x$ and $x = y - 4$
   $144 = 2y + 2(y - 4)$
   $144 = 2y + 2y - 8$
   $144 = 4y - 8$
   $152 = 4y$
   $38 = y$
   $x = y - 4 = (38) - 4 = 34$
   The dimensions are 34 feet by 38 feet.

2. Let $x = $ salary before the increase
   $y = $ amount of increase
   $x + y = 22,400$ and $y = 0.12x$
   $x + 0.12x = 22,400$
   $1.12x = 22,400$
   $x = 20,000$
   John's salary before the increase was $20,000.

3. Let $x = $ amount invested at 7%
   $y = $ amount invested at 9%
   $0.07x + 0.09y = 387$ and $y = 4x$
   $0.07x + 0.09(4x) = 387$
   $0.07x + 0.36x = 387$
   $0.43x = 387$
   $x = 900$
   $y = 4x = 4(900) = 3600$
   Linda has $900 invested at 7% and $3600 invested at 9%.

4. Let $x = $ grams of 50% solution
   $y = $ grams of 80% solution
   $100 = $ grams of 68% solution
   $0.50x + 0.80y = 0.68(100)$
   $x + y = 100$
   Multiply the first equation by 100.
   $50x + 80y = 6800$
   Multiply the second equation by -50.
   $-50x - 50y = -5000$
   Add the two equations.
   $$\begin{aligned} 50x + 80y &= 6800 \\ -50x - 50y &= -5000 \\ \hline 30y &= 1800 \\ y &= 60 \end{aligned}$$
   $x + y = 100$
   $x + (60) = 100$
   $x = 40$
   She should use 40 grams of 50% solution and 60 grams of 80% solution.

PROBLEM SET 7.4

1. The length of a rectangle is twice the width. If the perimeter is 132 meters, find the dimensions of the rectangle.

2. Robert has 200 feet of fencing. He wants to build a rectangular fence with length 10 feet greater than the width. What should be the dimensions of the fence?

3. The perimeter of a rectangle is 258 centimeters. If the length is 43 centimeters more than the width, find the width of the rectangle.

4. The length of a house is three times the width. If the perimeter of the house is 192 feet, find the length of the house.

5. Two angles are supplementary. One angle is 45° more than the other. Find the angles. (If the sum of two angles is 180°, they are supplementary.)

6. Two angles are supplementary. One angle is 20° less than three times the other. Find the angles.

7. Two angles are complementary. One angle is 6° less than twice the other. Find the angles. (If the sum of two angles is 90°, they are complementary.)

8. Two angles are complementary. Their difference is 28°. Find the angles.

9. The price of a home increased 12% to $78,400. What was the original price of the home?

10. Jean's salary was reduced by 6% to $4.70 per hour. What was her former salary?

11. The sale price of a truck was $8245 after the cost of it was reduced 15%. What was the original price of the truck?

12. Binh's salary was increased by 4% to $4.68 per hour. What was his original salary?

13. John left a sum of money in the bank at 7% simple interest for 1 year. At the end of the year the account contained $856. How much did John originally leave in the bank?

14. A store buys a coat and marks it up 20% so that the selling price is $96. What did the store pay for the coat?

15. Sharon invested a sum of money at 7% per year and $2500 less at 10% per year. How much did she invest at each rate if her annual interest was $600?

16. A total annual interest of $975 is received on two sums of money. How much is invested at each rate if $3000 more is invested at 9% than is invested at 6%?

17. Patricia earns $600 from her investments. She has some money invested at 7% and twice as much invested at 9%. How much does she have invested at each rate?

18. Ken pays 12% simple interest on one loan and 10% simple interest on another loan. If the loan at 10% is $1500 less than twice as much as the loan at 12% how much did he borrow at each rate if he paid $810 in interest?

19. A chemical technician combines a 40% acid solution with an 80% acid solution to obtain 12 liters of a 50% solution. How many liters of the 40% solution and how many liters of the 80% solution should she use?

20. Jim wants to mix a 30% alcohol solution with a 70% alcohol solution to get 100 milliliters of a 40% solution. How many milliliters of each solution should he use?

21. How many quarts of 55% antifreeze solution should be mixed with a 10% solution to get 12 quarts of a 25% solution?

22. How many liters of 15% salt solution should be added to a 50% salt solution to get 350 liters of a 30% solution?

23. Juan combines 6 liters of a 35% acid solution with a 50% acid solution to get a 45% solution. How many liters of 50% solution should he use?

24. Martha wants to make a 70% alcohol solution. How many milliliters of 90% solution should she combine with 15 milliliters of 55% solution?

25. The last step in solving a word problem is to check the answer in the _____ problem.

Checkup

The following problems provide a review of some of Sections 6.2 and 6.6. Some of these problems will help you with the next section.
Is the given ordered pair a solution of the equation?

26. $(-3, 1)$; $2x - y = -7$

27. $(4, -1)$; $-3x - y = -13$

28. $(-2, -5)$; $-4x + 2y = 2$

29. $(-4, -3)$; $-5x + 3y = 11$

Graph using intercepts.

30. $x + y = 4$

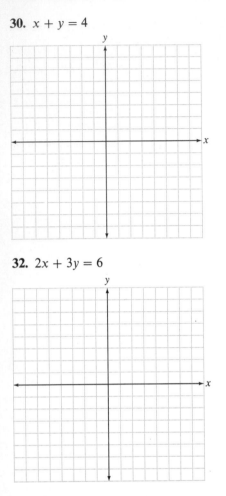

31. $2x + y = 6$

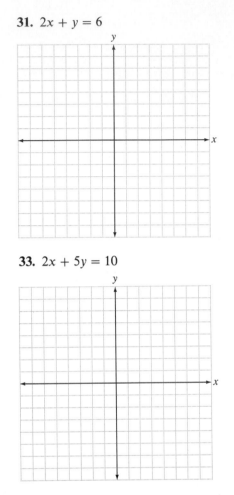

32. $2x + 3y = 6$

33. $2x + 5y = 10$

34. For the Jem Company the relationship between the number of rings sold and the profit is linear. If the profit is $500 when six rings are sold and $3000 when 11 rings are sold, write the equation relating profit y to units sold x. What is the profit when nine rings are sold?

35. Assume that the temperature f feet above the surface of the earth is $t°$ Celsius. Find the linear equation that relates f to t if the temperature at 6000 feet is 20° and the temperature at 3000 feet is 50°. At how many feet above the earth is the temperature 35°?

36. Find the linear equation that relates the book value V to the number of years t of depreciation of a new car that cost $14,000 and is being depreciated at $1500 per year. What is its book value after 6 years?

37. A real estate salesperson receives a salary of $1100 per month plus a 6% commission on sales. Find the amount of her sales if her wages were $5900.

7.5 LINEAR INEQUALITIES IN TWO VARIABLES

Linear inequalities occur in the practical applications of mathematics, especially in the allocation of limited resources.

Some examples of linear inequalities in two variables are

$$x + y < 3 \qquad y \leq 2x \qquad x - 3y > 2 \qquad y - 2x \geq 4$$

1 Solutions

There are many solutions to an inequality in two variables. The solutions are ordered pairs of real numbers.

EXAMPLE 1 Is the ordered pair a solution of $x - 3y > 1$?

a. (9, 2)

$$
\begin{array}{c|c}
\multicolumn{2}{c}{x - 3y > 1} \\
\hline
9 - 3(2) & 1 \\
9 - 6 & 1 \\
3 & 1
\end{array}
$$

Since $3 > 1$, (9, 2) is a solution.

b. (0, 0)

$$
\begin{array}{c|c}
\multicolumn{2}{c}{x - 3y > 1} \\
\hline
0 - 3(0) & 1 \\
0 - 0 & 1 \\
0 & 1
\end{array}
$$

Since $0 > 1$ is false, (0, 0) is not a solution. ∎

□ **DO EXERCISE 1.**

□ **Exercise 1** Is the ordered pair a solution of $2x - 5y < 3$?

a. (5, 1)

b. (0, 0)

2 Graphing Linear Inequalities in Two Variables

The following are steps in graphing a linear inequality in two variables.

1. Sketch the boundary line. This is a graph of the equation formed by replacing the inequality sign with an equal sign. The boundary line will be *solid* if the inequality sign is $\geq$ or $\leq$. This means that points on the boundary line are solutions. The boundary line will be *dashed* if the inequality sign is $>$ or $<$. This means that points on the line are not solutions.

2. Choose a test point that is not on the line and substitute its coordinates into the *inequality*. The point (0, 0) is an easy one to use if the boundary does not pass through it. If the inequality is true for the coordinates of the chosen point, shade the side of the boundary line containing it. If the inequality is false for the coordinates of the point, shade the other side of the line.

□ **Exercise 2** Graph.

a. $x + y \leq 5$

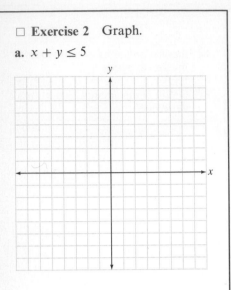

b. $2x - y > 4$

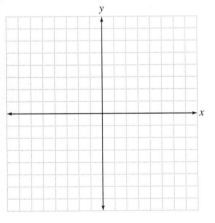

EXAMPLE 2 Graph $x + y \leq 3$.

1. $x + y = 3$ Replacing $\leq$ with $=$

Sketch the graph using the techniques of Chapter 6. The boundary line is solid because the inequality sign is $\leq$.

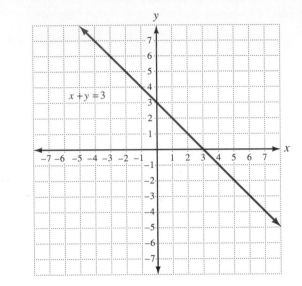

2. We chose the point $(0, 0)$. Hence $x = 0$ and $y = 0$. Substitute these values into the inequality.

$$
\begin{array}{c|c}
\multicolumn{2}{c}{x + y \leq 3} \\
\hline
0 + 0 & 3 \\
0 & 3
\end{array}
$$

$0 \leq 3$ is true. Shade the side of the boundary line that contains the point $(0, 0)$. All points on this side of the line are solutions.

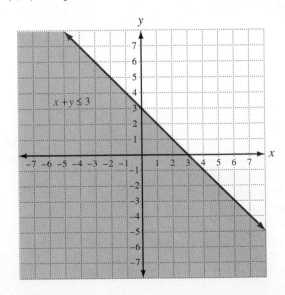

□ **DO EXERCISE 2.**

EXAMPLE 3   Graph $y < 3x$.

1. $y = 3x$ Replacing $<$ with $=$
   Sketch the graph. The boundary line is dashed because the inequality sign is $<$.

2. We must choose a point other than $(0, 0)$ since the boundary line passes through it. We chose the point $(1, 0)$. Then $x = 1$ and $y = 0$. Test these values in the inequality.

$0 < 3$ is true, so shade the side of the line containing $(1, 0)$.

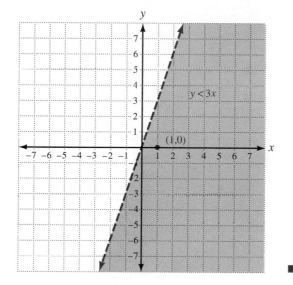

□ **DO EXERCISE 3.**

EXAMPLE 4   Graph $2x + 3y > 6$.

1. $2x + 3y = 6$ Replacing $>$ with $=$
   Sketch the graph.

2. We choose the point $(0, 0)$.

$$
\begin{array}{c|c}
\multicolumn{2}{c}{2x + 3y > 6} \\
\hline
2(0) + 3(0) & 6 \\
0 & 6
\end{array}
$$

□ **Exercise 3**   Graph.

a. $y < x$

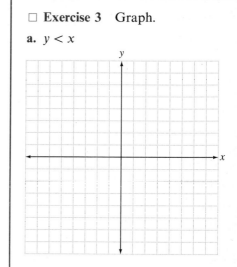

b. $y > 2x$

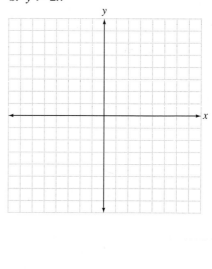

□ **Exercise 4** Graph.

a. $2x - 5y \leq 10$

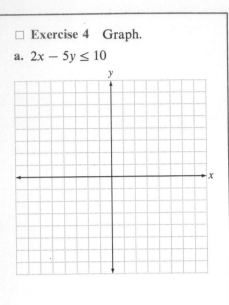

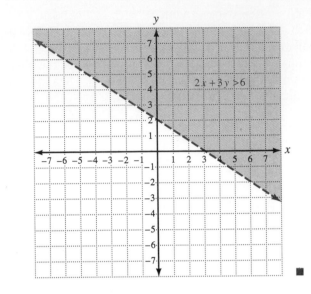

$2x + 3y > 6$

□ **DO EXERCISE 4.**

Answers to Exercises ————————————

1. a. No   **b.** Yes

b. $4x + 3y \geq 12$

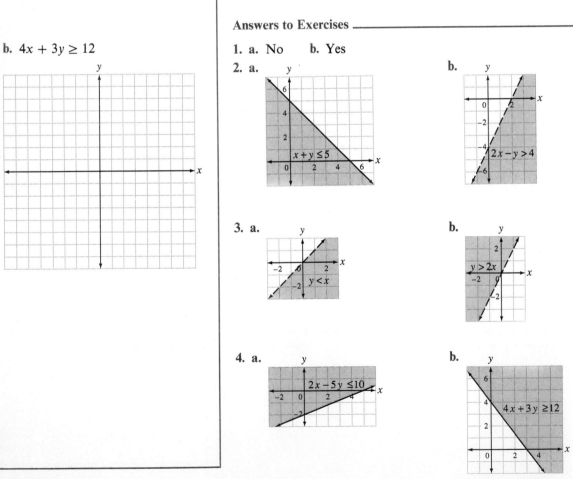

2. a.

$x + y \leq 5$

b.

$2x - y > 4$

3. a.

$y < x$

b.

$y > 2x$

4. a.

$2x - 5y \leq 10$

b.

$4x + 3y \geq 12$

PROBLEM SET 7.5

Graph.

1. $x - y < 2$

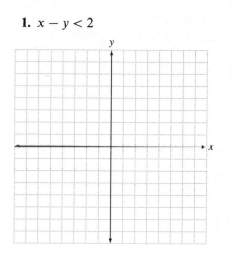

2. $x + y > 5$

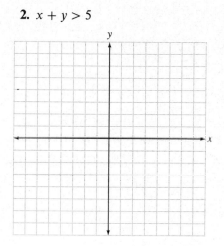

3. $x + 3y \geq 6$

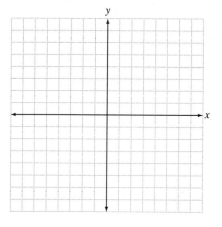

4. $2x - y \leq 4$

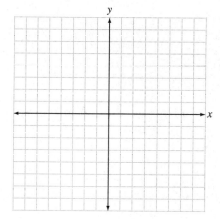

5. $y > 4x$

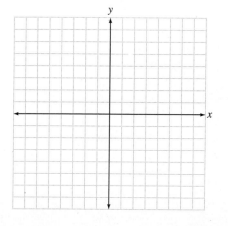

6. $y < 3x$

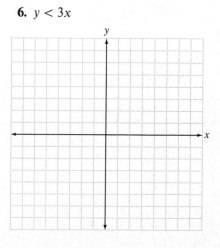

7. $3y \leq 4x$

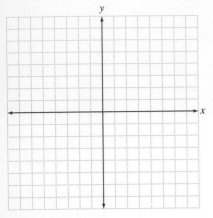

8. $2y \geq 5x$

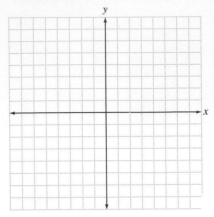

9. $3x - 2y < 6$

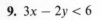
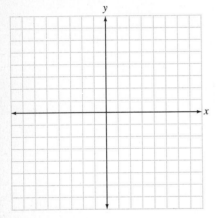

10. $5x + 2y > 10$

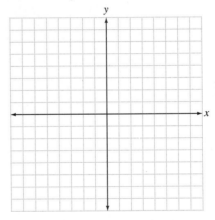

11. $2x + 7y > 14$

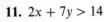
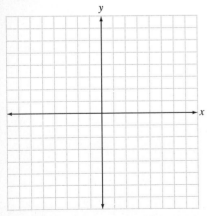

12. $3x - 5y < 15$

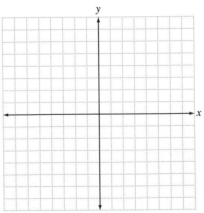

13. $y < 3x + 6$

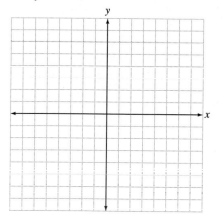

14. $y > 4x - 8$

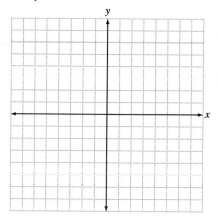

15. $y \geq -x + 4$

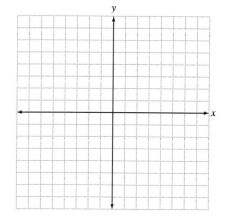

16. $y \leq -x + 2$

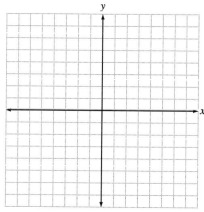

17. $y - \dfrac{1}{2}x \leq 2$

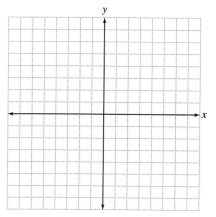

18. $y \geq \dfrac{1}{4}x + 4$

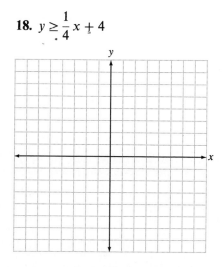

19. The solutions of an inequality in two variables are _____ _____ of real numbers.

20. For the graph of an inequality in two variables, the boundary line will be _____ if the inequality sign is $>$ or $<$.

21. To decide which side of the boundary line to _____, an easy test point to use is $(0, 0)$ if the boundary line does not pass through it.

22. If an inequality is _____ for the coordinates of the test point, shade the side of the boundary line that does not contain this point.

Checkup

The following problems provide a review of some of Sections 2.8 and 6.5.
Evaluate for the given value of the variable.

23. $3y + 5$, $y = -4$

24. $-2x - 7$, $x = 4$

25. $2x^2 + 3x - 4$, $x = -3$

26. $3x^2 - 2x + 5$, $x = -2$

Find an equation for the line with the following y-intercept and slope. Use the slope-intercept form.

27. $(0, 3)$, $m = -5$

28. $(0, -1)$, $m = 6$

29. $(0, -4)$, $m = -\dfrac{7}{3}$

30. $(0, 5)$, $m = -\dfrac{3}{4}$

Find an equation for the line through the given two points.

31. $(6, 4)$, $(3, 1)$

32. $(-5, 2)$, $(-3, 6)$

33. $(-1, -4)$, $(-5, -7)$

34. $(8, -4)$, $(-2, 7)$

CHAPTER 7 ADDITIONAL EXERCISES (OPTIONAL)

Section 7.1

Is the ordered pair $(-3, -6)$ a solution of the following pairs of equations?

1. $2x - 3y = 12$
$4x - y = -6$

2. $2x - 5y = 20$
$3x + 2y = -21$

3. $y + \dfrac{1}{3}x = 7$
$\dfrac{1}{2}y - 2x = 3$

4. $\dfrac{3}{4}x + \dfrac{1}{8}y = -3$
$4x - \dfrac{1}{2}y = -9$

Solve by graphing.

5. $y = -3x$
$2x - y = 5$

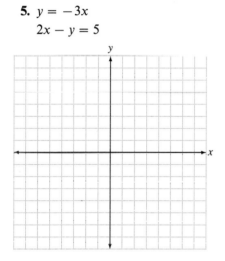

6. $3x + 2y = 6$
$y = x - 2$

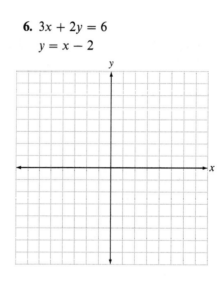

7. $3x + 2y = 4$
$2y = -3x - 2$

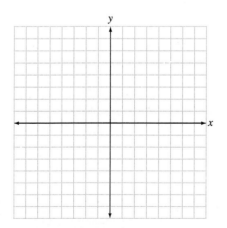

8. $x + 2y = 6$
$\dfrac{x}{2} + y = 3$

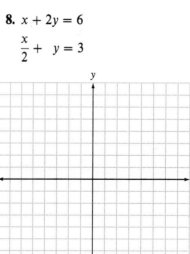

Section 7.2

Solve using the addition method.

9. $x - y = 3$
$x + y = 7$

10. $x + y = 2$
$x - y = -6$

11. $2x - 3y = 4$
$4x + y = -6$

12. $x + 3y = 8$
$2x + 4y = 10$

13. $3x + 2y = -5$
$5x - 3y = -21$

14. $5x - 4y = -1$
$-7x + 5y = 2$

15. $24x + 12y = 19$
$16x - 18y = -9$

16. $2.4x - 3.4y = 4.8$
$7.2x - 10.2y = 15.4$

17. $5.7x + 6.8y = 7.2$
$17.1x + 20.4y = 20.5$

Section 7.3

Solve using the substitution method.

18. $y = x - 5$
$2x + 3y = -10$

19. $x = y - 1$
$2x - y = 1$

20. $y = x - 1$
$5x - 2y = -1$

21. $x = y - 7$
$-3x - 2y = -4$

22. $4 + 3x - 3y = 34$
$4x = -y - 2 + 3x$

23. $3x + 5y = 7 + 4y$
$3x + 4 = 11 - y$

Solve by any method.

24. $3x + 2y = 9$
$2x + 5y = -5$

25. $4x - 3y = 10$
$y = x - 3$

26. $\frac{2}{3}x + \frac{1}{2}y = 6$

$\frac{1}{2}x - \frac{3}{4}y = 0$

27. $\frac{x}{2} - \frac{y}{4} = -4$

$\frac{2}{3}x + \frac{5}{4}y = 1$

Section 7.4

28. Two angles are supplementary. One angle is 65° more than the other. Find the angles.

29. Two angles are complementary. One angle is 9° less than twice the other. Find the angles.

30. The sale price of a car was \$12,320 after it was reduced 12%. What was the original price of the car?

31. Carlos's salary was increased by 8%, to \$8.10 per hour. What was his original salary?

32. If a sum of money is invested at $6\frac{1}{4}\%$ per year and \$2000 more than that amount is invested at $7\frac{3}{4}\%$ per year, how much is invested at each rate if the total annual interest is \$463?

33. A total annual interest of \$405 is received on two sums of money. How much is invested at each rate if twice as much is invested at 8.5% as is invested at 5.5%?

34. How many milliliters of pure alcohol should be added to 400 milliliters of a 35% alcohol solution to get a 60% solution?

35. How many quarts of water should be added to 18 quarts of a 75% salt solution to make a 40% solution?

Section 7.5

Graph.

36. $2x + y \leq 4$

37. $3x + 5y \geq 15$

38. $y > -3x$

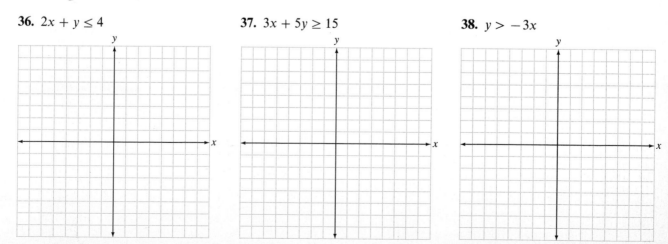

39. $y < -4x$

40. $4x - 3y \geq 2x - 6$

41. $x - 3y < 4 + 5y$

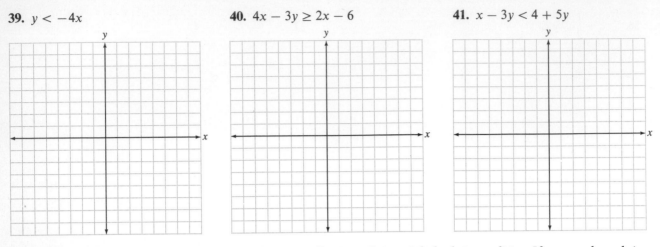

The solution of a system of two linear inequalities is all points that satisfy both inequalities. If we graph each inequality on the same axes, the solution is the overlap of the graphs of each inequality.
Graph the solution of the following systems of inequalities.

42. $x + y \leq 1$
$x - y \leq 1$

43. $x + 2y \geq 4$
$x - y \geq 1$

44. $3x + y > 6$
$2x - y > 8$

45. $4x + 5y < 20$
$x - 2y < 5$

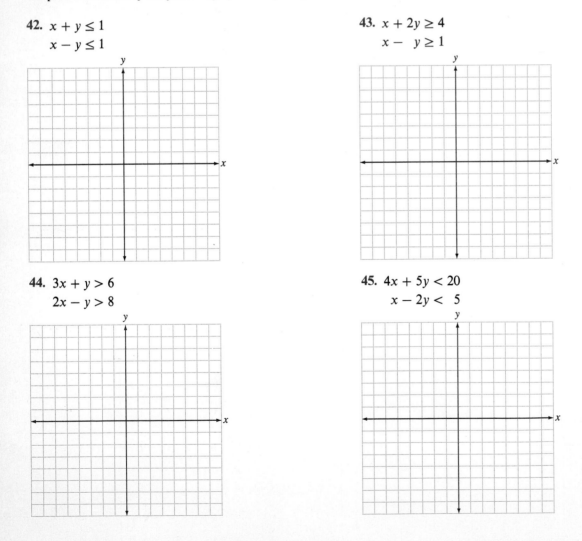

CHAPTER 7   PRACTICE TEST

1. Is $(-1, 4)$ a solution of the following pair of equations?

$$2x + y = 2$$
$$x - 2y = 0$$

1. _____

2. Solve by graphing:

$$x - y = 3$$
$$x - 2y = 0$$

2. _____

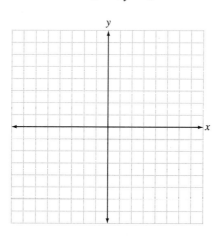

Solve using the addition method.

3. $x - 4y = -6$
   $x + 4y = 10$

3. _____

4. $3x + 5y = -1$
   $5x - 4y = -14$

4. _____

Solve using the substitution method.

5. $6x + 3y = 3$
   $x = y - 7$

5. _____

Solve by any method.

6. _____

6. $4x + 5y = 2$
$-4x - 3y = -6$

7. _____

7. $y = x + 5$
$2x - y = -9$

8. _____

8. Karen's salary was increased by 7% to $6.42 per hour. What was her former salary?

9. _____

9. How many liters of a 60% acid solution should be mixed with a 40% acid solution to get 12 liters of a 45% solution?

10. _____

10. Graph: $2x + 3y > 6$.

Rational Expressions

8.1 RATIONAL EXPRESSIONS

1 Identifying Rational Expressions

In Section 1.5 we explained that a rational number is a number that can be written as the quotient of two integers (with denominator not zero). The quotient of two polynomials is called a **rational expression**. Recall from Section 5.1 that a polynomial in a variable is an expression whose terms contain only whole-number powers of the variable.

EXAMPLE 1 Are the following rational expressions?

a. $\dfrac{x^2 + 3x + 1}{x - 2}$ Yes

b. $\dfrac{x^{-3} + x}{x}$ No; the numerator is not a polynomial since it contains an exponent of -3, which is not a whole number

c. $\dfrac{3}{4}$ Yes; $3 = 3x^0$ and $4 = 4x^0$; zero is a whole number

d. $\dfrac{\dfrac{1}{x}}{x^2 + 2}$ No; $1/x = x^{-1}$; the numerator is not a polynomial

e. $x - 3$ Yes; this is the same as $(x - 3)/1$

f. $\dfrac{3}{x + 5}$ Yes   ■

☐ DO EXERCISE 1.

☐ **Exercise 1**   Are the following rational expressions?

a. $\dfrac{x - 1}{x^2 + x - 3}$

b. $\dfrac{2}{x^{-2} + 4}$

c. $\dfrac{x + 1}{x + 3}$

d. $\dfrac{-7}{x - 4}$

361

□ **Exercise 2**   Evaluate for the given value of the variable.

□ **Exercise 2**   Evaluate for the given value of the variable.

a. $\dfrac{3}{2x}$; $x = 5$

b. $\dfrac{x^2 + 1}{2x + 3}$; $x = 3$

c. $\dfrac{2x^2 + x}{x + 5}$; $x = -3$

d. $\dfrac{3y}{2y^2 - y - 3}$; $y = -2$

2 **Evaluating a Rational Expression**

We may evaluate a given rational expression for a specific value of the variable.

EXAMPLE 2   Evaluate for the given value of the variable.

a. $\dfrac{y - 2}{3y + 4}$; $y = 1$

$$\frac{y - 2}{3y + 4} = \frac{(1) - 2}{3(1) + 4} = -\frac{1}{7}$$

b. $\dfrac{3}{x^2 + x - 12}$; $x = 4$

$$\frac{3}{x^2 + x - 12} = \frac{3}{4^2 + 4 - 12} = \frac{3}{8} \qquad \blacksquare$$

□ **DO EXERCISE 2.**

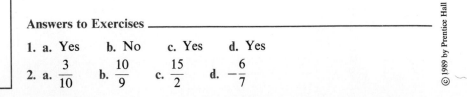

Answers to Exercises

1. **a.** Yes   **b.** No   **c.** Yes   **d.** Yes

2. **a.** $\dfrac{3}{10}$ **b.** $\dfrac{10}{9}$ **c.** $\dfrac{15}{2}$ **d.** $-\dfrac{6}{7}$

PROBLEM SET 8.1

A. *Are the following rational expressions?*

1. $\dfrac{x+4}{2x^2-3x+8}$ **2.** $\dfrac{5x^2-x}{4x^3+3}$ **3.** $\dfrac{3}{x^{-1}}$ **4.** $\dfrac{\frac{4}{x}}{x^2}$

5. $y \mid 2$ **6.** $\dfrac{x^{-2}+3}{4}$ **7.** $-\dfrac{7}{3}$ **8.** 4

B. *Evaluate for (a) $x = 1$; (b) $x = -2$.*

9. $\dfrac{5}{2x}$ **10.** $-\dfrac{3}{x}$

11. $\dfrac{x+1}{x-3}$ **12.** $\dfrac{3}{x+4}$

13. $\dfrac{4x^2-4}{3x+2}$ **14.** $\dfrac{x^2-1}{2-x}$

15. $\dfrac{2x}{3x^2+2}$ **16.** $\dfrac{x-2}{2x^2+1}$

17. $\dfrac{x}{2x^2-x-9}$ **18.** $\dfrac{-x+3}{3x^2+x-3}$

19. A polynomial in a variable is an expression whose terms contain only _____ -number powers of the variable.

20. The quotient of two _____ is a rational expression.

Checkup

The following problems provide a review of some of Sections 2.7 and 5.8. These problems will help you with the next section.
Divide and write with positive exponents.

21. $\dfrac{y^5}{y^2}$

22. $\dfrac{y^2}{y^7}$

23. $\dfrac{x^3}{x^8}$

24. $\dfrac{x^9}{x^2}$

25. $\dfrac{x}{x}$

26. $\dfrac{y^4}{y^4}$

Factor.

27. $x^3 - xy^2$

28. $2x^2 - 18$

29. $6x^2 - 5x - 6$

30. $12y^2 + 11y + 2$

31. $3x^2 - 24x + 48$

32. $x^4 + 4x^3 + 4x^2$

33. $3y^3 - 7y^2 + 2y$

34. $12x^2 + 8x - 32$

8.2 SIMPLIFYING RATIONAL EXPRESSIONS

1 Rational expressions are fractions. In Section 1.1 we studied simplifying fractions. We simplified a fraction by factoring the numerator and denominator into primes and then removing the 1.

There is a rule for rational expressions which says that if A/B is a rational expression and if C is any polynomial where $C \neq 0$, then

$$\frac{AC}{BC} = \frac{A}{B}$$

Notice that this is true since

$$\frac{AC}{BC} = \frac{A \cdot C}{B \cdot C} = \frac{A}{B} \cdot \frac{C}{C} = \frac{A}{B} \cdot 1 = \frac{A}{B} \qquad \text{Removing the 1}$$

If a fractional expression has the same numerator and denominator (other than zero), it is the same as 1.

$$\frac{-6}{-6} = 1 \qquad \frac{2x^2 - 1}{2x^2 - 1} = 1 \qquad \frac{x - 2}{x - 2} = 1$$

Hence an expression can be simplified as follows.

$$\frac{3(x - 2)}{2(x - 2)} = \frac{3}{2} \qquad \text{Since } AC/BC = A/B$$

Caution We must be careful to remove only factors that are 1. We may not remove terms that are 1. Remember that factors are separated by multiplication signs. Terms are separated by plus or minus signs. Notice that

$$\frac{12 + 4}{4} \neq 12 + \frac{4}{4} \neq 12$$

because

$$\frac{12 + 4}{4} = \frac{12}{4} + \frac{4}{4} = 3 + 1 = 4$$

However,

$$\frac{12 \cdot 4}{4} = 12 \cdot \frac{4}{4} = 12$$

Some rational expressions may be simplified using the laws of exponents from Section 2.7.

□ **Exercise 1** Simplify by writing with the least number of symbols. Write answers with positive exponents.

a. $\dfrac{12xy^2}{4x^2y}$

b. $\dfrac{y}{5y}$

c. $\dfrac{3x^2}{27x}$

d. $\dfrac{5x}{25x^4}$

EXAMPLE 1 Simplify by writing with the least number of symbols. Write answers with positive exponents.

a. $\dfrac{15x}{3x^2}$

$$\frac{15x}{3x^2} = 5x^{1-2} = 5x^{-1}$$

$$= 5 \cdot \frac{1}{x} = \frac{5}{x}$$

b. $\dfrac{8x}{x}$

In this problem it is easier to remove the 1 than to use the approach used in Example a.

$$\frac{8x}{x} = 8 \cdot \frac{x}{x} = 8$$

c. $\dfrac{3x^3}{15x}$

$$\frac{3x^3}{15x} = \frac{3}{3} \cdot \frac{x^3}{5x} \qquad \text{Factoring}$$

$$= \frac{x^3}{5x} \qquad \text{Removing } \tfrac{3}{3} \text{ or } 1$$

$$= \frac{x^{3-1}}{5} = \frac{x^2}{5} \qquad \blacksquare$$

□ **DO EXERCISE 1.**

To simplify most rational expressions, we must factor the numerator or denominator or both. We will review our general method for factoring discussed in Chapter 5.

General Method for Factoring

1. Factor out the greatest common factor first.
2. If there are *two terms* to be factored: Try to factor the expression as the difference of squares.
   If there are *three terms*: Check to see if the expression is a perfect square trinomial. If it is not, use the general method for factoring a trinomial.
   If there are *four terms*: Divide the expression into groups of two terms which have a common factor and factor completely.
   Some expressions do not factor.

EXAMPLE 2 Factor.

a. $6x^2 - 54$

　　1. Factor out the greatest common factor, 6.

$$6(x^2 - 9)$$

　　2. Factor the difference of two squares. Remember to keep the 6.

$$6(x + 3)(x - 3)$$
$$6x^2 - 54 = 6(x + 3)(x - 3)$$

b. $3x^3 + 24x^2 + 48x$

　　1. Factor out the greatest common factor, $3x$.

$$3x(x^2 + 8x + 16)$$

　　2. Is $x^2 + 8x + 16$ a perfect square trinomial? Yes.
　　Factor it. Remember that perfect square trinomials factor into
　　identical factors. Keep the greatest common factor, $3x$.

$$3x(x + 4)(x + 4)$$
$$3x^3 + 24x^2 + 48x = 3x(x + 4)(x + 4)$$

c. $6x^2 - 7x + 2$

　　1. There is no greatest common factor.
　　2. Factor $6x^2$ and 2.

Factors of $6x^2$	*Factors of* $+2$
$x, 6x$	$1, \quad 2$
$2x, 3x$	$-1, -2$

　　3. Arrange pairs of these factors so that the sum of the products
　　of the inner and outer terms is the same as the middle term,
　　$-7x$, of the original expression. Since the middle term is nega-
　　tive, we will use the negative factors of 2.

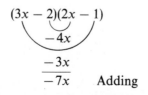

$$\frac{-3x}{-7x} \quad \text{Adding}$$

The sum of the products of the inner and outer terms is $-7x$.
The correct factorization is as follows.

$$6x^2 - 7x + 2 = (3x - 2)(2x - 1)$$

☐ **Exercise 2** Factor.

a. $3x^2 - 12$

b. $8x^2 + 2x - 3$

c. $4x^2 - 12x + 9$

d. $2ax + 2ay + x + y$

d. $xy + x + ay + a$

1. There is no greatest common factor.
2. The expression has four terms. Separate it into groups of two terms that have a common factor.

$$(xy + x) + (ay + a)$$

3. Factor out the greatest common factor from each binomial.

$$x(y + 1) + a(y + 1)$$

4. Factor out the common factor, $y + 1$.

$$(x + a)(y + 1)$$

The correct factorization is

$$xy + x + ay + a = (x + a)(y + 1) \quad \blacksquare$$

☐ **DO EXERCISE 2.**

Usually, we must factor the numerator and denominator before we can simplify.

EXAMPLE 3 Simplify by factoring and removing a 1.

a. $\dfrac{4x - 8}{3x - 6}$

$$\frac{4x - 8}{3x - 6} = \frac{4(x - 2)}{3(x - 2)} = \frac{4}{3} \cdot \frac{(x - 2)}{(x - 2)} = \frac{4}{3}$$ Removing the 1

b. $\dfrac{9x + 15}{3}$

$$\frac{9x + 15}{3} = \frac{3(3x + 5)}{3} = \frac{3}{3} \cdot (3x + 5) = 3x + 5$$

Notice that this expression could be simplified as

$$\frac{9x + 15}{3} = \frac{9x}{3} + \frac{15}{3} = 3x + 5$$

The 3 must be divided into *both* terms.

c. $\dfrac{4x^2 - 5x}{2x}$

$$\frac{4x^2 - 5x}{2x} = \frac{x(4x - 5)}{x \cdot 2} = \frac{x}{x} \cdot \frac{4x - 5}{2} = \frac{4x - 5}{2}$$ Removing x/x or 1

d. $\dfrac{x^2 - x - 6}{x^2 - 4}$

$$\frac{x^2 - x - 6}{x^2 - 4} = \frac{(x + 2)(x - 3)}{(x + 2)(x - 2)} = \frac{x - 3}{x - 2}$$ ∎

□ **DO EXERCISE 3.**

Sometimes it appears that we cannot simplify a rational expression when we can in fact simplify it by factoring a -1 out of the numerator. This method works when there are factors in the numerator and denominator that are opposites. The opposite of an expression has the opposite sign on each term from that of the original expression.

Expression	Opposite
3	-3
x	$-x$
$-y$	y
$x - y$	$-x + y$ or $y - x$
$x + y$	$-(x + y)$ or $-x - y$

□ **Exercise 3**   Simplify by factoring and removing a 1.

a. $\dfrac{5x + 5}{7x + 7}$

b. $\dfrac{2x + 3x^2}{4x}$

c. $\dfrac{4x + 8}{4}$

d. $\dfrac{6x + 2}{2}$

e. $\dfrac{x + 8}{x^2 + 7x - 8}$

f. $\dfrac{x^2 - 4x + 4}{x^2 + x - 6}$

□ **Exercise 4** Simplify by replacing opposites in the numerator and denominator with −1 in the numerator.

a. $\dfrac{y-x}{x-y}$

b. $\dfrac{x^2-1}{1-x}$

c. $\dfrac{3x-3y}{y-x}$

d. $\dfrac{x+y}{-x-y}$

EXAMPLE 4 Simplify by factoring a −1 out of the numerator and removing a 1.

$$\frac{x-y}{y-x}$$

We notice that $x-y$ and $y-x$ are opposites, so factor a −1 out of the numerator.

$$\frac{x-y}{y-x} = \frac{-1(-x+y)}{y-x} = \frac{-1(y-x)}{y-x} = -1$$

We may shorten our work by noticing that we may remove opposites by placing a −1 in the numerator or the denominator (not both). We choose to place the −1 in the numerator because we do not want the answer to have a negative sign in the denominator.

$$\frac{x-y}{y-x} = -1 \quad ■$$

□ **DO EXERCISE 4.**

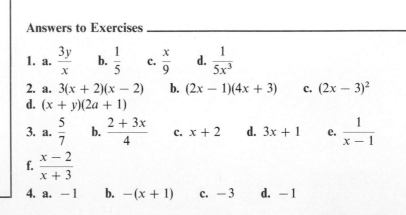

Answers to Exercises

1. a. $\dfrac{3y}{x}$ **b.** $\dfrac{1}{5}$ **c.** $\dfrac{x}{9}$ **d.** $\dfrac{1}{5x^3}$

2. a. $3(x+2)(x-2)$ **b.** $(2x-1)(4x+3)$ **c.** $(2x-3)^2$
d. $(x+y)(2a+1)$

3. a. $\dfrac{5}{7}$ **b.** $\dfrac{2+3x}{4}$ **c.** $x+2$ **d.** $3x+1$ **e.** $\dfrac{1}{x-1}$

f. $\dfrac{x-2}{x+3}$

4. a. -1 **b.** $-(x+1)$ **c.** -3 **d.** -1

PROBLEM SET 8.2

A. *Simplify by writing with the least number of symbols. Write answers with positive exponents.*

1. $\dfrac{x^2 y}{x}$

2. $\dfrac{xy}{y^2}$

3. $\dfrac{10x^2 y^2}{5x^3 y}$

4. $\dfrac{25x^3 y}{5xy^2}$

5. $\dfrac{30x^2}{6x^2}$

6. $\dfrac{9y}{18y}$

B. *Simplify by factoring and removing a 1.*

7. $\dfrac{6(x + 2)}{5(x + 2)}$

8. $\dfrac{x(x - 1)}{8(x - 1)}$

9. $\dfrac{4x + 12}{3x + 9}$

10. $\dfrac{5y - 10}{4y - 8}$

11. $\dfrac{2x + 4}{4x - 8}$

12. $\dfrac{10y - 10}{5y + 5}$

13. $\dfrac{3x^2 + x}{2x}$

14. $\dfrac{9x}{2x^2 - 4x}$

15. $\dfrac{x - y}{x^2 - y^2}$

16. $\dfrac{x + y}{(x + y)^2}$

17. $\dfrac{x^2 + 3x - 4}{x - 1}$

18. $\dfrac{2x + 4}{x^2 - x - 6}$

19. $\dfrac{x^2 - x - 6}{x^2 + x - 12}$

20. $\dfrac{6x^2 + 7x - 3}{3x^2 + 5x - 2}$

21. $\dfrac{9x^2 + 9x - 4}{6x^2 + 5x - 4}$

22. $\dfrac{2y^2 + 3y - 2}{y^2 + y - 2}$

C. *Simplify by replacing opposites in the numerator and denominator with -1 in the numerator.*

23. $\dfrac{-x + y}{y - x}$

24. $\dfrac{-y - x}{y + x}$

25. $\dfrac{5x - 5y}{y - x}$ $= \dfrac{5 \overset{-1}{(x - y)}}{(y - x)} = -5$

26. $\dfrac{-7x - 7y}{x + y}$

27. $\dfrac{x^2 - y^2}{y - x}$

28. $\dfrac{x^2 - y^2}{-x - y}$

29. To simplify a rational expression, _____ of the expression that are 1 may be removed.

30. To simplify most rational expressions, factor the _____ or _____ or both.

31. The first step in factoring is to factor out the _____ _____ _____ .

32. If there is a factor in the numerator of a rational expression and its opposite factor is in the denominator, the expression may be simplified by factoring a _____ _____ out of the numerator.

Checkup

The following problems provide a review of some of Section 7.1.
Decide if the given ordered pair is a solution of the system of equations.

33. $(5, -2)$;
$x - 3y = -1$
$2x + y = 8$

34. $(-4, 3)$;
$x + 5y = 11$
$2x + 7y = 13$

35. $(-3, -4)$;
$3x - 4y = 7$
$5x - 6y = 9$

36. $(2, 1)$;
$7x - 6y = 8$
$4x + 3y = 12$

Solve by graphing.

37. $y = 3x$
$2x - y = 1$

38. $x + y = 3$
$x - 2y = 6$

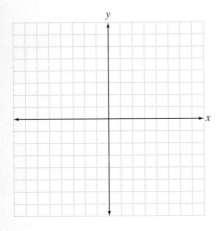

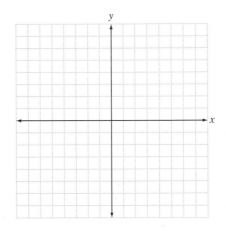

39. $y = 4x + 3$
$2y - 8x = 6$

40. $3x + 6y = 12$
$y = -\dfrac{1}{2}x + 6$

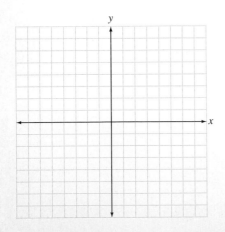

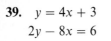

8.3 MULTIPLICATION AND DIVISION

1 Multiplication

When we multiply two fractions we multiply the numerators and multiply the denominators. Rational expressions are multiplied the same way.

> Rational expressions are multiplied by multiplying the numerators and multiplying the denominators.

EXAMPLE 1 Multiply $\dfrac{x^2}{x^3} \cdot \dfrac{y^2}{y}$

The multiplication problem may be worked using the rules for exponents.

$$\frac{x^2}{x^3} \cdot \frac{y^2}{y} = \frac{x^2 y^2}{x^3 y}$$

$$= x^{2-3} y^{2-1} = x^{-1} y = \frac{y}{x} \quad \blacksquare$$

□ **DO EXERCISE 1.**

EXAMPLE 2 Multiply.

$$\frac{2}{2x - 1} \cdot \frac{x + 4}{x - 2}$$

$$\frac{2}{2x - 1} \cdot \frac{x + 4}{x - 2} = \frac{2(x + 4)}{(2x - 1)(x - 2)}$$

It is often more useful to leave the answer in factored form. Integers, however, should be multiplied. ■

□ **DO EXERCISE 2.**

We make our work easier when we multiply fractions by factoring the numerators and denominators, indicating the multiplication and then simplifying, if possible.

EXAMPLE 3 Multiply.

a. $\dfrac{x^2 + 4x + 4}{7} \cdot \dfrac{3}{x^2 - 3x - 10}$

$$\frac{x^2 + 4x + 4}{7} \cdot \frac{3}{x^2 - 3x - 10}$$

$$= \frac{(x + 2)(x + 2)}{7} \cdot \frac{3}{(x + 2)(x - 5)} \quad \text{Factoring}$$

$$= \frac{(x + 2)(x + 2)(3)}{7(x + 2)(x - 5)}$$

$$= \frac{(x + 2)}{(x + 2)} \cdot \frac{(x + 2)3}{7(x - 5)} \quad \text{Factoring out } (x + 2)/(x + 2)$$

$$= \frac{3(x + 2)}{7(x - 5)} \quad \text{Removing the 1}$$

OBJECTIVES

1 *Multiply rational expressions*

2 *Divide rational expressions*

□ **Exercise 1** Multiply.

a. $\dfrac{3x}{y} \cdot \dfrac{y^3}{x^2}$

b. $\dfrac{5x^2}{y^3} \cdot \dfrac{xy}{4y}$

□ **Exercise 2** Multiply.

a. $\dfrac{-5}{2x + 5} \cdot \dfrac{-3}{x - 1}$

b. $\dfrac{y - 4}{3} \cdot \dfrac{y + 2}{y - 6}$

□ **Exercise 3**   Multiply.

a. $\dfrac{x^2 - 4}{x + 2} \cdot \dfrac{1}{x - 2}$

b. $\dfrac{4}{x^2 + 2x + 1} \cdot \dfrac{x + 1}{2}$

c. $\dfrac{x^2 + 2x - 15}{x^2 - 4} \cdot \dfrac{x + 2}{x - 3}$

d. $\dfrac{x^2 - 16}{x - 4} \cdot \dfrac{2}{2x + 8}$

□ **Exercise 4**   Multiply.

a. $\dfrac{7 - x}{3} \cdot \dfrac{2}{x - 7}$

b. $\dfrac{x^2 - 25}{5 - x} \cdot \dfrac{3}{4}$

b. $\dfrac{3}{x^2 - 9} \cdot \dfrac{x^2 + 2x - 3}{6}$

$$\dfrac{3}{x^2 - 9} \cdot \dfrac{x^2 + 2x - 3}{6} = \dfrac{3}{(x + 3)(x - 3)} \cdot \dfrac{(x + 3)(x - 1)}{3(2)}$$

$$= \dfrac{3(x + 3)(x - 1)}{(x + 3)(x - 3)(3)(2)}$$

$$= \dfrac{3(x + 3)}{3(x + 3)} \cdot \dfrac{(x - 1)}{(x - 3)(2)}$$

$$= \dfrac{x - 1}{2(x - 3)}$$

c. $\dfrac{x^2 + 4x - 5}{x + 5} \cdot \dfrac{x + 4}{2x^2 + 7x - 4}$

$$\dfrac{x^2 + 4x - 5}{x + 5} \cdot \dfrac{x + 4}{2x^2 + 7x - 4} = \dfrac{(x - 1)(x + 5)}{x + 5} \cdot \dfrac{(x + 4)}{(2x - 1)(x + 4)}$$

$$= \dfrac{(x - 1)(x + 5)(x + 4)}{(x + 5)(2x - 1)(x + 4)}$$

$$= \dfrac{(x + 5)(x + 4)}{(x + 5)(x + 4)} \cdot \dfrac{(x - 1)}{(2x - 1)}$$

$$= \dfrac{x - 1}{2x - 1} \quad \blacksquare$$

□ **DO EXERCISE 3.**

Remember that opposites may be simplified to give -1 in the numerator.

EXAMPLE 4   Multiply.

$$\dfrac{x^2 - 1}{3} \cdot \dfrac{7}{1 - x}$$

$$\dfrac{x^2 - 1}{3} \cdot \dfrac{7}{1 - x} = \dfrac{(x + 1)(x - 1)}{3} \cdot \dfrac{7}{1 - x}$$

$$= \dfrac{(x + 1)(x - 1)7}{3(1 - x)} = \dfrac{x - 1}{1 - x} \cdot \dfrac{(x + 1)7}{3}$$

$$= -1 \cdot \dfrac{(x + 1)7}{3} = \dfrac{-7(x + 1)}{3} \quad \blacksquare$$

□ **DO EXERCISE 4.**

2 Division

Recall that to divide two fractions, we invert the divisor and multiply. Division of rational expressions is done in the same way.

> To divide two rational expressions, invert the divisor and multiply the expressions.

EXAMPLE 5 Divide.

a. $\dfrac{3}{5} \div \dfrac{2}{3}$

$$\dfrac{3}{5} \div \dfrac{2}{3} = \dfrac{3}{5} \cdot \dfrac{3}{2}$$

$$= \dfrac{9}{10}$$

b. $\dfrac{x}{x+2} \div \dfrac{x}{x+3}$

$$\dfrac{x}{x+2} \div \dfrac{x}{x+3} = \dfrac{x}{x+2} \cdot \dfrac{x+3}{x}$$

$$= \dfrac{x(x+3)}{(x+2)x}$$

$$= \dfrac{x}{x} \cdot \dfrac{x+3}{x+2} = \dfrac{x+3}{x+2}$$

c. $\dfrac{x}{6} \div \dfrac{x}{3}$

$$\dfrac{x}{6} \div \dfrac{x}{3} = \dfrac{x}{6} \cdot \dfrac{3}{x}$$

$$= \dfrac{x(3)}{3(2)x} = \dfrac{3x}{3x} \cdot \dfrac{1}{2} = \dfrac{1}{2} \quad \blacksquare$$

□ **DO EXERCISE 5.**

Once we have inverted the divisor it may be necessary to factor the numerators and denominators of the rational expressions before completing the exercise.

EXAMPLE 6 Divide.

a. $\dfrac{x^2 - 16}{x - 3} \div \dfrac{x - 4}{x^2 - 9}$

$$\dfrac{x^2 - 16}{x - 3} \div \dfrac{x - 4}{x^2 - 9} = \dfrac{x^2 - 16}{x - 3} \cdot \dfrac{x^2 - 9}{x - 4} \qquad \text{Inverting the divisor}$$

$$= \dfrac{(x+4)(x-4)}{x-3} \cdot \dfrac{(x+3)(x-3)}{x-4}$$

$$= \dfrac{(x+4)(x-4)(x+3)(x-3)}{(x-3)(x-4)}$$

$$= \dfrac{(x-4)(x-3)}{(x-4)(x-3)} \cdot (x+4)(x+3)$$

$$= (x+4)(x+3)$$

□ **Exercise 5** Divide.

a. $\dfrac{3}{5} \div \dfrac{3}{7}$

b. $\dfrac{x+y}{4} \div \dfrac{x-y}{8}$

a. $\dfrac{x^2 - 1}{x - 2} \div \dfrac{x + 1}{x^2 + 3x - 10}$

b. $\dfrac{(x + 3)^2}{x - 6} \div \dfrac{x^2 - 9}{x^2 - 5x - 6}$

c. $\dfrac{2x - 3}{x^2 - 4} \div \dfrac{6x - 9}{x + 2}$

d. $\dfrac{5x - 10}{x - 2} \div \dfrac{x + 4}{x^2 + 2x - 8}$

b. $\dfrac{x^2 - x - 6}{x^2 + x - 12} \div \dfrac{x^2 + 2x - 3}{x^2 + 3x - 4}$

$$\dfrac{x^2 - x - 6}{x^2 + x - 12} \div \dfrac{x^2 + 2x - 3}{x^2 + 3x - 4} = \dfrac{x^2 - x - 6}{x^2 + x - 12} \cdot \dfrac{x^2 + 3x - 4}{x^2 + 2x - 3}$$

$$= \dfrac{(x - 3)(x + 2)}{(x + 4)(x - 3)} \cdot \dfrac{(x + 4)(x - 1)}{(x + 3)(x - 1)}$$

$$= \dfrac{(x - 3)(x + 2)(x + 4)(x - 1)}{(x + 4)(x - 3)(x + 3)(x - 1)}$$

$$= \dfrac{(x - 3)(x + 4)(x - 1)}{(x - 3)(x + 4)(x - 1)} \cdot \dfrac{x + 2}{x + 3}$$

$$= \dfrac{x + 2}{x + 3} \quad \blacksquare$$

□ **DO EXERCISE 6.**

DID YOU KNOW?

The Arabs used the nine digits and zero taught to them by the Hindus for their mathematical works. The Arabs introduced these numbers to the Western Europeans, who were still using the Roman numeral system, by the tenth century. Many Europeans fought against the use of the new numerals, and heavy fines (death) were imposed on those radicals who insisted on using the Hindu–Arabic numerals.

Answers to Exercises

1. a. $\dfrac{3y^2}{x}$ **b.** $\dfrac{5x^3}{4y^3}$

2. a. $\dfrac{15}{(2x + 5)(x - 1)}$ **b.** $\dfrac{(y - 4)(y + 2)}{3(y - 6)}$

3. a. 1 **b.** $\dfrac{2}{x + 1}$ **c.** $\dfrac{x + 5}{x - 2}$ **d.** 1

4. a. $-\dfrac{2}{3}$ **b.** $\dfrac{-3(x + 5)}{4}$

5. a. $\dfrac{7}{5}$ **b.** $\dfrac{2(x + y)}{x - y}$

6. a. $(x - 1)(x + 5)$ **b.** $\dfrac{(x + 3)(x + 1)}{x - 3}$ **c.** $\dfrac{1}{3(x - 2)}$

d. $5(x - 2)$

PROBLEM SET 8.3

A. *Multiply. Leave the answer in factored form except multiply integers. Write answers with positive exponents.*

1. $\dfrac{4x^2}{y^4} \cdot \dfrac{y}{x^3}$

2. $\dfrac{y^3}{4x} \cdot \dfrac{x^5}{y}$

3. $\dfrac{7x^2}{5x^6 y} \cdot \dfrac{y^2}{y}$

4. $\dfrac{3x}{2y^5} \cdot \dfrac{xy^7}{x^3 y}$

5. $\dfrac{x-2}{3} \cdot \dfrac{4}{x+3}$

6. $\dfrac{-7}{x+1} \cdot \dfrac{2}{x+5}$

7. $\dfrac{2x-3}{3} \cdot \dfrac{x+1}{x-4}$

8. $\dfrac{x-1}{x+1} \cdot \dfrac{2x-5}{2x+7}$

9. $\dfrac{x^2-25}{5} \cdot \dfrac{3}{x+5}$

10. $\dfrac{-7}{x^2-16} \cdot \dfrac{x+4}{8}$

11. $\dfrac{x^2+6x+9}{x+3} \cdot \dfrac{3x-6}{x-2}$

12. $\dfrac{x^2-4x+4}{5x-5} \cdot \dfrac{10}{x-2}$

13. $\dfrac{2x^2-3x-5}{x-2} \cdot \dfrac{x^2-4}{2x-5}$

14. $\dfrac{(x+4)^2}{x+4} \cdot \dfrac{3x^2+5x-2}{x+2}$

15. $\dfrac{y^2-2y}{y} \cdot \dfrac{y-1}{y^2+y-2}$

16. $\dfrac{y^2-4y}{y^2-y-12} \cdot \dfrac{y+3}{y^2}$

17. $\dfrac{y-8}{4} \cdot \dfrac{5}{8-y}$

18. $\dfrac{4-x}{3} \cdot \dfrac{2}{x-4}$

19. $\dfrac{x}{3-x} \cdot \dfrac{x^2-9}{2}$

20. $\dfrac{y^2}{1-x} \cdot \dfrac{x^2-1}{y}$

B. *Divide.*

21. $\dfrac{7}{8} \div \dfrac{3}{8}$

22. $\dfrac{5}{6} \div \dfrac{3}{4}$

23. $\dfrac{5}{x} \div \dfrac{7}{x}$

24. $\dfrac{y}{2} \div \dfrac{y}{9}$

25. $\dfrac{y^2}{x^3} \div \dfrac{y^3}{x^5}$

26. $\dfrac{x^2}{y^4} \div \dfrac{y^4}{x^3}$

27. $\dfrac{x^2}{2} \div x$

28. $\dfrac{y^4}{3} \div y^2$

29. $\dfrac{x+3}{x+4} \div \dfrac{x}{2}$

30. $\dfrac{y-8}{y-7} \div \dfrac{y+3}{y+4}$

31. $\dfrac{x+y}{2} \div \dfrac{(x+y)^2}{4}$

32. $\dfrac{x-4}{15} \div \dfrac{x-4}{5}$

33. $\dfrac{x+2}{x-2} \div \dfrac{2x}{x^2-4}$

34. $\dfrac{x^2-9}{2x+4} \div \dfrac{x-3}{x+2}$

35. $\dfrac{x+3}{x-3} \div \dfrac{x^2+5}{x^2-9}$

36. $\dfrac{x-y}{2y} \div \dfrac{x^2-y^2}{3y^4}$

37. $\dfrac{x^2+3x+2}{x^2+5x+4} \div \dfrac{x^2+5x+6}{x^2+10x+24}$

38. $\dfrac{16-y^2}{y^2+2y-8} \div \dfrac{y^2-2y-8}{4-y^2}$

39. $\dfrac{x^2+2x+1}{x^2+5x+4} \div \dfrac{x^2+2x+1}{x^2+3x+2}$

40. $\dfrac{2y^2+7y+3}{y^2-9} \div \dfrac{y^2-3y}{2y^2+11y+5}$

41. $\dfrac{2x^2-5x-12}{x^2-10x+24} \div x^2-9x+18$

42. $(y-2) \div \dfrac{y-2}{y^2+3y+8}$

43. Rational expressions are multiplied by multiplying the numerators and multiplying the _____.

44. If possible, rational expressions are factored and _____ before multiplying them.

45. If there is a factor in the numerator of a rational expression and its opposite factor is in the denominator, these factors may be simplified to give _____ _____ in the numerator.

46. To divide two rational expressions, _____ the divisor and multiply.

Checkup

The following problems provide a review of some of Sections 1.2 and 7.2. Some of these problems will help you with the next section.
Add or subtract as indicated.

47. $\dfrac{3}{5} + \dfrac{7}{10}$

48. $\dfrac{5}{8} - \dfrac{7}{24}$

49. $\dfrac{9}{16} - \dfrac{3}{24}$

50. $\dfrac{4}{21} + \dfrac{5}{36}$

Solve using the addition method.

51. $4x + y = 2$
$2x - 5y = 12$

52. $x - 3y = 4$
$3x - 7y = 8$

53. $3x - 4y = 6$
$2x - 7y = 17$

54. $4x - 5y = 11$
$3x + 2y = 14$

8.4 ADDITION AND SUBTRACTION WITH COMMON DENOMINATOR AND LOWEST COMMON DENOMINATOR

OBJECTIVES

1 *Add or subtract rational expressions with a common denominator*

2 *Find the lowest common denominator for rational expressions*

1 Addition of Rational Expressions with a Common Denominator

Recall that when we add fractions with a common denominator, we add only the numerators and use the common denominator. Addition of rational expressions with a common denominator is done the same way.

> To add rational expressions with the same denominator, add only the numerators and keep the common denominator.

EXAMPLE 1   Add.

a. $\dfrac{3}{5} + \dfrac{1}{5}$

$$\frac{3}{5} + \frac{1}{5} = \frac{3+1}{5} = \frac{4}{5}$$

b. $\dfrac{3}{x} + \dfrac{x+2}{x}$

$$\frac{3}{x} + \frac{x+2}{x} = \frac{3+x+2}{x} = \frac{5+x}{x}$$

c. $\dfrac{x^2+8}{3x-4} + \dfrac{3x^2-4x+2}{3x-4}$

$$\frac{x^2+8}{3x-4} + \frac{3x^2-4x+2}{3x-4} = \frac{(x^2+8)+(3x^2-4x+2)}{3x-4}$$
$$= \frac{4x^2-4x+10}{3x-4} \quad\blacksquare$$

□ **DO EXERCISE 1.**

Subtraction of Rational Expressions with a Common Denominator

Subtraction of rational expressions with a common denominator is done the same way as subtraction of fractions.

> To subtract two rational expressions with a common denominator, subtract the numerators and keep the common denominator.

□ **Exercise 1**   Add.

a. $\dfrac{4}{x} + \dfrac{3-x}{x}$

b. $\dfrac{2x^2+6x-3}{x+4} + \dfrac{x^2-8x}{x+4}$

a. $\dfrac{8}{x} - \dfrac{9}{x}$

b. $\dfrac{3}{y^2} - \dfrac{4y}{y^2}$

c. $\dfrac{7}{x+2} - \dfrac{3-x}{x+2}$

d. $\dfrac{-4x}{x-1} - \dfrac{6+x}{x-1}$

EXAMPLE 2 Subtract.

a. $\dfrac{3x+4}{x} - \dfrac{7}{x}$

$$\frac{3x+4}{x} - \frac{7}{x} = \frac{(3x+4) - 7}{x}$$

$$= \frac{3x - 3}{x}$$

We must be careful to subtract the entire numerator. Using parentheses around the second numerator helps us to remember to do this.

b. $\dfrac{5x}{x+4} - \dfrac{3x+2}{x+4}$

$$\frac{5x}{x+4} - \frac{3x+2}{x+4} = \frac{5x - (3x+2)}{x+4} \qquad \text{Using parentheses around the second numerator}$$

$$= \frac{5x - 3x - 2}{x+4}$$

$$= \frac{2x - 2}{x+4} \qquad \blacksquare$$

□ **DO EXERCISE 2.**

2 Lowest Common Denominator

Recall from Section 1.2 that to add fractions with different denominators we used multiplication by 1 to obtain fractions that have a common denominator. Recall also that our work was easier if we used the lowest common denominator to add the fractions. When we add rational expressions, our work is also easier if we use the lowest common denominator.

> To find the lowest common denominator, LCD, of two rational expressions, factor each denominator, factoring numbers into primes. Then use each factor in the LCD the greatest number of times it appears in either denominator.

EXAMPLE 3 Find the LCD for the rational expressions.

a. $\dfrac{3}{8x^2}, \quad \dfrac{1}{6x^4}$

  1. Factor the first denominator.
$$8x^2 = 2 \cdot 2 \cdot 2 \cdot x^2$$
  2. Factor the second denominator.
$$6x^4 = 2 \cdot 3 \cdot x^4$$
  3. Notice that the 2 occurs three times in the first denominator. Therefore, it must occur three times in the LCD. Similarly, 3

occurs once in the second denominator, so it must be in the LCD once. The factor x occurs four times in the second denominator. Therefore, it must appear four times in the LCD.

4. The LCD is $2 \cdot 2 \cdot 2 \cdot 3 \cdot x^4$, or $24x^4$.

b. $\dfrac{3}{t^2}, \ \dfrac{t+1}{3t^5+t^4}$

1. The factor t occurs twice in the first denominator.
2. Factor the second denominator.

$$3t^5 + t^4 = t^4(3t + 1)$$

3. The factor t occurs four times in the second denominator. Therefore, it must appear four times in the LCD. The LCD must also contain the factor $3t + 1$ since it appears once in the second denominator.
4. The LCD is $t^4(3t + 1)$.

c. $\dfrac{7}{24x^3yz^2}, \ \dfrac{3}{20x^2y^4z^3}$

1. Factor the first denominator.

$$24x^3yz^2 = 2 \cdot 2 \cdot 2 \cdot 3 \cdot x^3\,yz^2$$

2. Factor the second denominator.

$$20x^2y^4z^3 = 2 \cdot 2 \cdot 5 \cdot x^2\,y^4\,z^3$$

3. The LCD is $2 \cdot 2 \cdot 2 \cdot 3 \cdot 5 \cdot x^3y^4z^3$, or $120x^3y^4z^3$. ∎

☐ **DO EXERCISE 3.**

EXAMPLE 4 Find the LCD for the rational expressions.

a. $\dfrac{3}{x-2}, \ \dfrac{5}{2x-4}$

1. The first denominator, $x - 2$, cannot be factored.
2. Factor the second denominator.

$$2x - 4 = 2(x - 2)$$

3. We now have the following.

$$\dfrac{3}{x-2}, \ \dfrac{5}{\underset{2(x-2)}{2x-4}}$$

4. Notice that $x - 2$ occurs once in each denominator, so it must appear once in the LCD. The 2 appears in the second denominator, so it must also be included in the LCD.
5. The LCD is $(x - 2) \cdot 2$, or $2(x - 2)$.

☐ **Exercise 3** Find the LCD.

a. $\dfrac{4}{9x^3}, \ \dfrac{-1}{21x}$

b. $\dfrac{5}{y}, \ \dfrac{y-3}{y^4+2y}$

c. $\dfrac{-5}{15x^2}, \ \dfrac{2}{40x^4}$

d. $\dfrac{3}{12x^2yz}, \ \dfrac{1}{32xy^3z^4}$

□ **Exercise 4** Find the LCD for the rational expressions.

a. $\dfrac{2x}{3x - 9}$, $\dfrac{2}{x - 3}$

b. $\dfrac{-5}{x + 9}$, $\dfrac{x}{7}$, $\dfrac{x + 1}{x - 1}$

c. $\dfrac{4x}{x^2 - 5x + 6}$, $\dfrac{3x}{x - 2}$

d. $\dfrac{x + 4}{x^2 - 4}$, $\dfrac{x - 2}{x^2 + 4x + 4}$

b. $\dfrac{2x}{x + 2}$, $\dfrac{3}{x^2 + 4x + 4}$

1. The first denominator, $x + 2$, is not factorable.
2. Factor the second denominator.

$$x^2 + 4x + 4 = (x + 2)(x + 2,$$

3. We have the following.

$$\dfrac{2x}{x + 2}, \quad \dfrac{3}{x^2 + 4x + 4}$$
$$(x + 2)(x + 2)$$

4. The factor, $x + 2$, occurs twice in the second denominator, so it must appear twice in the LCD.
5. The LCD is $(x + 2)(x + 2)$.

c. $\dfrac{x}{2}$, $\dfrac{5}{2x - 5}$, $\dfrac{-3x}{x - 11}$

None of these denominators is factorable. The LCD is their product, $2(2x - 5)(x - 11)$.

d. $\dfrac{4}{x^2 - 1}$, $\dfrac{3}{x^2 + 2x + 1}$

1. Factor the first denominator.

$$x^2 - 1 = (x + 1)(x - 1)$$

2. Factor the second denominator.

$$x^2 + 2x + 1 = (x + 1)(x + 1)$$

3. The factor $x + 1$ occurs twice in the second denominator, so it must appear twice in the LCD. The factor $x - 1$ appears once in the first denominator, so it must also appear in the LCD.
4. The LCD is $(x + 1)(x + 1)(x - 1)$, or $(x + 1)^2(x - 1)$. ■

□ **DO EXERCISE 4.**

Answers to Exercises

1. **a.** $\dfrac{7 - x}{x}$ **b.** $\dfrac{3x^2 - 2x - 3}{x + 4}$

2. **a.** $-\dfrac{1}{x}$ **b.** $\dfrac{3 - 4y}{y^2}$ **c.** $\dfrac{4 + x}{x + 2}$ **d.** $\dfrac{-5x - 6}{x - 1}$

3. **a.** $63x^3$ **b.** $y(y^3 + 2)$ **c.** $120x^4$ **d.** $96x^2y^3z^4$

4. **a.** $3(x - 3)$ **b.** $7(x + 9)(x - 1)$ **c.** $(x - 3)(x - 2)$
   d. $(x + 2)^2(x - 2)$

PROBLEM SET 8.4

A. *Add.*

1. $\dfrac{3}{5} + \dfrac{8}{5}$

2. $\dfrac{5}{12} + \dfrac{7}{12}$

3. $\dfrac{5}{x} + \dfrac{9}{x}$

4. $-\dfrac{4}{y} + \dfrac{3}{y}$

5. $\dfrac{8}{x+1} + \dfrac{3}{x+1}$

6. $\dfrac{-2}{x-2} + \dfrac{8}{x-2}$

7. $\dfrac{x^2 + 2x}{3x - 4} + \dfrac{2x^2 - 3x - 7}{3x - 4}$

8. $\dfrac{x^2 - 3}{5x + 6} + \dfrac{2x^2 - x + 6}{5x + 6}$

B. *Subtract.*

9. $\dfrac{7}{9} - \dfrac{2}{9}$

10. $\dfrac{8}{x} - \dfrac{12}{x}$

11. $\dfrac{3}{y^2} - \dfrac{2y}{y^2}$

12. $\dfrac{-4x}{x+1} - \dfrac{3}{x+1}$

13. $\dfrac{7}{x-5} - \dfrac{2x-3}{x-5}$

14. $\dfrac{8x}{x-1} - \dfrac{8}{x-1}$

15. $\dfrac{x^2 - 3x}{4x + 7} - \dfrac{2x^2 + 4}{4x + 7}$

16. $\dfrac{5x - 9}{x^2 - 2} - \dfrac{x^2 - 3x}{x^2 - 2}$

C. *Do the operations indicated.*

17. $\dfrac{3}{x} + \dfrac{5}{x} - \dfrac{7}{x}$

18. $\dfrac{2y}{y+1} - \dfrac{3y}{y+1} + \dfrac{7}{y+1}$

19. $\dfrac{8 + 3x}{x + 4} - \dfrac{2x}{x + 4} - \dfrac{4}{x + 4}$

20. $\dfrac{22y}{2y + 5} + \dfrac{35}{2y + 5} - \dfrac{8y}{2y + 5}$

D. *Find the LCD for the rational expressions.*

21. $\dfrac{7}{2x^2}, \quad \dfrac{8}{6x^4}$

22. $\dfrac{-3}{5xy^2}, \quad \dfrac{2x}{10x^2y}$

23. $\dfrac{1}{16x^2y^3}, \quad \dfrac{3}{24xy^4}$

24. $\dfrac{5}{27xy^3}, \quad \dfrac{7}{36x^4y^2}$

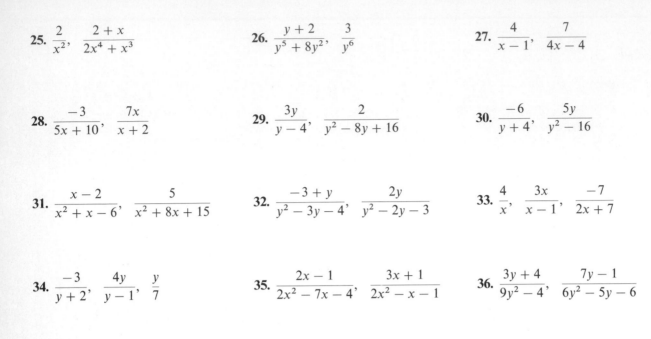

25. $\dfrac{2}{x^2}$, $\dfrac{2+x}{2x^4+x^3}$

26. $\dfrac{y+2}{y^5+8y^2}$, $\dfrac{3}{y^6}$

27. $\dfrac{4}{x-1}$, $\dfrac{7}{4x-4}$

28. $\dfrac{-3}{5x+10}$, $\dfrac{7x}{x+2}$

29. $\dfrac{3y}{y-4}$, $\dfrac{2}{y^2-8y+16}$

30. $\dfrac{-6}{y+4}$, $\dfrac{5y}{y^2-16}$

31. $\dfrac{x-2}{x^2+x-6}$, $\dfrac{5}{x^2+8x+15}$

32. $\dfrac{-3+y}{y^2-3y-4}$, $\dfrac{2y}{y^2-2y-3}$

33. $\dfrac{4}{x}$, $\dfrac{3x}{x-1}$, $\dfrac{-7}{2x+7}$

34. $\dfrac{-3}{y+2}$, $\dfrac{4y}{y-1}$, $\dfrac{y}{7}$

35. $\dfrac{2x-1}{2x^2-7x-4}$, $\dfrac{3x+1}{2x^2-x-1}$

36. $\dfrac{3y+4}{9y^2-4}$, $\dfrac{7y-1}{6y^2-5y-6}$

37. To add rational expressions with the same denominator, add only the _____ and keep the common denominator.

38. To subtract rational expressions with the same denominator, subtract the numerators and keep the _____ _____.

39. To add fractions with different denominators, multiply each fraction by _____ to obtain fractions that have a common denominator.

40. The first step in finding the lowest common denominator of two rational expressions is to _____ each denominator.

Checkup

The following problems provide a review of some of Section 7.3.
Solve by substitution.

41. $y = x + 1$
    $5x - 2y = 4$

42. $x = y + 6$
    $2x + 3y = 2$

43. $x - 2y = 4$
    $3x - 5y = 9$

44. $3x + y = 2$
    $4x - y = -9$

Solve by the substitution or the addition method.

45. $8x - 3y = 1$
    $7x + 3y = -16$

46. $2x - 7y = -4$
    $x + 7y = 19$

47. $3x - 2y = 14$
    $y = -4$

48. $2x + 3y = 8$
    $x = y + 4$

8.5 ADDITION AND SUBTRACTION WITH DIFFERENT DENOMINATORS

OBJECTIVES

1 *Multiply rational expressions by 1*

2 *Add or subtract rational expressions with different denominators*

3 *Combine terms in a rational expression using both addition and subtraction*

1 Multiplication by 1

Recall from Section 1.2 that the numerator and denominator of a fraction may be multiplied by the same number to give an equivalent fraction. This was called multiplying by 1. In the same way, a rational expression may be multiplied by 1 to give an equivalent rational expression. This allows us to add or subtract rational expressions with different denominators. We multiply each expression by 1 to give the same denominators. Then we may add or subtract the rational expressions.

The following are examples of multiplication by 1.

EXAMPLE 1 Multiply.

a. $\dfrac{3}{4} \cdot \dfrac{2}{2} = \dfrac{6}{8}$

Notice that the "1" is $\frac{2}{2}$, and $\frac{3}{4}$ is equivalent to $\frac{6}{8}$.

b. $\dfrac{5}{y} \cdot \dfrac{y}{y} = \dfrac{5y}{y^2}$

The "1" is y/y, and $5/y$ is equivalent to $5y/y^2$.

c. $\dfrac{x-4}{x+5} \cdot \dfrac{x+3}{x+3} = \dfrac{(x-4)(x+3)}{(x+5)(x+3)}$

Here the "1" is $(x+3)/(x+3)$. ∎

□ **DO EXERCISE 1.**

2 Addition of Rational Expressions with Different Denominators

To combine two rational expressions, they must have the same denominators. The denominator that we want them to have is the lowest common denominator (LCD). We multiply each expression by 1 to give it the LCD. Then combine the expressions.

□ **Exercise 1** Multiply.

a. $\dfrac{3}{x} \cdot \dfrac{2x}{2x}$

b. $\dfrac{2x+4}{x-5} \cdot \dfrac{x+1}{x+1}$

EXAMPLE 2 Add.

a. $\dfrac{3}{2y} + \dfrac{5}{6y}$

1. The LCD is $6y$. We want both expressions to have the same denominator of $6y$.
2. Multiply the first rational expression by $\frac{3}{3}$ to give it the LCD. It is not necessary to multiply the second expression by 1 since it already has the LCD.

$$\frac{3}{2y} \cdot \frac{3}{3} + \frac{5}{6y} = \frac{9}{6y} + \frac{5}{6y}$$

3. The expressions now have a common denominator, so we may add them and simplify.

$$\frac{9}{6y} + \frac{5}{6y} = \frac{9+5}{6y} = \frac{14}{6y} = \frac{7}{3y}$$

b. $\dfrac{4x}{x+2} + \dfrac{5}{x-1}$

1. Neither denominator is factorable, so the LCD is the product of the denominators.

$$\text{LCD} = (x+2)(x-1)$$

2. Multiply the first expression by $(x-1)/(x-1)$ to give it the LCD.
   Multiply the second expression by $(x+2)/(x+2)$ to give it the LCD.
   Add the two expressions.

$$\frac{4x}{x+2} \cdot \frac{x-1}{x-1} + \frac{5}{x-1} \cdot \frac{x+2}{x+2} = \frac{4x(x-1)}{(x+2)(x-1)} + \frac{5(x+2)}{(x+2)(x-1)}$$

$$= \frac{4x(x-1) + 5(x+2)}{(x+2)(x-1)}$$

$$= \frac{4x^2 - 4x + 5x + 10}{(x+2)(x-1)}$$

$$= \frac{4x^2 + x + 10}{(x+2)(x-1)}$$

The denominator may be left in factored form. Notice that this expression cannot be simplified. We will discuss an expression that must be simplified in Example 5.

c. $\dfrac{3}{x^2-1} + \dfrac{5}{x-1}$

1. Factor the first denominator.

$$x^2 - 1 = (x-1)(x+1)$$

2. The second denominator does not factor.

3. The original expression now looks like this:

$$\frac{3}{(x-1)(x+1)} + \frac{5}{x-1}$$

4. The LCD is $(x-1)(x+1)$. We want both expressions to have a denominator of $(x-1)(x+1)$.

5. The denominator of the first expression already is the LCD. Multiply the second expression in step 3 by 1 so that both fractions have the same denominator. Then add.

$$\frac{3}{(x-1)(x+1)} + \frac{5}{x-1} \cdot \boxed{\frac{x+1}{x+1}} = \frac{3}{(x-1)(x+1)} + \frac{5(x+1)}{(x-1)(x+1)}$$

$$= \frac{3 + 5(x+1)}{(x-1)(x+1)}$$

$$= \frac{3 + 5x + 5}{(x-1)(x+1)}$$

$$= \frac{8 + 5x}{(x-1)(x+1)}$$

d. $\dfrac{2}{x^2 + 5x + 6} + \dfrac{x-2}{x^2 + 2x - 3}$

1. Factor the first denominator.

$$x^2 + 5x + 6 = (x+3)(x+2)$$

2. Factor the second denominator.

$$x^2 + 2x - 3 = (x+3)(x-1)$$

3. The original expression now appears as follows:

$$\frac{2}{(x+3)(x+2)} + \frac{x-2}{(x+3)(x-1)}$$

4. The LCD is $(x+3)(x+2)(x-1)$.

5. Multiply each expression in step 3 by 1 to get the common denominator. Then add the expressions.

$$\frac{2}{(x+3)(x+2)} \cdot \boxed{\frac{x-1}{x-1}} + \frac{x-2}{(x+3)(x-1)} \cdot \boxed{\frac{x+2}{x+2}}$$

$$= \frac{2(x-1)}{(x+3)(x+2)(x-1)} + \frac{(x-2)(x+2)}{(x+3)(x+2)(x-1)}$$

$$= \frac{2x - 2 + x^2 - 4}{(x+3)(x+2)(x-1)}$$

$$= \frac{x^2 + 2x - 6}{(x+3)(x+2)(x-1)} \quad \blacksquare$$

☐ **DO EXERCISE 2.**

☐ **Exercise 2** Add.

a. $\dfrac{7}{5y} + \dfrac{4}{3y}$

b. $\dfrac{2x}{x-3} + \dfrac{x-4}{x+2}$

c. $\dfrac{4x}{x^2 - 9} + \dfrac{x}{x+3}$

d. $\dfrac{3}{x^2 + 2x + 1} + \dfrac{x}{x^2 + 5x + 4}$

□ **Exercise 3**   Subtract.

a. $\dfrac{6}{x+2} - \dfrac{5}{x}$

b. $\dfrac{7x}{(x+2)^2} - \dfrac{3}{x+2}$

c. $\dfrac{7}{x-4} - \dfrac{2}{x+3}$

d. $\dfrac{x+4}{x^2+2x+1} - \dfrac{x-1}{x^2+4x+3}$

Subtraction of Rational Expressions with Different Denominators

We use the same procedures for subtraction of rational expressions with different denominators as we use for addition of these expressions except that we subtract the numerators. Remember to subtract the *entire* numerator.

EXAMPLE 3   Subtract.

$$\frac{5x}{(x-1)^2} - \frac{x+1}{x-1}$$

1. The LCD is $(x-1)^2$.
2. Multiply the second expression by 1 to get the LCD. Then subtract.

$$\frac{5x}{(x-1)^2} - \frac{x+1}{x-1} \cdot \frac{x-1}{x-1} = \frac{5x}{(x-1)^2} - \frac{(x+1)(x-1)}{(x-1)^2}$$

$$= \frac{5x - (x+1)(x-1)}{(x-1)^2}$$

$$= \frac{5x - (x^2-1)}{(x-1)^2} \qquad \text{Use parentheses to subtract the entire.}$$

$$= \frac{5x - x^2 + 1}{(x-1)^2} \qquad \text{numerator}$$

$$= \frac{-x^2 + 5x + 1}{(x-1)^2} \qquad \blacksquare$$

□ **DO EXERCISE 3.**

Addition or Subraction when the Denominators Are Opposites

When two rational expressions have denominators that are opposites, we can make them the same by multiplying one of the rational expressions by 1. In this case we multiply by $-1/-1$.

EXAMPLE 4

a. Add $\dfrac{4x+2}{5} + \dfrac{x-3}{-5}$.

1. It is easier to multiply the second expression by $-1/-1$ than to multiply the first expression by $-1/-1$. Multiplying the second expression by $-1/-1$ gives us a positive LCD.

$$\frac{4x+2}{5} + \frac{x-3}{-5} \cdot \boxed{\frac{-1}{-1}} = \frac{4x+2}{5} + \frac{(x-3)(-1)}{(-5)(-1)}$$

$$= \frac{4x+2}{5} + \frac{(-1)(x-3)}{5}$$

$$= \frac{4x+2-x+3}{5}$$

$$= \frac{3x+5}{5}$$

b. Subtract $\dfrac{5x}{x-2} - \dfrac{2x}{2-x}$.

1. We may multiply the second expression by $-1/-1$ to get a common denominator. Then subtract.

$$\frac{5x}{x-2} - \frac{2x}{2-x} \cdot \boxed{\frac{-1}{-1}} = \frac{5x}{x-2} - \frac{2x(-1)}{(2-x)(-1)}$$

$$= \frac{5x}{x-2} - \frac{-2x}{-2+x}$$

$$= \frac{5x}{x-2} - \frac{-2x}{x-2}$$

$$= \frac{5x-(-2x)}{x-2} = \frac{7x}{x-2} \quad \blacksquare$$

☐ **DO EXERCISE 4.**

3 Combined Additions and Subtractions and Simplifying

Many exercises involve both addition and subtraction. It may be necessary to simplify the final result.

EXAMPLE 5 Combine the terms and simplify.

$$\frac{x+10}{x^2-2x} + \frac{2}{x} - \frac{6}{x-2}$$

1. Factor the first denominator.

$$x^2 - 2x = x(x-2)$$

2. The other denominators are not factorable. The LCD is

$$x(x-2)$$

3. Multiply by 1 to give all the rational expressions the LCD.

$$\frac{x+10}{x(x-2)} + \frac{2}{x} \cdot \boxed{\frac{x-2}{x-2}} - \frac{6}{x-2} \cdot \boxed{\frac{x}{x}}$$

$$= \frac{x+10}{x(x-2)} + \frac{2(x-2)}{x(x-2)} - \frac{6x}{(x-2)x}$$

$$= \frac{x+10}{x(x-2)} + \frac{2x-4}{x(x-2)} - \frac{6x}{x(x-2)}$$

$$= \frac{x+10+2x-4-6x}{x(x-2)}$$

$$= \frac{-3x+6}{x(x-2)}$$

$$= \frac{-3(x-2)}{x(x-2)} \qquad \text{Factor out } -3$$

$$= \frac{-3}{x} \qquad \text{Removing a one} \quad \blacksquare$$

☐ **DO EXERCISE 5.**

☐ **Exercise 4** Do the indicated operation.

a. $\dfrac{6y}{-3} - \dfrac{3y}{3}$

b. $\dfrac{4x}{x-3} - \dfrac{2x}{3-x}$

☐ **Exercise 5** Combine the terms and simplify.

a. $\dfrac{2}{x+3} - \dfrac{1}{x-3} + \dfrac{2x}{x^2-9}$

b. $\dfrac{2x}{x^2-1} + \dfrac{1}{x+1} - \dfrac{1}{x-1}$

When Do We Need a Common Denominator?

1. To add or subtract rational expressions find a common denominator.
2. To multiply or divide rational expressions, we do not need a common denominator.

Answers to Exercises

1. a. $\dfrac{6x}{2x^2}$ b. $\dfrac{(2x + 4)(x + 1)}{(x - 5)(x + 1)}$

2. a. $\dfrac{41}{15y}$ b. $\dfrac{3x^2 - 3x + 12}{(x - 3)(x + 2)}$ c. $\dfrac{x^2 + x}{(x + 3)(x - 3)}$

   d. $\dfrac{x^2 + 4x + 12}{(x + 1)^2(x + 4)}$

3. a. $\dfrac{x - 10}{(x + 2)x}$ b. $\dfrac{4x - 6}{(x + 2)^2}$ c. $\dfrac{5x + 29}{(x - 4)(x + 3)}$

   d. $\dfrac{7x + 13}{(x + 1)^2(x + 3)}$

4. a. $-3y$ b. $\dfrac{6x}{x - 3}$

5. a. $\dfrac{3}{x + 3}$ b. $\dfrac{2}{x + 1}$

PROBLEM SET 8.5

A. *Multiply. Leave the numerator and denominator in factored form. Do not simplify.*

1. $\dfrac{x-2}{x+4} \cdot \dfrac{3}{3}$

2. $\dfrac{2x+5}{x-3} \cdot \dfrac{6}{6}$

3. $\dfrac{x-1}{x+5} \cdot \dfrac{x+1}{x+1}$

4. $\dfrac{2x+3}{3x-4} \cdot \dfrac{x-4}{x-4}$

B. *Add. Simplify, if possible.*

5. $\dfrac{6}{4x} + \dfrac{5}{3x}$

6. $\dfrac{3}{x} + \dfrac{2}{x^2}$

7. $\dfrac{5}{x-2} + \dfrac{3}{x}$

8. $\dfrac{4}{y} + \dfrac{7}{y+1}$

9. $\dfrac{3y}{y+4} + \dfrac{y}{y^2-16}$

10. $\dfrac{x}{3x-6} + \dfrac{2x}{x-2}$

11. $\dfrac{4}{x+3} + \dfrac{2x}{2x+6}$

12. $\dfrac{3y}{y^2-9} + \dfrac{2}{y-3}$

13. $\dfrac{1}{x^2-x-2} + \dfrac{3}{x^2+2x+1}$

14. $\dfrac{3x}{x^2+3x-10} + \dfrac{2x}{x^2+x-6}$

15. $\dfrac{4}{x+1} + \dfrac{3}{(x+1)^2}$

16. $\dfrac{7}{(x-3)^2} + \dfrac{3}{x-3}$

17. $\dfrac{x}{x^2-1} + \dfrac{3x}{x^2-x}$

18. $\dfrac{4x-3}{x^2-16} + \dfrac{x-1}{3x-12}$

19. $\dfrac{x+2}{x-4} + \dfrac{x-1}{x+3}$

20. $\dfrac{2y}{3y-4} + \dfrac{5y}{4y-3}$

21. $\dfrac{5}{x-y} + \dfrac{3}{y-x}$

22. $\dfrac{4}{x-5} + \dfrac{3}{5-x}$

23. $\dfrac{x+2y}{x^2y} + \dfrac{3y}{xy^2}$

24. $\dfrac{x-y}{x^3y^2} + \dfrac{y}{xy^4}$

C. *Subtract. Simplify, if possible.*

25. $\dfrac{6}{x} - \dfrac{4}{x+5}$

26. $\dfrac{4}{x-3} - \dfrac{3}{x+2}$

27. $\dfrac{8y-4}{3x^2y} - \dfrac{3}{6xy^2}$

28. $\dfrac{5}{4x^2y^2} - \dfrac{3x}{3xy^2}$

29. $\dfrac{7}{x+1} - \dfrac{3}{x-1}$

30. $\dfrac{3y}{2y+3} - \dfrac{4y}{y+2}$

31. $\dfrac{1}{x-1} - \dfrac{2x}{x^2-1}$

32. $\dfrac{-3}{y^2-4} - \dfrac{3}{2y-4}$

33. $\dfrac{4}{2-x} - \dfrac{3}{x-2}$

34. $\dfrac{8}{7-y} - \dfrac{3}{y-7}$

35. $\dfrac{1}{x^2-1} - \dfrac{1}{x^2+3x+2}$

36. $\dfrac{x}{x^2-1} - \dfrac{x-1}{x^2+2x+1}$

D. *Combine the terms. Simplify, if possible.*

37. $\dfrac{3}{x} + \dfrac{2}{x^2} - \dfrac{5}{x^3}$

38. $\dfrac{4}{y^4} - \dfrac{3}{y^2} + \dfrac{2}{y}$

39. $\dfrac{4x}{x^2-1} - \dfrac{2}{x} - \dfrac{2}{x+1}$

40. $\dfrac{x+6}{4-x^2} - \dfrac{x+3}{x+2} + \dfrac{x-3}{2-x}$

41. $\dfrac{3}{x+2} + \dfrac{6}{x-2} + \dfrac{2x-5}{4-x^2}$

42. $\dfrac{3}{y^2-25} - \dfrac{1}{y+5} - \dfrac{y+1}{5-y}$

43. To combine two rational expressions, they must have the same _____.

44. To subtract one rational expression from another rational expression, remember to subtract the _____ numerator of the second rational expression.

45. If two rational expressions have denominators that are opposites, find a common denominator by _____ one of the expressions by $-1/-1$.

46. To multiply or divide rational expressions, a common _____ is not necessary.

Checkup

The following problems provide a review of some of Section 7.4.

47. Gagik has 128 feet of fencing. He wants to build a rectangular fence with length 10 feet greater than the width. What should be the width of the fence?

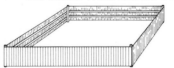

48. The length of a rectangle is four times the width. If the perimeter is 120 meters, find the length of the rectangle.

49. If a sum of money is invested at 9% per year and $1500 less is invested at 8% per year, how much is invested at 8% if the total annual interest is $645?

50. Bill earned $308 from his investments. He had some money invested at 6% and twice as much money invested at 8%. How much does he have invested at 6%?

51. A chemical technician wants to combine a 30% acid solution with a 60% acid solution to get 8 liters of a 40% solution. How much 30% solution should she use?

52. How many quarts of 55% antifreeze solution should Dean mix with a 20% solution to get 12 quarts of a 35% solution?

CHAPTER 8   ADDITIONAL EXERCISES (OPTIONAL)

Section 8.1

Are the following rational expressions?

1. $\dfrac{x^2 + 4}{x - 2}$

2. $\dfrac{3}{x^2 - 5}$

3. $\dfrac{x^2 + x + 2}{x^{-1}}$

4. $\dfrac{7}{x^2 + x^{-1} - 3}$

Evaluate for the given value of the variable.

5. $\dfrac{3x}{x - 4}$ for $x = -4$

6. $\dfrac{2x - 1}{3x + 2}$ for $x = -2$

7. $\dfrac{x^3 - 4}{x + 7}$ for $x = -2$

8. $\dfrac{x^2 + x - 3}{x^2 + 1}$ for $x = -1$

9. $\dfrac{-x^2 - 3}{7 + x}$ for $x = 4$

10. $\dfrac{-x^2 - 2x + 2}{4 - x}$ for $x = 6$

Section 8.2

Simplify. Write answers with positive exponents.

11. $\dfrac{-72x^6y^5}{8x^7y^9}$

12. $\dfrac{-54x^3y^9}{-6x^5y}$

13. $\dfrac{-108x^4y^3z^5}{-12x^6yz^8}$

14. $\dfrac{105a^7b^2c^6}{-35a^4b^5c^2}$

Simplify.

15. $\dfrac{x^2 - 9}{x + 3}$

16. $\dfrac{(3x - 2)^2}{3x - 2}$

17. $\dfrac{9x - 9y}{9y - 9x}$

18. $\dfrac{-7x - 7y}{7y + 7x}$

19. $\dfrac{21x^2 + 44x - 32}{14x^2 - 29x + 12}$

20. $\dfrac{30x^2 + 54x - 12}{10x^2 + 22x + 4}$

Section 8.3

Multiply. Write answers with positive exponents.

21. $\dfrac{6x^2}{y^4} \cdot \dfrac{y^7}{3x^5}$

22. $\dfrac{y^3}{12x^6} \cdot \dfrac{8x^9}{y^4}$

23. $\dfrac{34x^2}{21y^6} \cdot \dfrac{7x^2y^8}{-17x^5y}$

24. $\dfrac{-9x^7y^2}{144y} \cdot \dfrac{15y^3}{25x^4y^5}$

Multiply.

25. $\dfrac{y^3 - 2y^2}{2y^2 - 8} \cdot \dfrac{y^2 + 4y + 4}{y}$

26. $\dfrac{3x^2 - 6x - 24}{6x - 24} \cdot \dfrac{x + 6}{9x + 18}$

Divide. Write answers with positive exponents.

27. $\dfrac{25a^2b^3}{-5b^4} \div \dfrac{-35a^4b^5}{7a^2}$

28. $\dfrac{-64x^5y^4}{81x^7y} \div \dfrac{8x^6y}{9xy^3}$

Divide.

29. $\dfrac{2x^2 - 5x + 3}{3x^2 + 2x - 1} \div \dfrac{2x^2 - 7x + 6}{3x^2 + 11x - 4}$

30. $\dfrac{12 - 7x + x^2}{16 - x^2} \div \dfrac{9 - 6x + x^2}{8 - 2x - x^2}$

31. $\dfrac{y^2 - 8y + 16}{2y^2 - 5y - 12} \div \dfrac{4 - y}{2y + 3}$

32. $\dfrac{6 - x}{3x + 4} \div \dfrac{x^2 - 36}{(x + 6)^2}$

Section 8.4

Subtract.

33. $\dfrac{7x^2}{3x^2 + 2} - \dfrac{8x^2 + 4}{3x^2 + 2}$

34. $\dfrac{5x}{x^3 + 8} - \dfrac{3x - 7}{x^3 + 8}$

Find the lowest common denominator for the rational expressions.

35. $\dfrac{5}{18xy^3}, \quad \dfrac{7}{27x^4y^2}$

36. $\dfrac{1}{32x^4y^4}, \quad \dfrac{9}{24x^5y^2}$

37. $\dfrac{3}{x}, \quad \dfrac{7}{x + 4}$

38. $\dfrac{5}{x^2 + 2x + 3}, \quad \dfrac{3x}{x^2}$

39. $\dfrac{2y}{y-5}, \quad \dfrac{4}{y^2-7y+10}$

40. $\dfrac{-3}{2x+3}, \quad \dfrac{5x}{4x^2+12x+9}$

Section 8.5

Perform the operations indicated. Simplify, if possible.

41. $\dfrac{5}{6x^2} + \dfrac{7}{12x}$

42. $\dfrac{1}{9x} + \dfrac{4}{15x^4}$

43. $\dfrac{3y}{2y-5} - \dfrac{4}{y+3}$

44. $\dfrac{5x}{x-1} - \dfrac{3}{x+4}$

45. $\dfrac{4y}{y^2-25} + \dfrac{3}{y+5}$

46. $\dfrac{2}{x+4} - \dfrac{3x}{2x+8}$

47. $\dfrac{3}{4-x} - \dfrac{7}{x-4}$

48. $\dfrac{x-2}{x+3} + \dfrac{x-4}{x-3}$

49. $\dfrac{2}{x^2+6x+8} - \dfrac{3}{x^2-x-20}$

50. $\dfrac{7}{x^2-5x+6} + \dfrac{4}{x^2+2x-8}$

51. $\dfrac{x}{x^2-4} - \dfrac{x+2}{x^2-5x+6} + \dfrac{3}{2-x}$

52. $\dfrac{4}{15-8x+x^2} - \dfrac{2x+3}{x-5} + \dfrac{5}{(5-x)^2}$

CHAPTER 8   PRACTICE TEST

1. Is the following a rational expression?

$$\frac{x + 3}{x^2 + 2x + 4}$$

1. _____

2. Evaluate $(3x^2 - x)/(x^2 + 1)$ for $x = -2$.

2. _____

Simplify.

3. $\dfrac{3x - 9}{4x^2 - 7x - 15}$

3. _____

4. $\dfrac{4y - 4x}{x - y}$

4. _____

5. Multiply.

$$\frac{x^2 - 4}{x^2 + 3x - 10} \cdot \frac{8}{2x + 4}$$

5. _____

6. _____

6. Divide.

$$\frac{x^4}{x+y} \div \frac{x^2}{(x+y)^2}$$

Add.

7. _____

7. $\dfrac{5x}{x^2-16} + \dfrac{3}{x-4}$

8. _____

8. $\dfrac{7}{x-8} + \dfrac{3}{8-x}$

·9. _____

9. Subtract.

$$\frac{3}{x^2+4x-5} - \frac{2x}{x^2-25}$$

Are the following solutions of the pair of equations?

$$3x - 4y = 15$$
$$7x + 8y = -2$$

1. $(1, -3)$

2. $(2, -2)$

3. Solve by graphing.

$$2y = 3x + 1$$
$$2x + y = -3$$

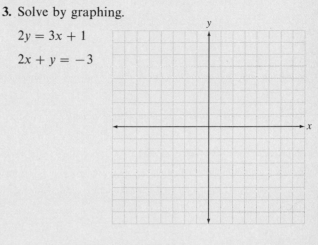

Solve using the addition method.

4. $x - y = 8$
    $x + y = 4$

5. $3x + 5y = -1$
    $2x - y = -5$

6. $7x - 2y = -20$
    $3x + 5y = 9$

Solve using the substitution method.

7. $9x - 2y = 1$
    $y = 4x$

8. $3x + 5y = 6$
    $3x \quad y = 12$

9. Two angles are supplementary. One angle is 15° more than twice the other. Find the angles.

10. Two angles are complementary. Onc angle is 38° less than the other. Find the angles.

11. Karen earned $139.50 from her investments. She has some money invested for 1 year at 7% interest and three times as much invested at 8% interest. How much does she have invested at each rate?

12. Greg wants to mix a 40% alcohol solution with a 65% alcohol solution to get 80 milliliters of a 55% solution. How many milliliters of each solution should he use?

Graph.

13. $x - 4y < 8$

14. $y \geq 5x$

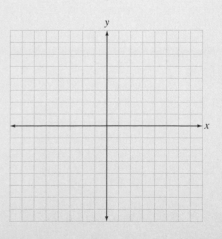

15. Which of the following are rational expressions?

(a) $\dfrac{x^2 + 2}{\frac{3}{4}x - 3}$ (b) $\dfrac{x^{-5} + 4}{x}$ (c) $\dfrac{x^2 + 2x + 9}{x - 4}$

Evaluate for the given value of the variable.

16. $\dfrac{3x - 1}{x^2 + x - 2}$ for $x = -3$.

17. $\dfrac{x^2 - 4x + 2}{x - 3}$ for $x = 5$

Multiply. Write answers with positive exponents.

18. $\dfrac{18x^3}{y^4} \cdot \dfrac{y^3}{6x^2}$

19. $\dfrac{3x - 6}{x + 4} \cdot \dfrac{x^2 - 16}{x^2 + 3x - 10}$

Divide. Write answers with positive exponents.

20. $\dfrac{4x^5}{y^2} \div \dfrac{2x^3}{y}$

21. $\dfrac{x - 8}{2x + 6} \div \dfrac{(x - 8)^2}{2}$

22. $\dfrac{x^2 - 9}{x^2 + 3x - 4} \div \dfrac{x^2 + 6x + 9}{x^2 + 2x - 8}$

Perform the operation indicated.

23. $\dfrac{7}{x + 4} - \dfrac{3}{x + 4}$

24. $\dfrac{x}{4x - 8} + \dfrac{3}{x - 2}$

25. $\dfrac{3y - 4}{y^2 - y} - \dfrac{2y}{y^2 - 1}$

26. $\dfrac{x + 3}{x^2 + 6x + 5} + \dfrac{4}{x^2 + 3x - 10}$

27. $\dfrac{6}{x + 3} - \dfrac{7x}{x + 5} + \dfrac{3}{(x + 5)^2}$

9

Rational Equations and Complex Fractions

9.1 SOLVING RATIONAL EQUATIONS

1 In Chapter 8 we studied rational expressions. In this chapter we solve rational equations. A **rational equation** is an equation with rational expressions. The following are examples of rational equations.

$$\frac{x}{4} = 7 \qquad \frac{x}{2} + \frac{2}{3} = 4 \qquad \frac{x}{x-3} = \frac{1}{x-3}$$

Recall that in Section 3.3 the first step in solving the first equation was to multiply both sides of the equation by 4. This removed the fraction so that we could solve the equation. We said that both sides of an equation may be multiplied by the same number. We use this fact to solve rational equations. The number that we multiply by is the lowest common denominator (LCD). Multiplying by the LCD removes the fractions.

> To solve a rational equation, first multiply both sides of the equation by the LCD.

EXAMPLE 1 Solve.

a. $\dfrac{x+1}{2} = \dfrac{x+2}{3}$

The LCD is 6 so we multiply both sides of the equation by 6.

$$6\left(\frac{x+1}{2}\right) = 6\left(\frac{x+2}{3}\right) \qquad \text{Use parentheses to multiply the entire numerator by 6.}$$

$$\frac{6}{2}(x+1) = \frac{6}{3}(x+2)$$

$$3(x+1) = 2(x+2) \qquad \text{Simplifying}$$

$$3x + 3 = 2x + 4 \qquad \text{Using a distributive law}$$

$$3x - 2x = 4 - 3$$

$$x = 1$$

401

Rational equation   an equation with rational expressions.

Meaningless rational expression   a rational expression is meaningless for any value of a variable which makes the denominator zero.

☐ **Exercise 1**   Solve.

a. $\dfrac{3}{4} + \dfrac{7}{8} = \dfrac{x}{2}$

b. $\dfrac{x+3}{x} = \dfrac{5}{4}$

c. $\dfrac{x-4}{3} = x + 2$

d. $\dfrac{5}{3y} + \dfrac{3}{y} = 1$

Check $\dfrac{x+1}{2} = \dfrac{x+2}{3}$

$$\dfrac{\dfrac{1+1}{2}}{1} \Bigg| \dfrac{\dfrac{1+2}{3}}{1}$$

The answer checks, so 1 is the solution.

b. $\dfrac{4}{x} + \dfrac{2}{3} = 1$

The LCD is $3x$. Multiply both sides of the equation by $3x$.

$$3x\left(\dfrac{4}{x} + \dfrac{2}{3}\right) = 3x(1)$$

$$3x\left(\dfrac{4}{x}\right) + 3x\left(\dfrac{2}{3}\right) = 3x(1)$$

$$12 + 2x = 3x \qquad \text{Simplifying}$$

$$12 = 3x - 2x$$

$$12 = x$$

Check $\dfrac{4}{x} + \dfrac{2}{3} = 1$

$$\dfrac{4}{12} + \dfrac{2}{3} \Bigg| 1$$

$$\dfrac{1}{3} + \dfrac{2}{3} \Bigg| 1$$

$$1 \Bigg| 1$$

The answer checks, so 12 is the solution.   ■

☐ **DO EXERCISE 1.**

We cannot have a rational number with denominator 0. We also cannot have a rational expression with denominator 0. Values of the variable that make the denominator 0 are said to make the **rational expression meaningless** or undefined.

EXAMPLE 2   Find the values that make the rational expression meaningless.

a. $\dfrac{3}{x}$ The expression is meaningless for $x = 0$.

b. $\dfrac{1}{x-3}$

To find the values of x that make the denominator 0, set the denominator equal to 0 and solve for x.

Let $x - 3 = 0$

$x = 3$

The value $x = 3$ makes the denominator zero. Therefore, the expression is meaningless for $x = 3$.

c. $\dfrac{x-1}{x^2+2}$

Are there values of x that make the denominator 0?

Let $x^2 + 2 = 0$

$$x^2 = -2$$

There are no values of x such that squaring it will give -2 and make the denominator 0. The denominator $x^2 + 2$ is always positive. Therefore, there are no values of x for which the expression is meaningless. ■

□ **DO EXERCISE 2.**

When we solve rational equations containing a variable in the denominator, we must always check our answers. Sometimes numbers that appear to be solutions are not solutions because they make a denominator of the original equation zero. When solving rational equations we should first determine what values of the variable will make a denominator zero.

EXAMPLE 3 Solve.

a. $\dfrac{x+1}{x-3} = \dfrac{4}{x-3} + 6$ (Original equation)

Notice that x *cannot equal 3* because this makes a denominator zero and we would have division by zero.

The LCD is $x - 3$, so multiply both sides of the equation by $x - 3$.

$$(x-3)\left(\dfrac{x+1}{x-3}\right) = (x-3)\left(\dfrac{4}{x-3} + 6\right)$$

$$(x-3)\left(\dfrac{x+1}{x-3}\right) = (x-3)\left(\dfrac{4}{x-3}\right) + (x-3)6$$

$$x + 1 = 4 + (x-3)6 \quad \text{Simplifying}$$

$$x + 1 = 4 + 6x - 18$$

$$x + 1 = 6x - 14$$

$$14 + 1 = 6x - x \qquad \text{Adding 14 and } -x \text{ to both sides}$$

$$15 = 5x$$

$$3 = x$$

Check $\dfrac{x+1}{x-3} = \dfrac{4}{x-3} + 6$

$$\dfrac{3+1}{3-3} \quad \bigg| \quad \dfrac{4}{3-3} + 6$$

$$\dfrac{4}{0} \quad \bigg| \quad \dfrac{4}{0} + 6$$

Remember that the value of $x = 3$ makes a denominator of the original equation zero, so 3 is not a solution. *There is no solution to this equation.*

□ **Exercise 2** Find the value that makes the rational expression meaningless.

a. $\dfrac{5}{4x}$

b. $\dfrac{3}{2x-8}$

c. $\dfrac{x-5}{x+5}$

d. $\dfrac{3x+4}{x^2+1}$

□ **Exercise 3**   Solve.

a. $\dfrac{x}{x-4} = \dfrac{4}{x-4} + 2$

b. $\dfrac{3}{x^2-1} = \dfrac{2}{x+1}$

---

b. $\dfrac{1}{x+3} + \dfrac{1}{x-3} = \dfrac{1}{x^2-9}$

Notice that x cannot equal 3 or -3 since this would make a zero denominator. The LCD is $x^2 - 9$ since it factors into $(x+3)(x-3)$. We multiply both sides of the equation by $(x+3)(x-3)$.

$$(x+3)(x-3)\left(\dfrac{1}{x+3} + \dfrac{1}{x-3}\right) = (x+3)(x-3)\left(\dfrac{1}{(x+3)(x-3)}\right)$$

$$(x+3)(x-3)\left(\dfrac{1}{x+3}\right) + (x+3)(x-3)\left(\dfrac{1}{x-3}\right)$$

$$= (x+3)(x-3)\left(\dfrac{1}{(x+3)(x-3)}\right)$$

$$x - 3 + x + 3 = 1 \qquad \text{Simplifying}$$

$$2x = 1$$

$$x = \dfrac{1}{2}$$

Check $\dfrac{1}{x+3} + \dfrac{1}{x-3} = \dfrac{1}{x^2-9}$

$$
\begin{array}{c|c}
\dfrac{1}{\frac{1}{2}+3} + \dfrac{1}{\frac{1}{2}-3} & \dfrac{1}{\left(\frac{1}{2}\right)^2 - 9} \\[2mm]
\dfrac{1}{\frac{1}{2}+3} + \dfrac{1}{\frac{1}{2}-3} & \dfrac{1}{\frac{1}{4}-9} \\[2mm]
\dfrac{1}{\frac{1}{2}+\frac{6}{2}} + \dfrac{1}{\frac{1}{2}-\frac{6}{2}} & \dfrac{1}{\frac{1}{4}-\frac{36}{4}} \\[2mm]
\dfrac{1}{\frac{7}{2}} + \dfrac{1}{\frac{-5}{2}} & \dfrac{1}{\frac{-35}{4}} \\[2mm]
\dfrac{2}{7} + \dfrac{-2}{5} & \dfrac{4}{-35} \\[2mm]
\dfrac{10}{35} - \dfrac{14}{35} & \dfrac{-4}{35} \\[2mm]
\dfrac{-4}{35} & \dfrac{-4}{35}
\end{array}
$$

The answer checks, so the solution is $\frac{1}{2}$. ■

□ **DO EXERCISE 3.**

404   Chapter 9   Rational Equations and Complex Fractions

It is important to understand the difference between adding or subtracting rational expressions and solving a rational equation. When we add or subtract rational expressions, we keep the lowest common denominator (LCD). When we solve rational equations, we multiply both sides of the equation by the LCD to remove the denominator.

$$\frac{x-7}{x+2}+\frac{1}{4} \qquad \text{is a rational expression}$$

$$\frac{x-7}{x+2}=\frac{1}{4} \qquad \text{is a rational equation}$$

EXAMPLE 4 Add or solve.

a. $\dfrac{x-7}{x+2}+\dfrac{1}{4}$

This is *not* an equation. The LCD is $4(x+2)$. *Multiply each expression by one to get the LCD.* Keep the LCD. *Add* the two expressions.

$$\frac{x-7}{x+2}\cdot\frac{4}{4}+\frac{1}{4}\cdot\frac{x+2}{x+2}=\frac{4(x-7)}{4(x+2)}+\frac{x+2}{4(x+2)}$$

$$=\frac{4x-28+x+2}{4(x+2)}$$

$$=\frac{5x-26}{4(x+2)}$$

The answer is $\dfrac{5x-26}{4(x+2)}$.

b. $\dfrac{x-7}{x+2}=\dfrac{1}{4}$

This is an equation. The LCD is $4(x+2)$. *Multiply each side of the equation by the LCD.* Simplify to remove the denominators. Then solve the equation.

$$4(x+2)\left(\frac{x-7}{x+2}\right)=4(x+2)\left(\frac{1}{4}\right)$$

$$4(x-7)=x+2 \qquad \text{Simplifying}$$

$$4x-28=x+2$$

$$3x=30$$

$$x=10$$

□ **Exercise 4** Add or subtract or solve.

a. $\dfrac{2y + 3}{y} = \dfrac{3}{2}$

b. $\dfrac{2y + 3}{y} - \dfrac{3}{2}$

Check $\quad \dfrac{x - 7}{x + 2} = \dfrac{1}{4}$

$$\begin{array}{c|c} \dfrac{10 - 7}{10 + 2} & \dfrac{1}{4} \\ \hline \dfrac{3}{12} & \dfrac{1}{4} \\ \dfrac{1}{4} & \dfrac{1}{4} \end{array}$$

The solution is 10. ■

□ **DO EXERCISE 4.**

Answers to Exercises

1. **a.** $\dfrac{13}{4}$ **b.** 12   **c.** -5 **d.** $\dfrac{14}{3}$

2. **a.** 0   **b.** 4   **c.** -5 **d.** None

3. **a.** No solution   **b.** $\dfrac{5}{2}$

4. **a.** -6 **b.** $\dfrac{y + 6}{2y}$

PROBLEM SET 9.1

A. *Find the value (if any) that makes the rational expression meaningless.*

1. $\dfrac{3}{7x}$

2. $\dfrac{-5}{2y}$

3. $\dfrac{x-2}{3x-9}$

4. $\dfrac{x+4}{2x+8}$

5. $\dfrac{3}{x^2+5}$

6. $\dfrac{-2}{3x^2+8}$

7. $\dfrac{x-2}{x+2}$

8. $\dfrac{x+1}{x-3}$

B. *Solve.*

9. $\dfrac{x}{2}-\dfrac{x}{4}=6$

10. $\dfrac{y-4}{4}=\dfrac{y+8}{16}$

11. $\dfrac{4}{x}+\dfrac{2}{3}=1$

12. $\dfrac{9}{x}=\dfrac{3}{4}$

13. $\dfrac{y}{y+2}=4$

14. $4+\dfrac{2}{x}=3$

15. $\dfrac{-3}{y+4}=\dfrac{2}{y-1}$

16. $\dfrac{6}{2y+1}-\dfrac{1}{y+4}=0$

17. $\dfrac{9}{x}=5-\dfrac{1}{x}$

18. $\dfrac{3}{y}+\dfrac{2}{y}=5$

19. $\dfrac{2x+8}{9}=\dfrac{10x+4}{27}$

20. $\dfrac{x}{3}+\dfrac{x}{9}=\dfrac{5}{6}$

21. $\dfrac{3x}{2}+x=5$

22. $\dfrac{1}{2}+\dfrac{2}{y}=1$

23. $\dfrac{x}{2}-\dfrac{x-1}{4}=\dfrac{5}{4}$

24. $\dfrac{8y}{5}-\dfrac{3y-4}{2}=\dfrac{5}{2}$

25. $\dfrac{y}{y-4}=\dfrac{2}{y-4}+5$

26. $\dfrac{x}{x-2}+4=\dfrac{2}{x-2}$

27. $\dfrac{4}{y-3}+\dfrac{2y}{y^2-9}=\dfrac{1}{y+3}$

28. $\dfrac{x}{x-3}=\dfrac{3}{x-3}+2$

29. $\dfrac{1}{x+5}-\dfrac{3}{x-5}=\dfrac{-10}{x^2-25}$

30. $\dfrac{2}{y-3}-\dfrac{12}{y^2-9}=\dfrac{3}{y+3}$

C. *Add or subtract or solve.*

31. $\dfrac{-7}{x}+\dfrac{3}{4}$

32. $\dfrac{2}{y}-\dfrac{5}{12}$

33. $\dfrac{-7}{x}=\dfrac{3}{4}$

34. $\dfrac{2}{y}=\dfrac{5}{12}$

35. $\dfrac{2}{y+1}=\dfrac{1}{y-2}$

36. $\dfrac{2}{x+3}=\dfrac{5}{x}$

37. $\dfrac{2}{y+1}-\dfrac{1}{y-2}$

38. $\dfrac{2}{x+3}+\dfrac{5}{x}$

39. $\dfrac{y}{8}-\dfrac{y}{12}=\dfrac{1}{8}$

40. $3+\dfrac{5}{x}=-6$

41. $\dfrac{y}{8}-\dfrac{y}{12}+\dfrac{1}{8}$

42. $3+\dfrac{5}{x}-6$

43. To solve a rational equation, first multiply both sides of the equation by the _____ _____ _____.

44. Values of a variable that make the denominator of a rational expression _____ are said to make the rational expression meaningless or undefined.

45. Possible solutions to rational equations that contain a _____ in the denominator must be checked.

46. To _____ or _____ rational expressions, keep the lowest common denominator.

Checkup

The following problems provide a review of some of Section 7.5. Graph.

47. $x + y \geq 3$

48. $x - y < 2$

49. $y < 4x$

50. $y \leq -2x$

51. $2x + 3y > 6$

52. $3x + 4y > 12$

53. $3x - 5y \leq 10$

54. $4x - 3y \geq 9$

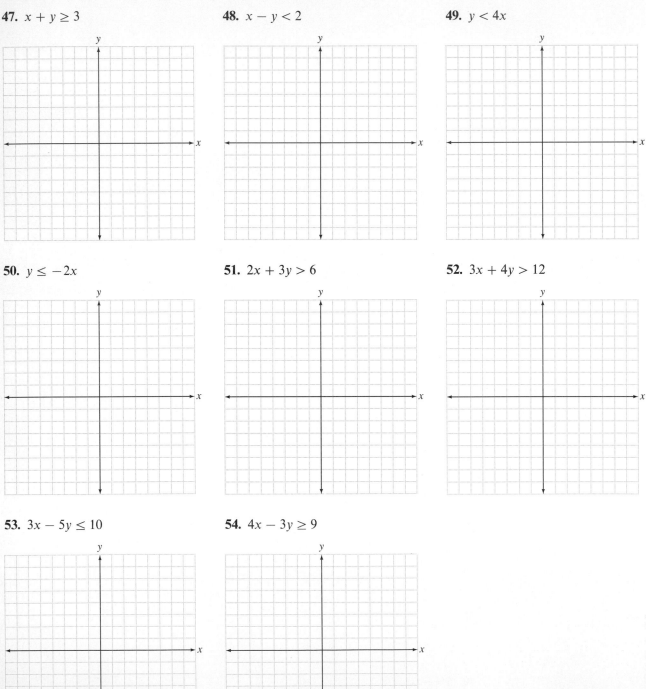

9.2 APPLIED PROBLEMS

1 Rational equations are used to solve many applied problems. Some examples and exercises follow.

Recall that the perimeter of a triangle is the distance around it. If a, b, and c are sides of a triangle, then the perimeter, P, is given by the formula

$$P = a + b + c$$

EXAMPLE 1 If one side of a triangle is one-third the perimeter, the second side is 8 meters, and the third side is two-fifths the perimeter, what is the perimeter of the triangle?

Variable   Let P = the perimeter

Then $\frac{1}{3} P$ = length of one side

8 – length of the second side (not a variable)

$\frac{2}{5} P$ = length of the third side

Drawing

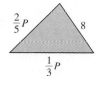

Equation $P = a + b + c$

$$P = \frac{1}{3} P + 8 + \frac{2}{5} P$$

Solve The LCD is 15. Multiply both sides of the equation by 15.

$$15(P) = 15 \left(\frac{1}{3} P + 8 + \frac{2}{5} P \right)$$

$$15P = 15 \left(\frac{1}{3} P \right) + 15(8) + 15 \left(\frac{2}{5} P \right)$$

$$15P = 5P + 120 + 6P$$

$$15P = 11P + 120$$

$$4P = 120$$

$$P = 30$$

The perimeter of the triangle is 30 meters.

Check The sum of the sides, $\frac{1}{3}P + 8 + \frac{2}{5}P = \frac{1}{3}(30) + 8 + \frac{2}{5}(30) = 10 + 8 + 12 = 30$, which is the perimeter. ■

☐ **DO EXERCISE 1.**

OBJECTIVE

1 *Use rational equations to solve applied problems, including work problems*

☐ **Exercise 1** If one side of a triangle is one-fourth the perimeter, the second side is 5 centimeters, and the third side is one-third the perimeter, what is the perimeter?

Rational equations are used to solve work problems.

> The amount of work done, W, is the rate of work, r, times the amount of time, t, spent on the job.
>
> $$W = rt$$

EXAMPLE 2 Cynthia can assemble an engine in 2 hours. Her friend Philip can do the job in 3 hours. How long does it take them both working together to assemble the engine?

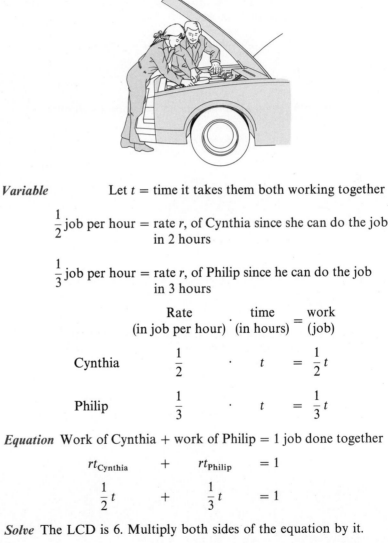

Variable   Let t = time it takes them both working together

$\frac{1}{2}$ job per hour = rate r, of Cynthia since she can do the job in 2 hours

$\frac{1}{3}$ job per hour = rate r, of Philip since he can do the job in 3 hours

	Rate (in job per hour)	$\cdot$	time (in hours)	=	work (job)
Cynthia	$\frac{1}{2}$	$\cdot$	t	=	$\frac{1}{2}t$
Philip	$\frac{1}{3}$	$\cdot$	t	=	$\frac{1}{3}t$

Equation Work of Cynthia + work of Philip = 1 job done together

$$rt_{\text{Cynthia}} \quad + \quad rt_{\text{Philip}} \quad = 1$$

$$\frac{1}{2}t \quad + \quad \frac{1}{3}t \quad = 1$$

Solve The LCD is 6. Multiply both sides of the equation by it.

$$6\left(\frac{1}{2}t + \frac{1}{3}t\right) = 6(1)$$

$$3t + 2t = 6$$

$$5t = 6$$

$$t = \frac{6}{5}$$

410

Working together it takes them $\frac{6}{5}$ hours or $1\frac{1}{5}$ hours to assemble the engine.

Check In $\frac{6}{5}$ hours Cynthia does $\frac{6}{5}(\frac{1}{2}) = \frac{3}{5}$ of the job. Philip does $\frac{6}{5}(\frac{1}{3}) = \frac{2}{5}$ of the job, so together they complete $\frac{3}{5} + \frac{2}{5} = 1$ job. ∎

☐ **DO EXERCISE 2.**

Some distance problems are worked using rational equations.

EXAMPLE 3   Jean traveled 120 miles in one direction. She made the return trip at double the speed and in 3 hours less time. Find her speed going to her destination.

Variable Let r = Jean's speed going to her destination. We know that distance = rate $\times$ time or $d = rt$. However, if we use the equation in this form, we will get two equations in two variables. Since we want to find r, it is easier to solve the equation $d = rt$ for t and eliminate t from the final equation. Since $d = rt$, $d/r = t$ or $d \div r = t$.

$$\left(\begin{array}{c}\text{Distance} \\ \text{in miles}\end{array}\right) \div \left(\begin{array}{c}\text{rate in} \\ \text{miles per} \\ \text{hour}\end{array}\right) = \left(\begin{array}{c}\text{time in} \\ \text{hours}\end{array}\right)$$

Going	120	$\div$	r	$=$	$\dfrac{120}{r}$	Use these facts in the equation
Return	120	$\div$	$2r$	$=$	$\dfrac{120}{2r}$	

Drawing

Equation Since Jean returns in 3 hours less time, we may subtract 3 from the longer time, which is the time going. The equation is

$$\frac{120}{r} - 3 = \frac{120}{2r}$$

Solve The LCD is $2r$. Multiply both sides of the equation by $2r$.

$$2r\left(\frac{120}{r} - 3\right) = 2r\left(\frac{120}{2r}\right)$$

$$240 - 6r = 120$$

$$-6r = -120$$

$$r = 20$$

Jean's speed going to her destination was 20 miles per hour.

Check Jean's speed returning was $2r = 2(20) = 40$ miles per hour. Her time going was $d/r = \frac{120}{20} = 6$ hours. Her time returning was $6 - 3 = 3$ hours. The distance of her return trip was $d = rt = 40(3) = 120$ miles, which is the same as her distance going to her destination. ∎

☐ **DO EXERCISE 3.**

☐ **Exercise 2**   A plumber can complete a job in 12 hours. Another plumber can do the same job in 16 hours. How long would it take the two plumbers to complete the job if they work together?

☐ **Exercise 3**   Joan drives 10 miles per hour faster than Jeff. If Joan travels 330 miles in the same time that Jeff travels 270 miles, find the rate of Joan.

411

1.
$$P = a + b + c$$

$$P = \frac{1}{4}P + 5 + \frac{1}{3}P$$

$$12(P) = 12\left(\frac{1}{4}P + 5 + \frac{1}{3}P\right)$$

$$12P = 12\left(\frac{1}{4}P\right) + 12(5) + 12\left(\frac{1}{3}P\right)$$

$$12P = 3P + 60 + 4P$$
$$12P = 7P + 60$$
$$5P = 60$$
$$P = 12 \qquad \text{The perimeter is 12 centimeters.}$$

2. Let t = time to do the job together

$\dfrac{1}{12}$ job per hour = rate of first plumber

$\dfrac{1}{16}$ job per hour = rate of second plumber

$$\begin{pmatrix} \text{Work of the} \\ \text{first plumber} \end{pmatrix} + \begin{pmatrix} \text{Work of the} \\ \text{second plumber} \end{pmatrix} = \begin{pmatrix} 1 \text{ job done} \\ \text{together} \end{pmatrix}$$

$$\frac{1}{12}t \quad + \quad \frac{1}{16}t \quad = \quad 1$$

LCD = 48

$$48\left(\frac{1}{12}t + \frac{1}{16}t\right) = 48(1)$$

$$48\left(\frac{1}{12}t\right) + 48\left(\frac{1}{16}t\right) = 48$$

$$4t + 3t = 48$$
$$7t = 48$$
$$t = \frac{48}{7}$$

They complete the job together in $\frac{48}{7}$ hours or $6\frac{6}{7}$ hours.

3. Let r = Joan's rate
$$d \div r = t$$

Joan $\qquad 330 \div r = \dfrac{330}{r}$

Jeff $\quad 270 \div (r - 10) = \dfrac{270}{r - 10}$

$$\frac{330}{r} = \frac{270}{r - 10}$$

The LCD is $r(r - 10)$.

$$r(r - 10)\frac{330}{r} = r(r - 10)\frac{270}{r - 10}$$

$$(r - 10)330 = 270r$$
$$330r - 3300 = 270r$$
$$60r = 3300$$
$$r = 55 \qquad \text{Joan's rate is 55 miles per hour.}$$

PROBLEM SET 9.2

1. If one side of a triangle is $\frac{1}{2}$ the perimeter, the second side is 3 meters, and the third side is $\frac{1}{5}$ the perimeter, find the perimeter of the triangle.

2. Find the dimensions of a rectangle with perimeter 56 centimeters if its width is five-ninths its length.

3. Five more than one-sixth a number is $\frac{2}{3}$. Find the number.

4. Eight more than $\frac{2}{3}$ a number is 5 more than $\frac{1}{6}$ the number. Find the number.

5. Susan can correct papers in 2 hours and Mark can correct them in 3 hours. How long will it take them to correct the papers together?

6. David can clean the house in 4 hours and Linda can clean it in 5 hours. How long does it take them to clean the house together?

7. A painter can paint a room in 5 hours and another painter can paint it in 6 hours. How long does it take them to paint the room together?

8. A cold water pipe can fill a swimming pool in 12 hours and a hot water pipe can fill the pool in 15 hours. How long will it take to fill the pool if both pipes are left open?

9. Kara can weed a garden in 6 hours and Kevin can weed it in 7 hours. How long does it take them to weed the garden together?

10. A plumber can do a certain job in 8 hours and another plumber can do the job in 12 hours. How long would it take them to do the job together?

11. Karen drives 20 miles per hour slower than Charles. If Karen travels 45 miles while Charles travels 63 miles, how fast is Charles going?

12. One train goes 10 miles per hour faster than another train. If the faster train goes 500 miles in the same time that the second train goes 450 miles, find the speed of each train.

13. The speed of a car is 15 kilometers per hour slower than the speed of a train. The car travels 130 kilometers in the time it takes the train to travel 160 kilometers. Find the speed of the car.

14. A train travels 40 miles per hour faster than a car. While the car goes 30 miles, the train travels 70 miles. Find the speed of the train.

15. Marilyn traveled 150 miles in one direction. She made the return trip at double the speed and it took 2 hours less time. Find her speed going to her destination.

16. A student traveled 180 miles by train. He returned on an express train at three times the speed in 3 hours less time. What was the speed of the express train?

17. The amount of work done on a job is the _____ of work times the amount of time spent on the job.

Checkup

The following problems provide a review of some of Section 8.1. Are the following rational expressions?

18. $\dfrac{x}{x^2 + 4}$

19. $\dfrac{x + 3}{x - 5}$

20. $\dfrac{x}{x^{-3} + 2}$

21. $\dfrac{x^3 - x + 4}{x^2 - 3}$

22. $\dfrac{1}{3x}$

23. $\dfrac{x^{-4} + 7}{3x}$

Evaluate for the given value of the variable.

24. $\dfrac{x}{3x - 4}, \quad x = 1$

25. $\dfrac{3x + 2}{x^2}, \quad x = -2$

26. $\dfrac{x - 2}{x^2 + 3x - 4}, \quad x = -3$

27. $\dfrac{-x + 5}{2x^2 - 4x + 1}, \quad x = 4$

Ratio of two quantities the quotient of two quantities.

Proportion formed by setting two ratios equal to one another.

Direct variation quantities are directly proportional. Either one quantity increases as the other quantity increases or one quantity decreases as the other quantity decreases.

9.3 RATIO, PROPORTION, AND VARIATION

Ratio

A **ratio** of two quantities is their quotient.

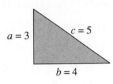

EXAMPLE 1 Consider the triangle shown.

a. The ratio of side a to side b is $\frac{3}{4}$. This ratio may also be written 3:4, which is read "3 to 4."

b. The ratio of side b to side a is $\frac{4}{3}$.

c. The ratio of side b to side c is $\frac{4}{5}$. ■

□ **DO EXERCISE 1.**

Proportion

> A **proportion** is formed by setting two ratios equal to each other.

The numerator or denominator of one of the ratios may be unknown. The following are examples of proportions.

$$\frac{4}{16} = \frac{1}{4} \qquad \frac{15}{x} = \frac{3}{1} \qquad \frac{y}{6} = \frac{1}{2}$$

1 Direct Variation

Quantities are directly proportional if one quantity increases as the other quantity increases or if one quantity decreases as the other quantity decreases. This is called **direct variation**. In the following example, as the number of hours worked goes up, the earnings go up. These quantities are directly proportional.

□ **Exercise 1** Find the ratios for the triangle shown.

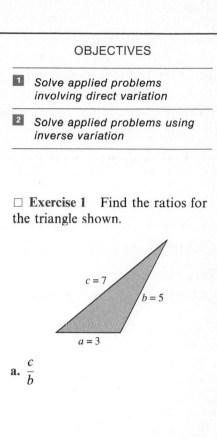

a. $\dfrac{c}{b}$

b. $\dfrac{a}{c}$

Inverse variation   quantities are inversely proportional. One quantity increases as the other quantity decreases.

□ **Exercise 2**   If 4 days at a hotel cost $225, find the cost to stay 7 days.

EXAMPLE 2   If Tracy works 35 hours, she earns $140. How much does she earn if she works 50 hours if her hourly rate does not change?

Variable   Let x = Tracy's earnings for working 50 hours. To solve the problem we write a proportion by setting two ratios equal to each other. The ratios are formed by comparing like quantities. We compare hours to hours and dollars to dollars. The ratios are

$$\frac{35 \text{ hours}}{50 \text{ hours}} \qquad \frac{\$140}{x}$$

Notice that 35 hours is related to $140, so both of these quantities should appear in either the numerators or the denominators of the proportion. We chose to put them in the numerators.

Equation $\dfrac{35}{50} = \dfrac{140}{x}$

Solve   The LCD is $50x$. Multiply both sides of the equation by it.

$$50x \left(\frac{35}{50}\right) = 50x \left(\frac{140}{x}\right)$$

$$35x = 7000$$

$$x = 200$$

Tracy earns $200 for working 50 hours.   ∎

The equation $35/140 = 50/x$ also works for the problem above. However, the type of comparison in Example 2 is necessary to work the inverse variation exercises of the next section correctly.

□ **DO EXERCISE 2.**

2 Inverse Variation

If quantities are inversely proportional, one quantity increases as the other quantity decreases. This is called **inverse variation**.

EXAMPLE 3   If a car takes 25 minutes to travel a certain distance at a rate of 45 miles per hour, how long does it take to travel the same distance at 60 miles per hour?

These quantities are inversely proportional. As the speed of the car increases, the time it takes to travel a given distance decreases.

Variable Let $x =$ the time it takes to travel the distance at 60 mph. The ratios are as follows.

$$\frac{25 \text{ minutes}}{x} \quad \text{and} \quad \frac{45 \text{ miles}}{60 \text{ miles}}$$

Since these quantities are inversely proportional, one of the ratios must be inverted to form the equation. When we invert the second ratio, we get the following.

Equation $\qquad \dfrac{25}{x} = \dfrac{60}{45}$

Solve $\qquad 45x\left(\dfrac{25}{x}\right) = 45x\left(\dfrac{60}{45}\right)$

$$1125 = 60x$$

$$18\frac{3}{4} = x$$

It takes the car $18\frac{3}{4}$ minutes to travel the distance at 60 mph. ■

□ **DO EXERCISE 3.**

We may know whether the quantities are directly or inversely proportional to each other. If the variation is direct, the quantities are directly proportional. If we have an inverse variation, the quantities are inversely proportional.

EXAMPLE 4 The number of rings a craftsman can make varies directly as the amount of time he works. If he can make 18 rings in 35 hours, how many rings can he make in 10 hours?

Variable Let $x =$ the number of rings he can make in 10 hours.

Equation $\qquad \dfrac{18}{x} = \dfrac{35}{10}$

$$10x\left(\dfrac{18}{x}\right) = 10x\left(\dfrac{35}{10}\right)$$

$$180 = 35x$$

$$5\frac{1}{7} = x$$

He can make $5\frac{1}{7}$ rings in 10 hours. ■

□ **DO EXERCISE 4.**

□ **Exercise 3** If 4 men can wash the windows on a building in 9 hours, how long will it take 7 men to do the job?

□ **Exercise 4** The pressure exerted by a liquid at a given point varies directly as the depth of the point beneath the surface of the liquid. If a liquid exerts a pressure of 300 pounds per square foot at a depth of 5 feet, what is the pressure at 30 feet?

☐ **Exercise 5** The length of a string and its fundamental frequency of vibration are inversely proportional. If a string of length 30 centimeters gives a fundamental frequency of 200 cycles per second, what is fundamental frequency of a string of length 12 centimeters?

EXAMPLE 5 The length of a rectangle of a given area varies inversely with the width. What is the width of a rectangle of length 6 centimeters if it has the same area as a rectangle with length 8 centimeters and width 3 centimeters?

Variable Let $x =$ the width of the rectangle

Equation

$$\frac{6}{8} = \frac{3}{x}$$

$$8x\left(\frac{6}{8}\right) = 8x\left(\frac{3}{x}\right)$$

$$6x = 24$$

$$x = 4$$

The width of the rectangle is 4 centimeters. ∎

☐ **DO EXERCISE 5.**

DID YOU KNOW?

If the sounds of two equally taut plucked strings harmonize, their lengths are related by numerical ratios. The Pythagoreans were the first to discover the relation between the length of a string and the pitch of the note made by touching the string. In about 540 B.C., the Pythagoreans were known for their religious brotherhood as well as for their school of learning.

Answers to Exercises

1. a. $\frac{7}{5}$ b. $\frac{3}{7}$

2. $393.75

3. $5\frac{1}{7}$ hours

4. 1800 lb/ft^2

5. 500 cycles/sec

PROBLEM SET 9.3

A. *Write the following ratios and simplify.*

1. 50 miles to 40 miles

2. 30 kilometers to 75 kilometers

3. 3 feet to 11 feet

4. 15 centimeters to 9 centimeters

B. *Solve.*

5. If six oranges cost $1.00, how much would 14 oranges cost?

6. If Jim works 20 hours, he earns $115. How much does he earn if he works 39 hours?

7. A jogger can run 4 miles in 45 minutes. How far can she jog in 30 minutes?

8. If a car travels 100 miles on 8 gallons of gasoline, how far will it go on 15 gallons?

9. If a car takes 20 minutes to travel a certain distance at 50 miles per hour, at what speed is it traveling if it covers the same distance in 25 minutes? (This is an inverse variation.)

10. If 2 students can weed the flower beds in 15 hours, how long will it take 5 students to do the job?

11. If it takes 3 pipes 4 hours to fill a tank, how many hours does it take 2 pipes of the same diameter to fill the tank?

12. If 2 people can clean a house in 7 hours, how many hours does it take 5 people to clean the house?

13. If a scientist knows that a 2.5-meter piece of steel rod weighs 21 kilograms, how much does a 6-meter piece of the rod weigh?

14. If a string of length 40 centimeters gives a fundamental frequency of 250 cycles per second, what length of string gives a fundamental frequency of 200 cycles per second?

15. If 2 people can type a document in 9 hours, how many hours will it take for 5 people to type the document?

16. If there are 7 grams of acid in 50 grams of solution, how many grams of acid are in 35 grams of solution?

17. The number of servings that can be obtained from a roast varies directly with the weight of the roast. If a 4-pound roast yields 9 servings, how many servings can be obtained from a 3-pound roast?

18. Hooke's law states that the force required to stretch a spring x inches beyond its natural length varies directly with x. If a 60-pound force stretches a spring 3 inches, how much force is required to stretch it 2 inches?

19. Assume that y varies directly as x. If $y = 12$ when $x = 20$, find y when $x = 30$.

20. Assume that y varies directly as x. If $y = 10$ when $x = 25$, find x when $y = 6$.

21. At a constant temperature the volume of a gas varies inversely as the pressure. If the pressure of a gas in a 200-cubic-foot container is 24 pounds per square inch, what is the pressure of the gas if it is expanded to 500 cubic feet?

22. If the voltage is constant, the current in an electrical circuit varies inversely as the resistance of the circuit. The current is 3.5 amperes when the resistance is 25 ohms. Find the current when the resistance is 20 ohms.

23. If y varies inversely with x, and $y = 30$ when $x = 14$, find y when $x = 28$.

24. If y varies inversely with x, and $x = 25$ when $y = 6$, find x when $y = 15$.

25. The number of pies a baker can make varies directly with the number of hours he works. If he can make 7 pies in 2 hours, how many pies can he make in 8 hours?

26. The fundamental frequency of vibration of a string varies inversely with the length of the string. A string of 15 cm gives a fundamental frequency of 475 cycles per second. What is the length of a string when it has a fundamental frequency of 300 cycles per second?

27. A ratio of two quantities is their _____.

28. A proportion is formed by setting two _____ equal to each other.

29. If two quantities are directly proportional, this is _____ variation.

30. Inverse variation occurs when two quantities are inversely _____.

Checkup

The following problems provide a review of some of Section 8.2. These problems will help you with the next section.
Simplify.

31. $\dfrac{15x^2y^2}{25x^3y}$

32. $\dfrac{16xy^3}{4xy^2}$

33. $\dfrac{3x^2 + 6x}{4x + 8}$

34. $\dfrac{7x}{14x^2 + 21x}$

35. $\dfrac{x^2 - 9}{2x^2 - x - 15}$

36. $\dfrac{(3x - 4)^2}{6x^2 - 5x - 4}$

37. $\dfrac{x^2 - 16}{4 - x}$

38. $\dfrac{y^2 - 25}{-y - 5}$

Complex fraction the numerator, denominator or both of these parts of the fraction contains fractions.

9.4 COMPLEX FRACTIONAL EXPRESSIONS

If the numerator or denominator of a fraction contains fractions, the fraction is called a **complex fraction**. Following are examples of *complex fractional expressions*.

$$\frac{\dfrac{2}{x}+3}{4} \qquad \frac{5}{\dfrac{x}{3}+\dfrac{3x}{2}} \qquad \frac{\dfrac{4}{x}+7}{\dfrac{8}{x}-\dfrac{x}{6}}$$

■ We want to simplify a complex fractional expression by removing all fractions in both the numerator and the denominator. A method of doing this is to add or subtract all terms in both the numerator and denominator. Then divide the numerator by the denominator.

EXAMPLE 1 Simplify by removing all fractions in the numerator and denominator of the complex fraction.

$$\frac{\dfrac{3}{x}+4}{\dfrac{5}{8}}$$

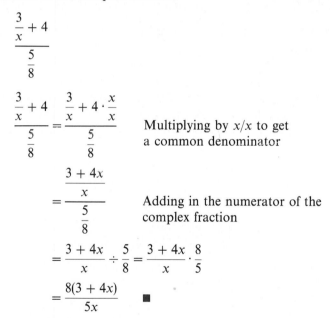

$$\frac{\dfrac{3}{x}+4}{\dfrac{5}{8}}=\frac{\dfrac{3}{x}+4\cdot\dfrac{x}{x}}{\dfrac{5}{8}}$$ Multiplying by x/x to get a common denominator

$$=\frac{\dfrac{3+4x}{x}}{\dfrac{5}{8}}$$ Adding in the numerator of the complex fraction

$$=\frac{3+4x}{x}\div\frac{5}{8}=\frac{3+4x}{x}\cdot\frac{8}{5}$$

$$=\frac{8(3+4x)}{5x} \qquad ■$$

☐ **DO EXERCISE 1.**

Sometimes there is a common denominator in both the numerator and denominator. In that case we can immediately simplify by dividing the numerator by the denominator.

OBJECTIVE

■ *Simplify complex fractional expressions*

☐ **Exercise 1** Simplify by removing all fractions in the numerator and denominator of the complex fraction.

a. $\dfrac{x+\dfrac{4}{x}}{\dfrac{3}{7}}$

b. $\dfrac{\dfrac{3}{y}}{\dfrac{1}{y}+5y}$

□ **Exercise 2** Simplify by removing all fractions in the numerator and denominator of the complex fraction.

a. $\dfrac{\dfrac{x}{y+1}}{\dfrac{x}{y}}$

b. $\dfrac{\dfrac{xy}{z}}{\dfrac{x^2y}{z^2}}$

□ **Exercise 3** Simplify by removing all fractions in the numerator and denominator of the complex fraction.

$$\dfrac{\dfrac{3y}{5}+\dfrac{y}{2}}{y+\dfrac{y}{5}}$$

EXAMPLE 2 Simplify by removing all fractions in the numerator and denominator of the complex fraction.

$$\dfrac{\dfrac{1}{y}}{\dfrac{1+y}{y}}$$

$$\dfrac{\dfrac{1}{y}}{\dfrac{1+y}{y}}=\dfrac{1}{y}\div\dfrac{1+y}{y}$$

$$=\dfrac{1}{y}\cdot\dfrac{y}{1+y}=\dfrac{y\cdot 1}{y(1+y)}$$

$$=\dfrac{1}{1+y}\quad\blacksquare$$

□ **DO EXERCISE 2.**

EXAMPLE 3 Simplify by removing all fractions in the numerator and denominator of the complex fraction.

$$\dfrac{\dfrac{3}{x}+\dfrac{1}{3x}}{x+\dfrac{x}{3}}$$

$$\dfrac{\dfrac{3}{x}+\dfrac{1}{3x}}{x+\dfrac{x}{3}}=\dfrac{\dfrac{3}{x}\cdot\dfrac{3}{3}+\dfrac{1}{3x}}{x\cdot\dfrac{3}{3}+\dfrac{x}{3}}=\dfrac{\dfrac{9+1}{3x}}{\dfrac{3x+x}{3}}=\dfrac{\dfrac{10}{3x}}{\dfrac{4x}{3}}$$

$$=\dfrac{10}{3x}\div\dfrac{4x}{3}=\dfrac{10}{3x}\cdot\dfrac{3}{4x}=\dfrac{2\cdot 5}{3\cdot x}\cdot\dfrac{3}{2\cdot 2\cdot x}$$

$$=\dfrac{2\cdot 5\cdot 3}{2\cdot 2\cdot 3\cdot x\cdot x}=\dfrac{2\cdot 3\cdot 5}{2\cdot 3\cdot 2\cdot x^2}$$

$$=\dfrac{5}{2x^2}\quad\blacksquare$$

□ **DO EXERCISE 3.**

© 1989 by Prentice Hall

EXAMPLE 4 Simplify by removing all fractions in the numerator and denominator of the complex fraction.

$$\frac{8 + \dfrac{4}{x}}{\dfrac{2x + 1}{3}}$$

$$\frac{8 + \dfrac{4}{x}}{\dfrac{2x + 1}{3}} = \frac{8 \cdot \dfrac{x}{x} + \dfrac{4}{x}}{\dfrac{2x + 1}{3}} = \frac{\dfrac{8x + 4}{x}}{\dfrac{2x + 1}{3}}$$

$$= \frac{8x + 4}{x} \div \frac{2x + 1}{3} = \frac{4(2x + 1)}{x} \cdot \frac{3}{2x + 1} = \frac{4 \cdot 3(2x + 1)}{x(2x + 1)}$$

$$= \frac{12}{x} \quad \blacksquare$$

☐ **DO EXERCISE 4.**

Complex fractions may also be simplified by multiplying the numerator and denominator by the lowest common denominator of the denominators.

☐ **Exercise 4** Simplify by removing all fractions in the numerator and denominator of the complex fraction.

$$\frac{7 + \dfrac{7}{x}}{1 + \dfrac{1}{x}}$$

EXAMPLE 5 Simplify by removing all fractions in the numerator and denominator of the complex fraction.

$$\frac{8 + \dfrac{4}{x}}{\dfrac{2x + 1}{3}}$$

The LCD is $3x$, so multiply numerator and denominator by $3x$.

$$\frac{8 + \dfrac{4}{x}}{\dfrac{2x + 1}{3}} = \frac{3x\left(8 + \dfrac{4}{x}\right)}{3x\left(\dfrac{2x + 1}{3}\right)} \qquad \text{Multiplying by } 3x$$

$$= \frac{3x(8) + 3x\left(\dfrac{4}{x}\right)}{3x\left(\dfrac{2x + 1}{3}\right)}$$

$$= \frac{3x(8) + 3(4)}{x(2x + 1)}$$

$$= \frac{24x + 12}{x(2x + 1)}$$

$$= \frac{12(2x + 1)}{x(2x + 1)} \qquad \text{Factoring}$$

$$= \frac{12}{x}$$

This is the same answer that we got to this exercise when we simplified it using the method described in Example 4. ■

Answers to Exercises _____

1. a. $\dfrac{7(x^2 + 4)}{3x}$ **b.** $\dfrac{3}{1 + 5y^2}$

2. a. $\dfrac{y}{y + 1}$ **b.** $\dfrac{z}{x}$

3. $\dfrac{11}{12}$

4. 7

PROBLEM SET 9.4

Simplify by removing all fractions from the numerator and denominator of the complex fraction.

1. $\dfrac{\dfrac{1}{x} - 2}{\dfrac{3}{5}}$

2. $\dfrac{\dfrac{5}{9}}{\dfrac{2}{y} + 3}$

3. $\dfrac{\dfrac{3}{x} + 1}{\dfrac{4 + x}{5}}$

4. $\dfrac{x + \dfrac{x}{2}}{\dfrac{x + 2}{x}}$

5. $\dfrac{\dfrac{x + 1}{x}}{\dfrac{x + 1}{y}}$

6. $\dfrac{\dfrac{3}{x}}{\dfrac{5 - x}{x}}$

7. $\dfrac{\dfrac{4}{x + 2}}{\dfrac{12}{x}}$

8. $\dfrac{\dfrac{5}{y}}{\dfrac{3}{y - 1}}$

9. $\dfrac{\dfrac{4}{x} + \dfrac{1}{6x}}{2x + \dfrac{x}{2}}$

10. $\dfrac{\dfrac{5}{x} - \dfrac{1}{2x}}{\dfrac{x}{4} - x}$

11. $\dfrac{1 + \dfrac{1}{y}}{3 + \dfrac{3}{y}}$

12. $\dfrac{\dfrac{5}{x} - 5}{\dfrac{1}{x} - 1}$

13. $\dfrac{x + \dfrac{1}{y}}{\dfrac{1}{x} + y}$

14. $\dfrac{\dfrac{x + y}{x}}{\dfrac{1}{x} + \dfrac{1}{y}}$

15. $\dfrac{\dfrac{x}{x + y}}{\dfrac{x}{x^2 - y^2}}$

16. $\dfrac{\dfrac{x}{x+1}}{\dfrac{3}{x^2-1}}$

17. $\dfrac{\dfrac{2}{x+y}}{\dfrac{4}{x^2-y^2}}$

18. $\dfrac{\dfrac{3}{x-2}}{\dfrac{6}{x^2-x-2}}$

19. If the numerator or denominator of a fraction contains fractions, the fraction is called a _____ fraction.

20. If there is a _____ _____ for the numerator and denominator of a complex fraction, the fraction can be simplified by dividing the numerator by the denominator.

21. Complex fractions may be simplified by _____ the numerator and denominator by the lowest common denominator of the denominators.

Checkup

The following problems provide a review of some of Section 8.3.
Write answers with positive exponents.
Multiply.

22. $\dfrac{5x^2}{y^2} \cdot \dfrac{y^3}{25x^4}$

23. $\dfrac{3x}{x^2y^5} \cdot \dfrac{x^3y^2}{9x}$

24. $\dfrac{3x+9}{3x+12} \cdot \dfrac{x+4}{2x^2+11x+15}$

25. $\dfrac{2x-3}{6x^2-13x+6} \cdot \dfrac{6x-4}{5x+10}$

Divide.

26. $\dfrac{x^2}{6y^3} \div \dfrac{y^3}{3x^4}$

27. $\dfrac{7x}{3y} \div \dfrac{x^5}{14y^2}$

28. $\dfrac{x^2-4}{3x^2+x-14} \div \dfrac{x+2}{6x+12}$

29. $\dfrac{x-1}{2x+8} \div \dfrac{2x^2-2}{x^2+3x-4}$

30. $\dfrac{x^2+x-20}{2x^2+13x+15} \div 3x^2-8x-16$

31. $(3y-3) \div \dfrac{6y^2-5y-1}{12y^2+8y+1}$

CHAPTER 9   ADDITIONAL EXERCISES (OPTIONAL)

Section 9.1

Find the value that makes the rational expression meaningless.

1. $\dfrac{4}{2x}$
2. $\dfrac{-7}{5y}$
3. $\dfrac{7}{2x + 3}$
4. $\dfrac{8x}{4x - 9}$
5. $\dfrac{5x + 2}{3x^2 + 4}$
6. $\dfrac{7x - 8}{6x^2 + 3}$

Solve.

7. $\dfrac{8}{x} = 7 - \dfrac{6}{x}$

8. $\dfrac{4}{y} + \dfrac{6}{y} = 2$

9. $\dfrac{-2}{y - 3} = \dfrac{6}{y - 15}$

10. $\dfrac{3}{2x - 1} - \dfrac{6}{x - 8} = 0$

11. $\dfrac{2}{x + 4} + \dfrac{1}{x^2 - 16} = \dfrac{-1}{x - 4}$

12. $\dfrac{1}{x - 3} - \dfrac{4}{x - 4} = \dfrac{x}{x^2 - 7x + 12}$

Combine terms or solve.

13. $\dfrac{3}{x - 2} + \dfrac{4x}{x^2 - 4} - \dfrac{2}{x + 2}$

14. $\dfrac{2}{x - 3} - \dfrac{3}{x + 3} = \dfrac{12}{x^2 - 9}$

15. $\dfrac{3}{x^2 + 5x + 6} = \dfrac{5}{x^2 + 2x - 3} - \dfrac{1}{x^2 + x - 2}$

16. $\dfrac{4}{y^2 - 3y + 2} - \dfrac{5}{y^2 - 4y + 3} = \dfrac{-4}{y^2 - 5y + 6}$

Section 9.2

17. Find the dimensions of a rectangle with perimeter 144 meters if its length is $\frac{7}{2}$ of its width.

18. The reciprocal of 3 more than a number is twice the reciprocal of the number. Find the number.

19. Laura can clean the carpets in 2 hours and Sarah can clean them in 3 hours. How long does it take them to clean them together?

20. Sean can wash the windows in 2 hours and Matt can wash them in 4 hours. How long does it take them to wash them together?

21. It takes two painters 3 hours to paint a room together. If one painter can paint the room alone in 5 hours, how long does it take the other painter to paint the room alone?

22. Ken and John can mow the lawn together in 2 hours. If it takes Ken 3 hours to mow the lawn alone, how long does it take John to mow it alone?

23. Susan traveled 140 miles in one direction. She made the return trip at double the speed and in 2 hours less time. Find Susan's speed returning from her destination.

24. Greg took a 120-mile scenic train ride. He returned by express train at three times the speed in 2 hours less time. What was the speed of the express train?

25. Assume that y varies directly as x. If $y = 12$ when $x = 15$, find y when $x = 40$.

26. Assume that y varies directly as x. If $y = 9$ when $x = 21$, find y when $x = 28$.

27. A paint store mixes red paint with blue paint by using 3 parts of red paint to 2 parts of blue paint. If they use 5 quarts of blue paint, how much red paint should they use?

28. If y varies inversely with x and $y = 20$ when $x = 14$, find y when $x = 35$.

29. If y varies inversely with x and $x = 32$ when $y = 4$, find x when $y = 8$.

30. If Juan drives 45 miles per hour, he gets to work in 20 minutes. How fast must he drive to get to work in 15 minutes?

31. A 12-tooth gear with a speed of 450 revolutions per minute (rpm) is in mesh with a gear with a speed of 675 rpm. How many teeth must the gear with the speed of 675 rpm have? (This is an inverse variation.)

Section 9.4

Simplify.

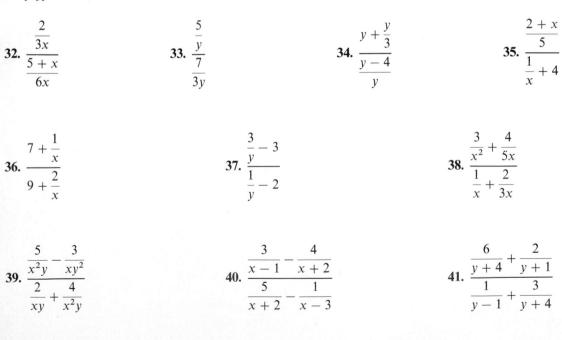

32. $\dfrac{\dfrac{2}{3x}}{\dfrac{5+x}{6x}}$

33. $\dfrac{\dfrac{5}{y}}{\dfrac{7}{3y}}$

34. $\dfrac{y + \dfrac{y}{3}}{\dfrac{y-4}{y}}$

35. $\dfrac{\dfrac{2+x}{5}}{\dfrac{1}{x} + 4}$

36. $\dfrac{7 + \dfrac{1}{x}}{9 + \dfrac{2}{x}}$

37. $\dfrac{\dfrac{3}{y} - 3}{\dfrac{1}{y} - 2}$

38. $\dfrac{\dfrac{3}{x^2} + \dfrac{4}{5x}}{\dfrac{1}{x} + \dfrac{2}{3x}}$

39. $\dfrac{\dfrac{5}{x^2 y} - \dfrac{3}{xy^2}}{\dfrac{2}{xy} + \dfrac{4}{x^2 y}}$

40. $\dfrac{\dfrac{3}{x-1} - \dfrac{4}{x+2}}{\dfrac{5}{x+2} - \dfrac{1}{x-3}}$

41. $\dfrac{\dfrac{6}{y+4} + \dfrac{2}{y+1}}{\dfrac{1}{y-1} + \dfrac{3}{y+4}}$

CHAPTER 9   PRACTICE TEST

Solve.

1. $\dfrac{5}{x} = \dfrac{15}{4 - x}$

1. _____

2. $\dfrac{x}{x - 2} = \dfrac{2}{x - 2} + 2$

2. _____

3. Add.

$$\frac{3}{x} + \frac{5}{x + 4}$$

3. _____

4. Joan can mow the lawn in 3 hours. Jean can mow the same lawn in 2 hours. How long will it take them to mow the lawn together?

4. _____

5. One car travels 20 miles per hour faster than another car. If the slower car travels 150 miles in the same time the faster car travels 200 miles, find the speed of the slower car.

5. _____

6. _____

6. If Ken works 26 hours, he earns $117. How much does he earn if he works 18 hours?

7. _____

7. Assume that y varies inversely as x. If $y = 16$ when $x = 3$, find y when $x = 15$.

Simplify.

8. _____

8. $\dfrac{\dfrac{5}{x}}{\dfrac{3+x}{x+1}}$

9. _____

9. $\dfrac{\dfrac{x-y}{x}}{\dfrac{1}{y}-\dfrac{1}{x}}$

Radical Expressions

10

10.1 SQUARE ROOTS

When we square a number, we multiply it by itself. Hence the square of 6 is

$$6 \cdot 6 = 36$$

We want to reverse the process. We want to find a number that when multiplied by itself gives 36. One answer is 6. It is called a square root of 36.

> The number c is a square root of b if $c^2 = b$.

1 Principal Square Roots

> The symbol $\sqrt{}$ is called a **radical sign** and is used to represent the positive square root of a number.

This is called the **principal square root**. We write

$$\sqrt{36} = 6$$

If we want to write a negative number whose square is 36, we use $-\sqrt{36}$ or -6. The number under the radical sign is called the **radicand**. The entire expression, including radical sign and radicand, is called a **radical**. For example, $\sqrt{7}$, $\sqrt{x}$, and $\sqrt{x^2 - 2}$ are radicals.

EXAMPLE 1   Find the square root.

a. $\sqrt{16} = 4$ Since $4 \cdot 4 = 16$

b. $-\sqrt{16} = -4$ Since $\sqrt{16} = 4$

c. $\sqrt{64} = 8$ Since $8 \cdot 8 = 64$

d. $-\sqrt{100} = -10$ Since $\sqrt{100} = 10$

e. $\sqrt{0} = 0$ Since $0 \cdot 0 = 0$ ■

□ DO EXERCISE 1.

1 *Find principal square roots of numbers*

2 *Identify rational and irrational numbers in radical, decimal, or fractional form*

3 *Use a table to approximate square roots of numbers*

□ **Exercise 1**   Find the square root.

a. $\sqrt{9}$ **b.** $-\sqrt{9}$

c. $\sqrt{25}$ **d.** $-\sqrt{4}$

e. $-\sqrt{49}$ **f.** $\sqrt{100}$

□ **Exercise 2**   Are the following square roots defined?

a. $\sqrt{-81}$

b. $-\sqrt{81}$

c. $\sqrt{-64}$

d. $-\sqrt{9}$

□ **Exercise 3**   Find the square root.

a. $\sqrt{y^2}$

b. $\sqrt{(3x)^2}$

c. $\sqrt{a^2 b^2}$

d. $\sqrt{(ab)^2}$

e. $\sqrt{(x+1)^2}$

f. $\sqrt{(-7)^2}$

Negative numbers do not have square roots in the real number system. We can not find any real number which multiplied times itself gives a negative number. For purposes of this book, we say that square roots of negative numbers are not defined. They will be defined in a more advanced mathematics course.

EXAMPLE 2   Are the following square roots defined?

a. $\sqrt{-16}$ No; this is the square root of a negative number

b. $-\sqrt{36}$ Yes; since $\sqrt{36} = 6$, $-\sqrt{36} = -6$; notice that the negative sign is not under the radical sign

c. $\sqrt{-100}$ No   ■

□ **DO EXERCISE 2.**

We may also find square roots of variables. Consider the problem

$$\sqrt{x^2} = ?$$

The radical sign indicates that we must find the positive square root which means that x is not an acceptable answer because x may be negative. For example,

$$\sqrt{(-2)^2} = \sqrt{4} = 2 \neq -2$$

We use the absolute value sign to indicate the positive answer. Hence

$$\sqrt{x^2} = |x|$$

EXAMPLE 3   Find the square root.

a. $\sqrt{(-2)^2} = |-2| = 2$

b. $\sqrt{x^2 y^2} = |xy|$

c. $\sqrt{x^2 + 4x + 4} = \sqrt{(x+2)^2} = |x+2|$

The absolute value of a product is the product of the absolute values.

$$|x \cdot y| = |x| \cdot |y| \qquad \text{for any numbers } x \text{ and } y$$

d. $\sqrt{(5y)^2} = |5y| = |5||y| = 5|y|$ ■

□ **DO EXERCISE 3.**

2 Rational and Irrational Numbers

Whole numbers such as 1, 4, 9, and 16 that have integers for square roots are called **perfect squares**. However, most whole numbers are not perfect squares. Hence their square roots are not integers, they are *rational numbers*. For example, $\sqrt{5}$ is not a rational number. We say that $\sqrt{5}$ is irrational.

An **irrational number** is a number that cannot be written in the form a/b, where a and b are integers and b is not zero.

EXAMPLE 4   Are the following numbers rational or irrational?

a. $\sqrt{7}$ Irrational; 7 is not a perfect square

b. $\sqrt{36}$ Rational; $\sqrt{36} = 6$, which is a rational number

c. $\sqrt{10}$ Irrational; 10 is not a perfect square

d. $\sqrt{49}$ Rational; $\sqrt{49} = 7$, which is a rational number and 49 is a perfect square   ■

□ **DO EXERCISE 4.**

When we write a rational number as a decimal number, the decimal number either repeats or terminates. When we write an irrational number as a decimal number, it does not repeat and it does not terminate.

EXAMPLE 5   Are the following rational or irrational?

a. $\dfrac{3}{4} = 0.75$ Rational; the decimal number terminates

b. $\sqrt{6} = 2.4494897\ldots$ Irrational; the decimal number does not repeat and it does not terminate

c. $\dfrac{1}{3} = 0.333333\ldots$ Rational; this a repeating decimal number

d. $\sqrt{17} = 4.1231056\ldots$ Irrational; the decimal number does not repeat and it does not terminate

e. $\sqrt{16} = 4$ Rational   ■

□ **DO EXERCISE 5.**

Square roots of numbers that are *not* perfect squares or ratios of perfect squares are irrational and π is also irrational. If we plot all the rational numbers and all the irrational numbers on the number line, we fill up the number line. There is a point on the number line for each rational and irrational number. These make up the real numbers.

> The set of real numbers is composed of the rational numbers and the irrational numbers.

Recall the following chart from Chapter 1, which shows the relationship of the subsets of the real numbers. The natural numbers are a subset of the whole numbers. Similarly, the whole numbers are a subset of the integers, the integers are a subset of the rational numbers, and the rational numbers are a subset of the real numbers. The irrational numbers are also a subset of the real numbers.

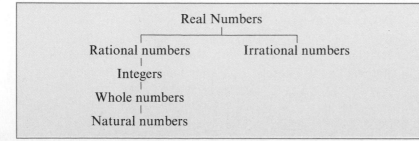

□ **Exercise 4**   Are the following numbers rational or irrational?

a. $\sqrt{3}$ **b.** $\sqrt{16}$

c. $\sqrt{64}$ **d.** $\sqrt{27}$

□ **Exercise 5**   Are the following rational or irrational?

a. $\dfrac{7}{8} = 0.875$

b. $\dfrac{1}{9} = 0.1111\ldots$

c. $0.707106\ldots$
does not repeat

d. $3.316624\ldots$
does not repeat

□ **Exercise 6** Use the square root table to find the following.

a. $\sqrt{8}$

b. $-\sqrt{19}$

c. $-\sqrt{75}$

d. $\sqrt{50}$

3 Approximating Square Roots

We know that since 3 is not a perfect square, its square root is not a rational number. However, we can approximate the irrational number which is its square root. We do this by using the $\sqrt{}$ key on our calculators or by referring to the square root table in the back of the book.

EXAMPLE 6 Use the square root table to find the following.

a. $\sqrt{3} \approx 1.732$ Rounded to three decimal places

b. $-\sqrt{27} \approx -5.196$ Rounded to three decimal places ■

□ **DO EXERCISE 6.**

DID YOU KNOW?

The Pythagoreans based their mathematical and religious philosophy on the properties of whole numbers. When the Pythagoreans discovered that not all numbers were whole numbers, they tried to keep the discovery secret. According to legend, at least one member of the brotherhood was banished and mourned as dead after revealing this secret.

Answers to Exercises _____

1. **a.** 3   **b.** −3   **c.** 5   **d.** −2   **e.** −7   **f.** 10
2. **a.** No   **b.** Yes   **c.** No   **d.** Yes
3. **a.** $|y|$ **b.** $3|x|$ **c.** $|ab|$ **d.** $|ab|$ **e.** $|x + 1|$ **f.** 7
4. **a.** Irrational   **b.** Rational   **c.** Rational   **d.** Irrational
5. **a.** Rational   **b.** Rational   **c.** Irrational   **d.** Irrational
6. **a.** 2.828   **b.** −4.359   **c.** −8.660   **d.** 7.071

PROBLEM SET 10.1

A. *Find the square roots.*

1. $\sqrt{4} = 2$

2. $\sqrt{16} = 4$

3. $-\sqrt{64} = -8$

4. $-\sqrt{81} = -9$

5. $\sqrt{1} = 1$

6. $\sqrt{100} = 10$

7. $\sqrt{36} = 6$

8. $-\sqrt{1} = -1$

9. $\sqrt{0} = 0$

10. $\sqrt{144} = 12$

11. $\sqrt{225} = 15$

12. $\sqrt{169} = 13$

B. *Are the following square roots defined?*

13. $\sqrt{100}$ Yes.

14. $-\sqrt{81}$ yes

15. $\sqrt{-4}$ No

16. $-\sqrt{9}$ No yes

17. $\sqrt{-16}$ yes

18. $\sqrt{-25}$ No

19. $\sqrt{1}$ yes

20. $\sqrt{-1}$ No

C. *Find the square roots.*

21. $\sqrt{x^2} = x$

22. $\sqrt{b^2} = b$

23. $\sqrt{16x^2} = 4x$

24. $\sqrt{25a^2} = 5a$

25. $\sqrt{(7ab)^2} = 7|ab|$

26. $\sqrt{(3xy)^2} = 3|xy|$

27. $\sqrt{(-6b)^2} =$

28. $\sqrt{(-xy)^2}$

29. $\sqrt{(x-3)^2}$

30. $\sqrt{(x+2)^2}$

31. $\sqrt{x^2 + 6x + 9}$

32. $\sqrt{x^2 - 8x + 16}$

D. *Are the following rational or irrational numbers?*

33. $\sqrt{8}$

34. $\sqrt{6}$

35. $\sqrt{25}$

36. $\sqrt{81}$

37. $-\sqrt{10}$

38. $-\sqrt{12}$

39. $\sqrt{75}$

40. $-\sqrt{25}$

41. $\dfrac{3}{5} = 0.6$

42. $\dfrac{5}{9} = 0.55555\ldots$

43. 4

44. -7

45. $5.2323\ldots$

46. 3.7

47. $0.447213\ldots$ (does not repeat)

48. $3.316624\ldots$ (does not repeat)

E. *Use the square root table to find the following.*

49. $\sqrt{6}$ **50.** $\sqrt{10}$ **51.** $-\sqrt{15}$ **52.** $-\sqrt{21}$

53. $\sqrt{98}$ **54.** $\sqrt{79}$ **55.** $\sqrt{23}$ **56.** $\sqrt{35}$

In clear weather, the miles that a person can see the view at an altitude A is given by $1.22\sqrt{A}$.

57. How far can a person see at an altitude of 4 miles?

58. At an altitude of 9 miles, how far can the view be seen?

59. The symbol $\sqrt{}$ is used to represent the _____ square root of a number.

60. The expression under the radical sign is called the _____.

61. Whole numbers that have integers for square roots are called _____ _____.

62. A number that cannot be written in the form a/b, where a and b are integers and $b \neq 0$, is an _____ number.

63. A rational number written as a decimal number either repeats or _____.

64. The set of real numbers is composed of the _____ numbers and the irrational numbers.

Checkup

The following problems provide a review of some of Sections 2.7 and 8.4. Some of these problems will help you with the next section.

Simplify by raising to powers.

65. $(x^4)^2$ **66.** $(x^5)^2$ **67.** $(x^3)^2$ **68.** $(x^7)^2$

Add or subtract as indicated.

69. $\dfrac{x^2 - 4x}{2x + 1} + \dfrac{3x^2 + 2x - 7}{2x + 1}$

70. $\dfrac{-7x + 3}{x^2 - 4} + \dfrac{2x^2 - 5x - 8}{x^2 - 4}$

71. $\dfrac{y^2 - 4y + 9}{x^2 + 3} - \dfrac{4y^2 - 7y - 16}{x^2 + 3}$

72. $\dfrac{3y^2 + 9y - 10}{y - 2} - \dfrac{7y^2 + 5y + 6}{y - 2}$

10.2 SIMPLIFYING RADICALS

1 If there is a number under the radical sign that has factors that are perfect squares, the radical is not in simplified form. We will see in the next section that we may need to simplify radicals in order to add or subtract them. Sometimes we simplify a large radical in order to approximate it by using the square root table. How do we simplify radicals? We remove all perfect square factors from the number under the radical (radicand). First we notice the following.

1. $\sqrt{16 \cdot 4} = \sqrt{64} = 8$ and $\sqrt{16} \cdot \sqrt{4} = 4 \cdot 2 = 8$

2. $\sqrt{25 \cdot 4} = \sqrt{100} = 10$ and $\sqrt{25} \cdot \sqrt{4} = 5 \cdot 2 = 10$

3. $\sqrt{(-9) \cdot (-4)} = \sqrt{36} = 6$ but $\sqrt{(-9)} \cdot \sqrt{(-4)}$ is not defined since square roots of negative numbers are not defined

From the examples above it appears that $\sqrt{x \cdot y} = \sqrt{x} \cdot \sqrt{y}$ if x and y are not negative. This is true and we have the following rule.

$$\sqrt{x \cdot y} = \sqrt{x} \cdot \sqrt{y} \qquad \text{if } x \text{ and } y \text{ are not negative}$$

This rule allows us to simplify radicals. It is helpful to memorize the following table of principal square roots.

Table of Principal Square Roots

Perfect Square	Principal Square Root
1	1
4	2
9	3
16	4
25	5
36	6
49	7
64	8
81	9
100	10
121	11
144	12

EXAMPLE 1 Simplify by removing the perfect square from the radicand.

a. $\sqrt{27}$

We find the largest perfect square in our principal square root table which divides evenly into 27. The perfect square is 9. 27 divided by 9 is 3. We factor 27 into $9 \cdot 3$.

$$\sqrt{27} = \sqrt{9 \cdot 3}$$

We also know that

$$\sqrt{9 \cdot 3} = \sqrt{9} \cdot \sqrt{3} \qquad \text{From our previous rule}$$

This allows us to simplify the radical since $\sqrt{9} = 3$.

$$\sqrt{9} \cdot \sqrt{3} = 3 \cdot \sqrt{3}$$

This is read "three times the square root of 3," but it is usually written $3\sqrt{3}$. Hence $\sqrt{27} = 3\sqrt{3}$.

b. $\sqrt{200}$

The largest perfect square in the principal square root table which divides evenly into 200 is 100. We factor 200 into $100 \cdot 2$ and then remove the perfect square from under the radical sign.

$$\sqrt{200} = \sqrt{100 \cdot 2}$$
$$= \sqrt{100} \cdot \sqrt{2}$$
$$= 10\sqrt{2} \qquad \blacksquare$$

□ **DO EXERCISE 1.**

We may also simplify radicals with variables under the radical sign. It is again helpful to make a table of principal square roots. Notice that the table continues on indefinitely.

Additional Table of Principal Square Roots

Perfect Square	Principal Square Root
x^2	$|x|$
x^4	$|x^2| = x^2$
x^6	$|x^3|$
x^8	$|x^4| = x^4$
x^{10}	$|x^5|$
x^{12}	$|x^6| = x^6$
x^{14}	$|x^7|$
x^{16}	$|x^8| = x^8$ Note $\sqrt{x^{16}} \neq x^4$

Since the even powers of x are not negative, we do not need to use the absolute value signs on them. For example, as shown above, $|x^2| = x^2$.

EXAMPLE 2 Simplify by removing the perfect squares from the radicand.

a. $\sqrt{x^9}$

Find the largest perfect square in the table that divides evenly into x^9. It is x^8. Factor x^9 into $x^8 \cdot x$. Then remove the perfect square from under the radical sign.

$$\sqrt{x^9} = \sqrt{x^8 \cdot x}$$
$$= \sqrt{x^8} \cdot \sqrt{x}$$
$$= x^4 \sqrt{x}$$

Notice that we may find that $\sqrt{x^8} = x^4$ in the principal square root table.

b. $\sqrt{24x^2}$

The x^2 is already a perfect square. The largest perfect square that divides evenly into 24 is 4. Factor 24 into $4 \cdot 6$. Then simplify.

$$\sqrt{24x^2} = \sqrt{4 \cdot 6 \, x^2}$$
$$= \sqrt{4 \cdot x^2 \cdot 6} \qquad \text{Rearranging terms}$$
$$= \sqrt{4} \cdot \sqrt{x^2} \cdot \sqrt{6}$$
$$= 2 \, |x| \, \sqrt{6}$$

c. $\sqrt{x^{11}}$

The largest perfect square that divides into x^{11} is x^{10}. Factor x^{11} into $x^{10} \cdot x$. Then simplify.

$$\sqrt{x^{11}} = \sqrt{x^{10} \cdot x}$$
$$= \sqrt{x^{10}} \cdot \sqrt{x}$$
$$= |x|^5 \sqrt{x}$$

d. $\sqrt{36x}$

$$\sqrt{36x} = \sqrt{36 \cdot x}$$
$$= \sqrt{36} \cdot \sqrt{x}$$
$$= 6 \sqrt{x} \quad \blacksquare$$

☐ **DO EXERCISE 2.**

We may approximate square roots of large numbers by factoring the number into numbers that appear in the square root table in the back of the book. We make our work easier by removing any perfect squares from under the radical sign.

☐ **Exercise 2** Simplify by removing the perfect squares from the radicand.

a. $\sqrt{x^7}$

b. $\sqrt{x^{15}}$

c. $\sqrt{64x^3}$

d. $\sqrt{20x^6}$

□ **Exercise 3** Use the square root table to find the following to three decimal places.

a. $\sqrt{360}$

b. $\sqrt{105}$

EXAMPLE 3 Use the square root table to find the following.

a. $\sqrt{250}$

We factor out the largest square that divides evenly into 250. It is 25. Then remove the perfect square, 25, from under the radical sign. We use the square root table to approximate the square root.

$$\sqrt{250} = \sqrt{25 \cdot 10}$$
$$= \sqrt{25} \cdot \sqrt{10}$$
$$= 5\sqrt{10}$$
$$\approx 5(3.162) \qquad \text{From the square root table, } \sqrt{10} \approx 3.162$$
$$\approx 15.810$$

b. $\sqrt{130}$

There are no perfect squares that divide into 130 evenly. Factor 130 into two numbers whose square roots appear in the table. Let us choose $65 \cdot 2$. Then we may approximate the solution using the square root table.

$$\sqrt{130} = \sqrt{65 \cdot 2}$$
$$= \sqrt{65} \cdot \sqrt{2}$$
$$\approx (8.062)(1.414)$$
$$\approx 11.400 \qquad \text{Rounded to three decimal places} \qquad ■$$

□ **DO EXERCISE 3.**

DID YOU KNOW?

Members of the Pythagorean brotherhood were angry about the discovery of irrational numbers. So, according to legend, they took the unfortunate member who made the discovery out to sea and threw him overboard far from shore.

Answers to Exercises

1. **a.** $3\sqrt{5}$ **b.** $2\sqrt{6}$ **c.** $5\sqrt{6}$ **d.** $6\sqrt{3}$

2. **a.** $|x^3|\sqrt{x}$ **b.** $|x^7|\sqrt{x}$ **c.** $8|x|\sqrt{x}$ **d.** $2|x^3|\sqrt{5}$

3. **a.** 18.972 **b.** 10.248

PROBLEM SET 10.2

A. *Simplify by removing all perfect squares from the radicand.*

1. $\sqrt{18}$ **2.** $\sqrt{20}$ **3.** $\sqrt{75}$

4. $\sqrt{50}$ **5.** $\sqrt{12}$ **6.** $\sqrt{32}$

7. $\sqrt{147}$ **8.** $\sqrt{245}$ **9.** $\sqrt{125}$

10. $\sqrt{180}$ **11.** $\sqrt{64x}$ **12.** $\sqrt{25x}$

13. $\sqrt{4x^2}$ **14.** $\sqrt{9x^2}$ **15.** $\sqrt{48x}$

16. $\sqrt{72x}$ **17.** $\sqrt{80x^2}$ **18.** $\sqrt{125x^2}$

19. $\sqrt{49y^2}$ **20.** $\sqrt{81y^2}$ **21.** $\sqrt{x^7}$

22. $\sqrt{x^{17}}$ **23.** $\sqrt{x^8}$ **24.** $\sqrt{x^{10}}$

25. $\sqrt{100x^7}$ **26.** $\sqrt{64x^9}$ **27.** $\sqrt{150x^4}$

28. $\sqrt{108x^8}$ **29.** $\sqrt{112x^3y^2}$ **30.** $\sqrt{147x^6y^5}$

B. *Use the square root table to find the following.*

31. $\sqrt{150}$
32. $\sqrt{160}$
33. $\sqrt{252}$

34. $\sqrt{343}$
35. $\sqrt{285}$
36. $\sqrt{265}$

37. If x and y are not _____,
$\sqrt{x \cdot y} = \sqrt{x} \cdot \sqrt{y}$.

38. Since principal square roots that are even powers of the variable are not negative, it is not necessary to use _____ _____ signs on them.

39. Square roots of large numbers that are not in the square root table may be approximated by _____ the numbers into numbers that are in the square root table.

Checkup

The following problems provide a review of some of Section 8.5.
Add.

40. $\dfrac{7}{2x-4} + \dfrac{3x}{x-2}$

41. $\dfrac{5x}{x^2-x} + \dfrac{6}{x-1}$

42. $\dfrac{2}{y-4} + \dfrac{3}{y^2-16}$

43. $\dfrac{7y}{y+3} + \dfrac{9y}{(y+3)^2}$

Subtract.

44. $\dfrac{7}{x+3} - \dfrac{5}{x-2}$

45. $\dfrac{3x}{x} - \dfrac{9x}{x-7}$

46. $\dfrac{3x+2}{(x-3)^2} - \dfrac{8x}{x^2-3x}$

47. $\dfrac{4}{x^2-36} - \dfrac{7+5x}{3x+18}$

48. $\dfrac{5}{2x^2+7x-4} - \dfrac{3x+1}{6x^2+x-2}$

49. $\dfrac{4y-8}{4y^2+7y+3} - \dfrac{9}{5y^2+3y-2}$

10.3 MULTIPLYING, ADDING, AND SUBTRACTING RADICALS

1 Multiplying Radicals

We stated in Section 10.2 that $\sqrt{x \cdot y} = \sqrt{x} \cdot \sqrt{y}$ if x and y are not negative. Since an equation is a balance, it is reversible and we have

$$\sqrt{x} \cdot \sqrt{y} = \sqrt{x \cdot y} \qquad \text{if } x \text{ and } y \text{ are not negative}$$

This rule allows us to multiply radicals.

Since many of our rules for radicals apply only to nonnegative real numbers, we often assume that all variables under the radical sign represent positive number.

EXAMPLE 1 Multiply.

a. $\sqrt{3} \cdot \sqrt{7} = \sqrt{3 \cdot 7} = \sqrt{21}$

b. $\sqrt{5}\sqrt{5} = \sqrt{5 \cdot 5} = \sqrt{25} = 5$

Notice that we may omit the multiplication sign when we multiply two radicals and that $\sqrt{5}\sqrt{5} = 5$. The result is the number under the radical when the radicals multiplied are identical.

c. $\sqrt{7}\sqrt{7} = 7$

d. $\sqrt{6}\sqrt{x} = \sqrt{6x}$ Assume that x is positive ∎

□ **DO EXERCISE 1.**

After we have completed the multiplication, we can often simplify the result by removing the largest perfect square that divides into the number under the radical sign.

EXAMPLE 2 Multiply and simplify. Assume that all variables under the radical sign are positive.

a. $\sqrt{3}\sqrt{8} = \sqrt{3 \cdot 8}$
$= \sqrt{24}$
$= \sqrt{4 \cdot 6}$ 4 is the largest square in 24
$= 2\sqrt{6}$

b. $\sqrt{2x}\sqrt{5x} = \sqrt{2x \cdot 5x}$
$= \sqrt{10x^2}$
$= x\sqrt{10}$

Notice that an absolute value sign is not necessary since the expressions under the radical sign are positive. ∎

□ **DO EXERCISE 2.**

□ **Exercise 1** Multiply. Assume that the variables under the radical sign are positive.

a. $\sqrt{2}\sqrt{7}$ **b.** $\sqrt{4}\sqrt{4}$

c. $\sqrt{3}\sqrt{5}$ **d.** $\sqrt{3x}\sqrt{3}$

□ **Exercise 2** Multiply and simplify. Assume that all variables under the radical sign are positive.

a. $\sqrt{3}\sqrt{6}$ **b.** $\sqrt{30}\sqrt{10}$

c. $\sqrt{5x^2}\sqrt{3x^4}$ **d.** $\sqrt{6x}\sqrt{2x^4}$

□ **Exercise 3** Add or subtract.

a. $4\sqrt{3} - 2\sqrt{3}$

b. $9\sqrt{2} + 6\sqrt{2}$

□ **Exercise 4** Add or subtract.

a. $\sqrt{6} - \sqrt{24}$

b. $4\sqrt{3} - \sqrt{12} + 4\sqrt{27}$

□ **Exercise 5** Add or subtract. Assume that the variables under the radical sign are positive.

a. $\sqrt{16x} - \sqrt{25x}$

b. $\sqrt{36y} + \sqrt{4y} - \sqrt{y}$

c. $\sqrt{36x^5} + \sqrt{x^5}$

d. $3\sqrt{x^9} - \sqrt{49x^9}$

2 **Addition and Subtraction of Radicals**

Recall that the sum of 5 and x is $5 + x$. Similarly, the sum of 5 and $\sqrt{3}$ is $5 + \sqrt{3}$. Also, the sum of $6x$ and $2x$ is $8x$. We can combine terms with variables if the variables are identical and raised to the same power. Similarly, we can combine terms containing square roots if the numbers under the radical sign are *identical*.

EXAMPLE 3 Add or subtract.

a. $8\sqrt{2} + 7\sqrt{2} = 15\sqrt{2}$

b. $7\sqrt{5} - 4\sqrt{5} = 3\sqrt{5}$ ■

□ **DO EXERCISE 3.**

Sometimes it appears that we cannot add or subtract radicals because the numbers under the radical sign are different. It is often possible to make them the same by removing a perfect square from under the radical sign.

EXAMPLE 4 Add or subtract.

a.
$$\begin{aligned}
\sqrt{3} + \sqrt{27} &= \sqrt{3} + \sqrt{9 \cdot 3} \\
&= \sqrt{3} + \sqrt{9}\sqrt{3} \\
&= \sqrt{3} + 3\sqrt{3} \qquad \text{Recall that } \sqrt{3} = 1\sqrt{3} \\
&= 4\sqrt{3}
\end{aligned}$$

b.
$$\begin{aligned}
5\sqrt{7} - 3\sqrt{28} + 3\sqrt{63} &= 5\sqrt{7} - 3\sqrt{4 \cdot 7} + 3\sqrt{9 \cdot 7} \\
&= 5\sqrt{7} - 3\sqrt{4}\sqrt{7} + 3\sqrt{9}\sqrt{7} \\
&= 5\sqrt{7} - 3 \cdot 2\sqrt{7} + 3 \cdot 3\sqrt{7} \\
&= 5\sqrt{7} - 6\sqrt{7} + 9\sqrt{7} \\
&= 8\sqrt{7} \qquad ■
\end{aligned}$$

□ **DO EXERCISE 4.**

EXAMPLE 5 Add or subtract. Assume that the variables under the radical sign are positive.

a.
$$\begin{aligned}
\sqrt{4x} - 3\sqrt{x} &= \sqrt{4}\sqrt{x} - 3\sqrt{x} \\
&= 2\sqrt{x} - 3\sqrt{x} \\
&= -\sqrt{x}
\end{aligned}$$

b.
$$\begin{aligned}
\sqrt{9x^5} - 5\sqrt{x^5} &= \sqrt{9 \cdot x^4 \cdot x} - 5\sqrt{x^4 \cdot x} \\
&= 3x^2\sqrt{x} - 5x^2\sqrt{x} \\
&= -2x^2\sqrt{x} \qquad ■
\end{aligned}$$

□ **DO EXERCISE 5.**

Answers to Exercises _____

1. a. $\sqrt{14}$ **b.** 4 **c.** $\sqrt{15}$ **d.** $3\sqrt{x}$

2. a. $3\sqrt{2}$ **b.** $10\sqrt{3}$ **c.** $x^3\sqrt{15}$ **d.** $2x^2\sqrt{3x}$

3. a. $2\sqrt{3}$ **b.** $15\sqrt{2}$ **4. a.** $-\sqrt{6}$ **b.** $14\sqrt{3}$

5. a. $-\sqrt{x}$ **b.** $7\sqrt{y}$ **c.** $7x^2\sqrt{x}$ **d.** $-4x^4\sqrt{x}$

PROBLEM SET 10.3

A. *Multiply. Assume that the variables are positive.*

1. $\sqrt{2}\sqrt{5}$

2. $\sqrt{2}\sqrt{3}$

3. $\sqrt{5}\sqrt{5}$

4. $\sqrt{7}\sqrt{7}$

5. $\sqrt{3}\sqrt{11}$

6. $\sqrt{5}\sqrt{7}$

7. $\sqrt{7}\sqrt{x}$

8. $\sqrt{2x}\sqrt{3}$

9. $\sqrt{5}\sqrt{14x}$

10. $\sqrt{17}\sqrt{x}$

11. $\sqrt{5}\sqrt{y}$

12. $\sqrt{33}\sqrt{y}$

B. *Multiply and simplify. Assume that the variables are positive.*

13. $\sqrt{15}\sqrt{3}$

14. $\sqrt{2}\sqrt{24}$

15. $\sqrt{18}\sqrt{2}$

16. $\sqrt{5}\sqrt{10}$

17. $\sqrt{6}\sqrt{3}$

18. $\sqrt{6}\sqrt{3}$

19. $\sqrt{x}\sqrt{x}$

20. $\sqrt{2y}\sqrt{2y}$

21. $\sqrt{2x}\sqrt{10x}$

22. $\sqrt{6x}\sqrt{2x}$

23. $\sqrt{5x}\sqrt{15x^2}$

24. $\sqrt{5x^3}\sqrt{20x^3}$

25. $\sqrt{3xy}\sqrt{6xy^2}$

26. $\sqrt{3x^3y}\sqrt{20x^2y}$

C. *Add or subtract. Assume that the variables are positive.*

27. $9\sqrt{3}+6\sqrt{3}$

28. $8\sqrt{2}+4\sqrt{2}$

29. $8\sqrt{17}-3\sqrt{17}$

30. $3\sqrt{10}-5\sqrt{10}$

31. $2\sqrt{20}+6\sqrt{5}$

32. $7\sqrt{3}+2\sqrt{12}$

33. $4\sqrt{27} - \sqrt{3}$

34. $\sqrt{75} - 6\sqrt{3}$

35. $\sqrt{45} + \sqrt{125}$

36. $\sqrt{80} + \sqrt{180}$

37. $\sqrt{48} - \sqrt{27}$

38. $\sqrt{90} - \sqrt{40}$

39. $4\sqrt{24} - \sqrt{54} + 5\sqrt{20}$

40. $4\sqrt{2} - 3\sqrt{27} - 2\sqrt{12}$

41. $\sqrt{16x} + 5\sqrt{x}$

42. $\sqrt{36x} - \sqrt{100x}$

43. $\sqrt{64x^3} - \sqrt{x^3}$

44. $2\sqrt{x^7} - \sqrt{4x^7}$

45. $\sqrt{9x + 9} + \sqrt{x + 1}$

46. $\sqrt{x + 4} + \sqrt{x^3 + 4x^2}$

47. Terms containing square roots may be combined if the radicands are _____.

48. It is sometimes possible to make two radicands the same by removing a _____ _____ from one or both radicands.

Checkup

The following problems provide a review of some of Section 9.1.
Solve.

49. $\dfrac{2}{x + 4} = \dfrac{4}{x + 9}$

50. $\dfrac{3}{2x - 1} = \dfrac{6}{x - 8}$

51. $5 + \dfrac{3}{x} = 2$

52. $\dfrac{5}{4}x + x = 9$

53. $\dfrac{3}{y^2 - 9} = \dfrac{6}{y + 3}$

54. $\dfrac{7x}{(x + 2)^2} = \dfrac{3}{x + 2}$

55. $\dfrac{1}{x + 3} - \dfrac{2}{x - 4} = \dfrac{-8x}{x^2 - x - 12}$

56. $\dfrac{4}{y - 1} + \dfrac{2}{y + 3} = \dfrac{3y}{y^2 + 2y - 3}$

Rationalizing the denominator One process of removing the radicals from the denominator.

10.4 RATIONALIZING THE DENOMINATOR

■ Denominator Is a Perfect Square

We may not want to have a fraction under the radical sign or a radical in the denominator of a fractional expression. The usual operation of removing the fraction from under the radical sign or the radical from the denominator is called **rationalizing the denominator**. If the number under the radical sign in the denominator is a perfect square, we do not need the special technique of rationalizing the denominator. Consider the following.

1. $\sqrt{\dfrac{16}{9}} = \dfrac{4}{3}$ because $\dfrac{4}{3} \cdot \dfrac{4}{3} = \dfrac{16}{9}$

2. $\sqrt{\dfrac{4}{25}} = \dfrac{2}{5}$ because $\dfrac{2}{5} \cdot \dfrac{2}{5} = \dfrac{4}{25}$

If we continue to work examples like those above, we would come to the following true conclusion.

$$\sqrt{\dfrac{x}{y}} = \dfrac{\sqrt{x}}{\sqrt{y}} \qquad \text{where } x \text{ and } y \text{ are not negative and } y \text{ is not zero.}$$

EXAMPLE 1 Simplify by removing all perfect squares from the radicand.

a. $\sqrt{\dfrac{1}{4}} = \dfrac{\sqrt{1}}{\sqrt{4}} = \dfrac{1}{2}$

b. $\sqrt{\dfrac{5}{9}} = \dfrac{\sqrt{5}}{\sqrt{9}} = \dfrac{\sqrt{5}}{3}$ ■

☐ **DO EXERCISE 1.**

OBJECTIVES

■ Simplify fractions with radicals when the radical in the denominator is a perfect square

■ Simplify fractions with radicals in the denominator by rationalizing the denominator.

■ Divide radicals

■ Use a table to approximate square roots of fractions when there is a radical in both the numerator and the denominator

☐ **Exercise 1** Simplify by removing all perfect squares from the radicand.

a. $\sqrt{\dfrac{1}{36}}$

b. $\sqrt{\dfrac{49}{16}}$

c. $\sqrt{\dfrac{7}{4}}$

d. $\sqrt{\dfrac{3}{25}}$

a. $\sqrt{\dfrac{1}{7}}$

b. $\sqrt{\dfrac{5}{3}}$

c. $\dfrac{\sqrt{11}}{\sqrt{12}}$

d. $\dfrac{\sqrt{3}}{\sqrt{32}}$

☑ Rationalizing the Denominator

When the radicand in the denominator is not a perfect square, we can *rationalize the denominator* by multiplying the numerator and denominator of the fraction by the same number (multiplying by 1).

EXAMPLE 2 Rationalize the denominator.

a.

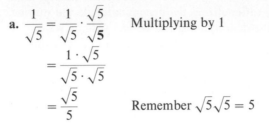

$$\dfrac{1}{\sqrt{5}} = \dfrac{1}{\sqrt{5}} \cdot \dfrac{\sqrt{5}}{\sqrt{5}} \qquad \text{Multiplying by 1}$$

$$= \dfrac{1 \cdot \sqrt{5}}{\sqrt{5} \cdot \sqrt{5}}$$

$$= \dfrac{\sqrt{5}}{5} \qquad \text{Remember } \sqrt{5}\sqrt{5} = 5$$

Notice that we cannot divide numbers under the radical sign by numbers that are not under the radical sign. The answer is $\sqrt{5}/5$.

b.

$$\sqrt{\dfrac{5}{6}} = \dfrac{\sqrt{5}}{\sqrt{6}}$$

$$= \dfrac{\sqrt{5}}{\sqrt{6}} \cdot \dfrac{\sqrt{6}}{\sqrt{6}} \qquad \text{Multiplying by 1}$$

$$= \dfrac{\sqrt{5} \cdot \sqrt{6}}{\sqrt{6} \cdot \sqrt{6}}$$

$$= \dfrac{\sqrt{30}}{6} \qquad \text{Since } \sqrt{6}\sqrt{6} = 6$$

c.

$$\dfrac{\sqrt{7}}{\sqrt{20}} = \dfrac{\sqrt{7}}{\sqrt{20}} \cdot \dfrac{\sqrt{5}}{\sqrt{5}}$$

$$= \dfrac{\sqrt{35}}{\sqrt{100}}$$

$$= \dfrac{\sqrt{35}}{10}$$

The "1" that we multiplied by was $\sqrt{5}/\sqrt{5}$ rather than $\sqrt{20}/\sqrt{20}$ because we use a number that gives us the smallest possible square in the denominator. ■

□ **DO EXERCISE 2.**

3 Division

We said earlier in this section that $\sqrt{x/y} = \sqrt{x}/\sqrt{y}$ if x and y are not negative and y is not zero. Since the equation is reversible, we have the following.

> $$\frac{\sqrt{x}}{\sqrt{y}} = \sqrt{\frac{x}{y}} \qquad \text{if } x \text{ and } y \text{ are not negative and } y \text{ is not zero.}$$

This rule allows us to divide radicals.

EXAMPLE 3 Divide and simplify. Assume that the variables under the radical sign are positive.

a. $\dfrac{\sqrt{24}}{\sqrt{6}} = \sqrt{\dfrac{24}{6}} = \sqrt{4} = 2$

b. $\dfrac{\sqrt{5}}{\sqrt{10}} = \sqrt{\dfrac{5}{10}} = \sqrt{\dfrac{1}{2}}$ Simplifying

$\qquad = \dfrac{\sqrt{1}}{\sqrt{2}} = \dfrac{\sqrt{1}}{\sqrt{2}} \cdot \dfrac{\sqrt{2}}{\sqrt{2}} = \dfrac{\sqrt{2}}{2}$

Notice that by simplifying first, we can multiply by a smaller number to rationalize the denominator.

c. $\dfrac{\sqrt{12x^5}}{\sqrt{6x^4}} = \sqrt{\dfrac{12x^5}{6x^4}} = \sqrt{2x}$ ■

☐ **DO EXERCISE 3.**

☐ **Exercise 3**   Divide and simplify. Assume that the variables under the radical sign are positive.

a. $\dfrac{\sqrt{8}}{\sqrt{2}}$

b. $\dfrac{\sqrt{2}}{\sqrt{8}}$

c. $\dfrac{\sqrt{3}}{\sqrt{9}}$

d. $\dfrac{\sqrt{25x^3}}{\sqrt{5x^2}}$

□ **Exercise 4** Use the square root table to find the following to three decimal places.

a. $\dfrac{\sqrt{7}}{\sqrt{8}}$

b. $\dfrac{\sqrt{10}}{\sqrt{5}}$

---

4 **Approximating Square Roots**

The procedures described above make it easier to approximate square roots of fractions using the square root table.

EXAMPLE 4 Use the square root table to find the following to three decimal places.

a. $\dfrac{\sqrt{4}}{\sqrt{24}} = \sqrt{\dfrac{4}{24}} = \sqrt{\dfrac{1}{6}} = \dfrac{\sqrt{1}}{\sqrt{6}}$

$= \dfrac{\sqrt{1}}{\sqrt{6}} \cdot \dfrac{\sqrt{6}}{\sqrt{6}} = \dfrac{\sqrt{6}}{6} \approx \dfrac{2.449}{6} \approx 0.408$

Notice that it is easier to find $\sqrt{6}$ and divide it by 6 than to find the square roots of both 4 and 24 and then divide.

b. $\dfrac{\sqrt{5}}{\sqrt{7}} = \dfrac{\sqrt{5}}{\sqrt{7}} \cdot \dfrac{\sqrt{7}}{\sqrt{7}}$

$= \dfrac{\sqrt{35}}{7} \approx \dfrac{5.916}{7} \approx 0.845$ ■

□ **DO EXERCISE 4.**

Answers to Exercises

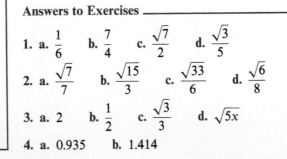

1. a. $\dfrac{1}{6}$ b. $\dfrac{7}{4}$ c. $\dfrac{\sqrt{7}}{2}$ d. $\dfrac{\sqrt{3}}{5}$

2. a. $\dfrac{\sqrt{7}}{7}$ b. $\dfrac{\sqrt{15}}{3}$ c. $\dfrac{\sqrt{33}}{6}$ d. $\dfrac{\sqrt{6}}{8}$

3. a. 2   b. $\dfrac{1}{2}$ c. $\dfrac{\sqrt{3}}{3}$ d. $\sqrt{5x}$

4. a. 0.935   b. 1.414

PROBLEM SET 10.4

Assume that the variables under the radical sign are positive.

A. *Simplify by removing perfect squares from the radicand.*

1. $\sqrt{\dfrac{4}{9}}$ **2.** $\sqrt{\dfrac{16}{25}}$ **3.** $-\sqrt{\dfrac{4}{81}}$ **4.** $-\sqrt{\dfrac{9}{49}}$

5. $\sqrt{\dfrac{1}{64}}$ **6.** $\sqrt{\dfrac{1}{100}}$ **7.** $\sqrt{\dfrac{81}{144}}$ **8.** $\sqrt{\dfrac{64}{225}}$

9. $\sqrt{\dfrac{25}{x^2}}$ **10.** $\sqrt{\dfrac{64}{y^2}}$ **11.** $\sqrt{\dfrac{x^2}{y^2}}$ **12.** $\dfrac{\sqrt{y}}{20}$

B. *Rationalize the denominator.*

13. $\sqrt{\dfrac{1}{5}}$ **14.** $\sqrt{\dfrac{1}{8}}$ **15.** $\sqrt{\dfrac{2}{3}}$ **16.** $\sqrt{\dfrac{3}{5}}$

17. $\sqrt{\dfrac{2}{7}}$ **18.** $\sqrt{\dfrac{5}{8}}$ **19.** $\sqrt{\dfrac{3}{2}}$ **20.** $\sqrt{\dfrac{1}{2}}$

21. $\dfrac{\sqrt{5}}{\sqrt{7}}$ **22.** $\dfrac{\sqrt{6}}{\sqrt{5}}$ **23.** $\dfrac{\sqrt{10}}{\sqrt{20}}$ **24.** $\dfrac{\sqrt{5}}{\sqrt{12}}$

25. $\dfrac{\sqrt{7}}{\sqrt{y}}$ **26.** $\dfrac{\sqrt{5}}{\sqrt{x}}$ **27.** $\dfrac{\sqrt{2x}}{\sqrt{7}}$ **28.** $\dfrac{\sqrt{3y}}{\sqrt{11}}$

C. *Divide and simplify. Then, if the denominator contains a radical, rationalize it.*

29. $\dfrac{\sqrt{27}}{\sqrt{3}}$ **30.** $\dfrac{\sqrt{8}}{\sqrt{2}}$ **31.** $\dfrac{\sqrt{100}}{\sqrt{4}}$ **32.** $\dfrac{\sqrt{72}}{\sqrt{2}}$

33. $\dfrac{\sqrt{35}}{\sqrt{7}}$ **34.** $\dfrac{\sqrt{40}}{\sqrt{20}}$ **35.** $\dfrac{\sqrt{32}}{\sqrt{4}}$ **36.** $\dfrac{\sqrt{24}}{\sqrt{3}}$

37. $\dfrac{\sqrt{36x}}{\sqrt{9x}}$

38. $\dfrac{\sqrt{27y}}{\sqrt{3y}}$

39. $\dfrac{\sqrt{2x^4}}{\sqrt{3x^2}}$

40. $\dfrac{\sqrt{7x^6}}{\sqrt{8x^4}}$

41. $\dfrac{\sqrt{x^4}}{\sqrt{x^5}}$

42. $\dfrac{\sqrt{x^7}}{\sqrt{x^8}}$

43. $\dfrac{\sqrt{16a^5}}{\sqrt{a^4}}$

44. $\dfrac{\sqrt{a^7}}{\sqrt{25a^5}}$

D. *Use the square root table to find the following to three decimal places.*

45. $\dfrac{\sqrt{5}}{\sqrt{8}}$

46. $\dfrac{\sqrt{4}}{\sqrt{3}}$

47. $\dfrac{\sqrt{2}}{\sqrt{5}}$

48. $\dfrac{\sqrt{5}}{\sqrt{6}}$

49. $\dfrac{\sqrt{200}}{\sqrt{5}}$

50. $\dfrac{\sqrt{150}}{\sqrt{5}}$

51. $\dfrac{\sqrt{72}}{\sqrt{6}}$

52. $\dfrac{\sqrt{84}}{\sqrt{7}}$

53. A denominator of a fraction may be rationalized by multiplying the fraction by a particular _____.

54. If x and y are not _____ and y is not zero, $\sqrt{x}/\sqrt{y} = \sqrt{x/y}$.

Checkup

The following problems provide a review of some of Section 9.2.

55. If one side of a triangle is $\frac{1}{4}$ the perimeter, the second side is 15 centimeters, and the third side is $\frac{1}{8}$ the perimeter, find the perimeter of the triangle.

56. Find the dimensions of a rectangle with perimeter 60 feet if its width is $\frac{3}{7}$ its length.

57. Three more than $\frac{3}{5}$ a number is -3. Find the number.

58. Six less than $\frac{3}{4}$ a number is $\frac{1}{2}$ the number. Find the number.

59. Sharon can clean the house in 3 hours and Dean can clean it in 4 hours. How long does it take them to clean the house together?

60. Kiet can correct papers in 3 hours and Binh can correct them in 5 hours. How long does it take them to correct the papers together?

61. A train travels 30 miles per hour faster than a car. The car goes 20 miles in the same amount of time as the train travels 60 miles. Find the speed of the car.

62. Carlos drives 10 miles per hour slower than Maria. If Carlos travels 27 miles while Maria travels 36 miles, what is the speed of Carlos's car?

Hypotenuse   the longest side of a right triangle.

Legs   the two shorter sides of a right triangle.

Pythagorean theorem   in a right triangle, $a^2 + b^2 = c^2$ where a and b are the lengths of the legs and c is the length of the hypotenuse.

10.5   THE PYTHAGOREAN THEOREM AND APPLIED PROBLEMS

OBJECTIVES

1 *Find the length of the side of a right triangle when the lengths of two sides are given*

2 *Solve word problems using right triangles*

A right triangle is a triangle with a right angle. The **Pythagorean theorem** allows us to find the length of a side of a *right* triangle if we know the lengths of the other two sides. In a right triangle the longest side is called the **hypotenuse** and the other two sides are called the **legs**. The hypotenuse is usually designated by the letter c and the two legs are labeled a and b.

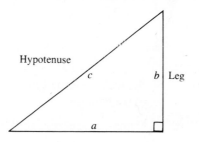

The Pythagorean theorem may be stated as follows: In a right triangle the sum of the squares of the legs equals the square of the hypotenuse. The theorem is usually written as a formula.

Pythagorean Theorem

In a right triangle,

$$a^2 + b^2 = c^2$$

where a and b are the lengths of the legs and c is the length of the hypotenuse.

---

1 **Using the Pythagorean Theorem**

We may use the theorem to find the length of one side of a triangle if we are given the lengths of the other two sides.

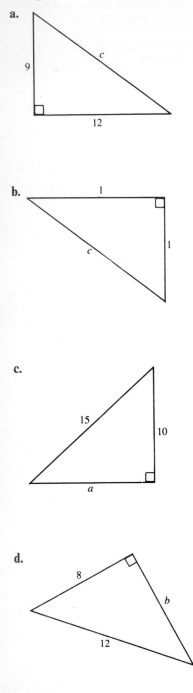

□ **Exercise 1** Find the lengths of the sides of the following right triangles.

a.

b.

c.

d.

EXAMPLE 1

a. Find the length of the hypotenuse of the triangle shown.

$$a^2 + b^2 = c^2$$
$$5^2 + 6^2 = c^2$$
$$25 + 36 = c^2$$
$$61 = c^2$$

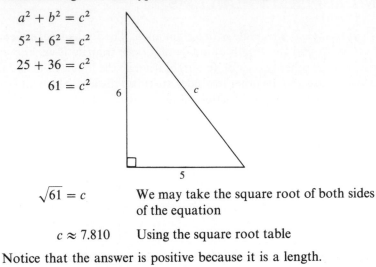

$\sqrt{61} = c$ We may take the square root of both sides of the equation

$c \approx 7.810$ Using the square root table

Notice that the answer is positive because it is a length.

b. Find the length of the leg of the triangle shown.

$$a^2 + b^2 = c^2$$
$$a^2 + 7^2 = 10^2$$
$$a^2 + 49 = 100$$
$$a^2 = 100 - 49$$
$$a^2 = 51$$
$$a = \sqrt{51}$$

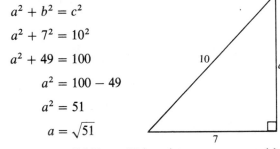

$a \approx 7.141$ Using the square root table

c. Find the length of the leg of the triangle shown.

$$a^2 + b^2 = c^2$$
$$3^2 + b^2 = 9^2$$
$$9 + b^2 = 81$$
$$b^2 = 81 - 9$$
$$b^2 = 72$$
$$b = \sqrt{72}$$
$$b = \sqrt{36 \cdot 2}$$
$$b = \sqrt{36} \cdot \sqrt{2}$$
$$b = 6\sqrt{2}$$

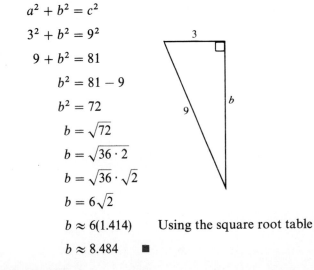

$b \approx 6(1.414)$ Using the square root table

$b \approx 8.484$ ■

□ **DO EXERCISE 1.**

2 Applied Problems

We can use the Pythagorean theorem to solve applied problems.

EXAMPLE 2 Two boats leave from the same point at the same time, one traveling due east at 10 miles per hour and the other traveling due north at 8 miles per hour. How far are they apart after 1 hour? We make a drawing showing the right triangle.

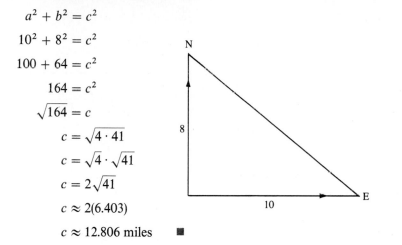

$$a^2 + b^2 = c^2$$

$$10^2 + 8^2 = c^2$$

$$100 + 64 = c^2$$

$$164 = c^2$$

$$\sqrt{164} = c$$

$$c = \sqrt{4 \cdot 41}$$

$$c = \sqrt{4} \cdot \sqrt{41}$$

$$c = 2\sqrt{41}$$

$$c \approx 2(6.403)$$

$$c \approx 12.806 \text{ miles} \quad \blacksquare$$

☐ **DO EXERCISE 2.**

☐ **Exercise 2** A ladder 14 feet long is leaning against a wall. The bottom of the ladder is 2 feet from the wall. How high on the wall is the top of the ladder?

DID YOU KNOW?

James Abram Garfield invented an original proof of the Pythagorean theorem in 1876, while he was a congressman. Five years later he became the president of the United States.

Answers to Exercises

1. a. 15   **b.** 1.414   **c.** 11.180   **d.** 8.944

2. 13.416 feet

PROBLEM SET 10.5

A. *Find the length of the side not given.*

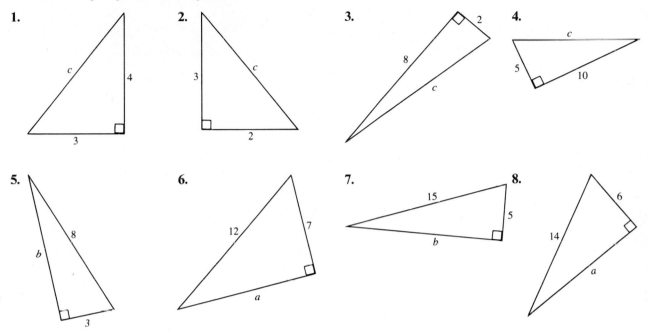

1.   **2.**   **3.**   **4.**

5.   **6.**   **7.**   **8.**

B. *Find the length of the third side of each right triangle if c is the hypotenuse.*

9. $a = 7,\ \ b = 3$ **10.** $a = 6,\ \ b = 7$ **11.** $a = 12,\ \ c = 15$ **12.** $b = 4,\ \ c = 8$

13. $b = 9,\ c = 15$ **14.** $a = 5,\ \ c = 12$ **15.** $a = 3,\ \ b = 4$ **16.** $b = 7,\ \ c = 9$

C. *Solve.*

17. Two boats leave the same point at the same time, one traveling due south at 11 miles per hour and the other traveling due west at 12 miles per hour. How far are they apart after 1 hour?

18. A ladder 12 feet long is leaning against a wall. The bottom of the ladder is 4 feet from the wall. How high on the wall is the top of the ladder?

19. What is the length of the diagonal of a square with sides 5 centimeters long?

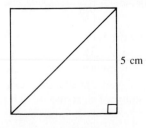

5 cm

20. A baseball diamond is a 90-foot square. How far is it from home plate to second base?

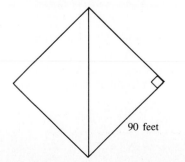

90 feet

10.5   The Pythagorean Theorem and Applied Problems   **457**

21. The tip of the shadow of a street light 15 feet high is at a point 12 feet from the base of the light. How far is it from the tip of the shadow to the top of the light pole?

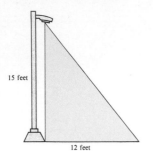

15 feet

12 feet

22. Cheryl is flying a balloon at a height of 20 feet. If the horizontal distance from the balloon to Cheryl is 8 feet, how far is she from the balloon?

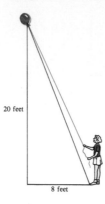

20 feet

8 feet

23. The longest side of a right triangle is called the _____.

24. In a right triangle, the sum of the squares of the _____ equals the square of the hypotenuse.

Checkup

The following problems provide a review of some of Section 9.3. Solve.

25. If 8 apples cost 98 cents, how much will 14 apples cost?

26. If a car travels 70 miles on 4 gallons of gas, how far will it go on 20 gallons?

27. If 2 people can paint a room in 5 hours, how long does it take 3 people to paint the room?

28. If a train takes 25 minutes to travel a certain distance at 60 miles per hour, how long does it take it to travel the same distance at 75 miles per hour?

29. The current in an electrical circuit varies inversely as the resistance of the circuit if the voltage is constant. The current is 4 amperes when the resistance is 20 ohms. Find the resistance when the current is 5 amperes.

30. The cost of staying at an inn varies directly with the length of stay. If a 6-day stay costs $276, how much does it cost to stay 10 days?

31. The number of servings that can be obtained from a turkey varies directly with the weight of the turkey. If a 14-pound turkey yields 8 servings, how many servings can be obtained from a 24-pound turkey?

32. The length of a string and its fundamental frequency of vibration are inversely proportional. If a string of length 20 centimeters gives a fundamental frequency of 300 cycles per second, what is the fundamental frequency of a string of length 32 centimeters?

10.6 EQUATIONS WITH RADICALS

1 In our study of business and the physical sciences, we often find equations with radicals. An example of this type of equation is

$$\sqrt{x} - 2 = 5$$

We need a new technique to remove the radical sign. The new method is to square both sides of the equation, because if both sides of an equation are squared, all solutions of the original equation are solutions of the squared equation. For example, if

$$x = 3$$

$$\text{then } x^2 = 9$$

☐ **Exercise 1** Solve $\sqrt{x} - 2 = 3$.

The second equation has two solutions, $x = 3$ and $x = -3$. The first equation has only one solution, $x = 3$. The solution, $x = 3$, is a solution to both equations.

The fact that we can square both sides of an equation may be stated as follows:

> If $x = y$, then $x^2 = y^2$.

We must check all solutions in the original equation because as noted above, some solutions to the squared equation may not be solutions to the original equation.

EXAMPLE 1 Solve $\sqrt{x} - 2 = 5$ (Original equation).

We get the radical alone on one side of the equation, if possible. Otherwise, when we square both sides, we get another radical.

$$\sqrt{x} - 2 = 5$$

$$\sqrt{x} = 7 \qquad \text{Adding 2 to both sides of the equation to isolate the radical}$$

$$(\sqrt{x})^2 = 7^2 \qquad \text{Squaring both sides}$$

$$x = 49$$

Check $\sqrt{x} - 2 = 5$

$$\begin{array}{c|c} \sqrt{49} - 2 & 5 \\ 7 - 2 & 5 \\ 5 & 5 \end{array}$$

The answer checks, so the solution is 49. ■

☐ **DO EXERCISE 1.**

□ **Exercise 2**   Solve $\sqrt{x} = -1$.

□ **Exercise 3**   Solve $2\sqrt{x} = \sqrt{x + 12}$.

EXAMPLE 2   Solve $\sqrt{y} = -2$.

$$\sqrt{y} = -2$$
$$(\sqrt{y})^2 = (-2)^2 \qquad \text{Squaring both sides}$$
$$y = 4$$

Check
$$\begin{array}{c|c} \sqrt{y} = -2 & \\ \hline \sqrt{4} & -2 \\ 2 & -2 \end{array}$$

The answer does not check. There is no solution to this equation.   ■

□ **DO EXERCISE 2.**

EXAMPLE 3   Solve $\sqrt{5x - 9} = 2\sqrt{x}$.

$$\sqrt{5x - 9} = 2\sqrt{x}$$
$$(\sqrt{5x - 9})^2 = (2\sqrt{x})^2 \qquad \text{Squaring both sides}$$
$$(\sqrt{5x - 9})^2 = 2^2(\sqrt{x})^2 \qquad \begin{array}{l}\text{Notice that the 2 must be squared since}\\ (ab)^n = a^n b^n\end{array}$$
$$5x - 9 = 4x$$
$$-9 = -x \qquad \text{Adding } -5x \text{ to both sides}$$
$$x = 9$$

Check
$$\begin{array}{c|c} \sqrt{5x - 9} = 2\sqrt{x} & \\ \hline \sqrt{5(9) - 9} & 2\sqrt{9} \\ \sqrt{45 - 9} & 2\sqrt{9} \\ \sqrt{36} & 2(3) \\ 6 & 6 \end{array}$$

The answer checks, so 9 is the solution.   ■

□ **DO EXERCISE 3.**

Answers to Exercises _____

1. 25
2. No solution
3. 4

PROBLEM SET 10.6

Solve.

1. $\sqrt{x} = 5$

2. $\sqrt{y} = 7$

3. $\sqrt{x} = \dfrac{2}{3}$

4. $\sqrt{x} = \dfrac{1}{9}$

5. $\sqrt{x - 1} = 4$

6. $\sqrt{y - 3} = 8$

7. $\sqrt{x + 6} = 11$

8. $\sqrt{y + 2} = 10$

9. $\sqrt{x + 5} = 5$

10. $\sqrt{x - 3} = 2$

11. $\sqrt{3x + 1} = 5$

12. $\sqrt{2x + 7} = 5$

13. $\sqrt{x} = -4$

14. $\sqrt{y} = -5$

15. $\sqrt{y} = -7$

16. $\sqrt{x} = -9$

17. $3\sqrt{x} = \sqrt{8x + 16}$

18. $2\sqrt{y} = \sqrt{3y + 9}$

19. $2\sqrt{3x} = 6$

20. $4\sqrt{5y} = 8$

21. $\sqrt{2x + 9} = \sqrt{x + 5}$

22. $\sqrt{4y - 4} = \sqrt{y - 1}$

If an automobile traveling at a rate r hits a tree, the force of the impact is the same as the force with which it would hit the ground when falling from a building s feet high where

$$r = 5.42\sqrt{s}$$

23. How fast is a car traveling if it hits a tree with the same force as if it had been dropped from a building 65 feet high?

24. What is the speed of an automobile if it hits a tree with the same force as if it had been dropped from a building 210 feet high?

25. If a car traveling 90 miles per hour hits a tree, the force of impact is the same as if it fell from a building how many feet high?

26. If a car moving 25 miles per hour hits a tree, the force of impact is the same as if it fell from a building how many feet high?

27. If we square both sides of an equation, some solutions of the squared equation may not be solutions of the _____ equation.

28. All possible solutions of an equation containing radicals must be _____ in the original equation.

Checkup

The following problems provide a review of some of Section 9.4.
Simplify.

29. $\dfrac{\dfrac{3}{x}}{\dfrac{5+x}{x}}$

30. $\dfrac{\dfrac{7}{2+y}}{\dfrac{y}{2+y}}$

31. $\dfrac{x + \dfrac{1}{y}}{\dfrac{1}{x} + \dfrac{3}{x^2}}$

32. $\dfrac{5 + \dfrac{1}{2x}}{\dfrac{3}{2x} - \dfrac{7}{3x}}$

33. $\dfrac{\dfrac{x}{x+1}}{\dfrac{4}{2x^2 - x - 3}}$

34. $\dfrac{\dfrac{5}{x-4}}{\dfrac{7}{x^2 - 16}}$

CHAPTER 10   ADDITIONAL EXERCISES (OPTIONAL)

Section 10.1

Are the following roots defined?

1. $-\sqrt{25}$ **2.** $\sqrt{-36}$ **3.** $\sqrt{-144}$ **4.** $-\sqrt{9}$

Find the square root.

5. $-\sqrt{225}$ **6.** $\sqrt{169}$ **7.** $\sqrt{x^2 + 6x + 9}$ **8.** $\sqrt{(-5)^2}$

Are the following numbers rational or irrational?

9. $\sqrt{\dfrac{81}{121}}$ **10.** $\sqrt{\dfrac{17}{31}}$ **11.** 0.394 **12.** $0.333\ldots$ (repeats)

Section 10.2

Simplify.

13. $\sqrt{72}$ **14.** $\sqrt{80}$ **15.** $\sqrt{54x^5}$ **16.** $\sqrt{108y^7}$

17. $\sqrt{96x^9y^3}$ **18.** $\sqrt{147a^5b^3}$ **19.** $\sqrt{320a^4b^7}$ **20.** $\sqrt{252x^{15}y^8}$

21. $\sqrt{(x-4)^3}$ **22.** $\sqrt{(x+8)^5}$

Section 10.3

Assume that the variables are positive. Multiply and simplify, if possible.

23. $\sqrt{37}\sqrt{37}$ **24.** $\sqrt{91}\sqrt{91}$ **25.** $\sqrt{2}\sqrt{12}$

26. $\sqrt{6}\sqrt{8}$ **27.** $\sqrt{7x^3y}\sqrt{14xy^4}$ **28.** $\sqrt{5a^3b^4}\sqrt{35a^2b^2}$

Perform the following operations.

29. $5\sqrt{3} - 4\sqrt{27}$ **30.** $2\sqrt{96} + 3\sqrt{6}$

31. $6\sqrt{128} + 3\sqrt{162} - 4\sqrt{32}$ **32.** $5\sqrt{252} - 3\sqrt{243} - 6\sqrt{448}$

Section 10.4

Assume that the variables under the radical sign are positive.
Simplify.

33. $\sqrt{\dfrac{81}{196}}$

34. $\sqrt{\dfrac{289}{625}}$

Rationalize the denominator.

35. $\sqrt{\dfrac{3}{7}}$

36. $\sqrt{\dfrac{3}{5}}$

37. $\dfrac{\sqrt{3}}{\sqrt{x^3}}$

38. $\dfrac{\sqrt{6}}{\sqrt{y^5}}$

Divide and simplify.

39. $\dfrac{\sqrt{60}}{\sqrt{3}}$

40. $\dfrac{\sqrt{128}}{\sqrt{2}}$

41. $\dfrac{\sqrt{1575x^8}}{\sqrt{7x^2}}$

42. $\dfrac{\sqrt{46}}{\sqrt{230}}$

Use the square root table to find the following to three decimal places.

43. $\dfrac{\sqrt{7}}{\sqrt{6}}$

44. $\dfrac{\sqrt{10}}{\sqrt{3}}$

Section 10.5

Find the length of the third side of each right triangle if c is the hypotenuse.

45. $a = 6, \quad b = 9$

46. $a = 2, \quad b = 8$

47. $a = 3, \quad c = 7$

48. $b = 4, \quad c = 10$

49. What is the length of the diagonal of a rectangle with width 5 meters and length 8 meters?

50. The dimensions of a flower garden are 9 feet by 12 feet. What is the length of the diagonal path across the garden?

Section 10.6

Solve.

51. $\sqrt{x + 3} = 7$

52. $\sqrt{y - 4} = 5$

53. $\sqrt{2x - 5} = 5$

54. $\sqrt{3x + 1} = -4$

55. $3\sqrt{y + 2} = 9\sqrt{y - 6}$

56. $5\sqrt{x - 4} = 2\sqrt{x + 17}$

57. $\sqrt{3x - 2} = \sqrt{x + 6}$

58. $\sqrt{x + 4} = -\sqrt{2x - 8}$

CHAPTER 10   PRACTICE TEST

Find the square root.

1. $\sqrt{100}$ **2.** $-\sqrt{64}$

3. $\sqrt{x^2}$ **4.** $\sqrt{36y^2}$

5. $\sqrt{(xy)^2}$

Are the following rational or irrational numbers?

6. $\sqrt{10}$ **7.** $\sqrt{16}$

8. $\sqrt{50}$

Simplify.

9. $\sqrt{12}$ **10.** $\sqrt{75x^2}$

Use the square root table to find the following to three decimal places.

11. $\sqrt{245}$

For the rest of the test, assume that the variables under the radical sign are positive.
Multiply and simplify, if possible.

12. $\sqrt{5}\sqrt{2}$ **13.** $\sqrt{2}\sqrt{24}$

14. $\sqrt{2x}\sqrt{32xy}$

1. _____

2. _____

3. _____

4. _____

5. _____

6. _____

7. _____

8. _____

9. _____

10. _____

11. _____

12. _____

13. _____

14. _____

15. _____

16. _____

17. _____

18. _____

19. _____

20. _____

21. _____

22. _____

23. _____

24. _____

25. _____

26. _____

27. _____

28. _____

29. _____

Add or subtract.

15. $6\sqrt{2} - 4\sqrt{2}$

16. $3\sqrt{90} + 2\sqrt{40}$

17. $\sqrt{25x} - 5\sqrt{x}$

Simplify.

18. $-\sqrt{\dfrac{25}{36}}$

19. $\sqrt{\dfrac{y^2}{4}}$

Rationalize the denominator.

20. $\sqrt{\dfrac{1}{3}}$

21. $\sqrt{\dfrac{3}{x}}$

22. $\dfrac{\sqrt{7}}{\sqrt{8}}$

Divide and simplify.

23. $\dfrac{\sqrt{50}}{\sqrt{2}}$

24. $\dfrac{\sqrt{2x^4}}{\sqrt{5x^2}}$

Use the square root table to find the following to three decimal places.

25. $\dfrac{\sqrt{5}}{\sqrt{6}}$

26. $\dfrac{\sqrt{84}}{\sqrt{2}}$

Find the third side of the right triangle.

27. $a = 6,\quad b = 5,\quad c = ?$

Solve.

28. $\sqrt{5x + 1} = 6$

29. $\sqrt{x} = -10$

CUMULATIVE REVIEW   CHAPTERS 9 AND 10

Find the value of x that makes the rational expression meaningless.

1. $\dfrac{2x + 5}{x}$

2. $\dfrac{3x + 4}{x - 5}$

Solve.

3. $\dfrac{x}{4} - 3 = \dfrac{5}{8}$

4. $\dfrac{5}{2x + 3} = \dfrac{10}{x + 15}$

5. $\dfrac{5}{x} + 2 = \dfrac{19}{7}$

6. $\dfrac{3}{x - 4} - \dfrac{2}{x + 4} = \dfrac{x}{x^2 - 16}$

Perform the operation indicated.

7. $\dfrac{3}{x - 7} + \dfrac{x}{x^2 - 49}$

8. $\dfrac{3}{y + 1} + \dfrac{7}{2y - 3}$

9. Find the dimensions of a rectangle with a perimeter of 72 inches if its length is $\frac{7}{5}$ its width.

10. Jon can mow the lawn in 2 hours and Ken can mow it in 3 hours. How long does it take them to mow the lawn together?

11. Janice drives 20 miles per hour faster than Greg. If Janice travels 126 miles in the same time that Greg travels 90 miles, what is the speed of Janice's car?

12. If a baker can make 9 pies in 2 hours, how many pies can he make in 5 hours?

13. If 3 students can paint a room in 5 hours, how long will it take 2 students to do the job?

14. The weight of a steel rod varies directly with the length of the rod. If a 3.5-meter rod weighs 29.4 kilograms, find the weight of a 2-meter piece of the rod.

15. At a constant temperature the volume of a gas varies inversely as the pressure. If the pressure of a gas in a 300-cubic-foot container is 16 pounds per square inch, what is the pressure of the gas if it is expanded to 450 cubic feet?

Simplify.

16. $\dfrac{\dfrac{x + 2}{5}}{\dfrac{3x}{15}}$

17. $\dfrac{\dfrac{x}{x - 2}}{\dfrac{4x}{x^2 - 4}}$

Find the following roots.

18. $\sqrt{121}$

19. $\sqrt{64a^2}$

20. $\sqrt{x^2 + 10x + 25}$

Simplify.

21. $\sqrt{24}$

22. $\sqrt{54x^2}$

23. $\sqrt{81x^7}$

Multiply and simplify. Assume that the variables are positive.

24. $\sqrt{3}\sqrt{15}$

25. $\sqrt{7}\sqrt{7}$

26. $\sqrt{5x^2}\sqrt{15x^3}$

Simplify and combine like terms. Assume that the variables are positive.

27. $3\sqrt{28} - 2\sqrt{63}$

28. $5\sqrt{40} - 3\sqrt{48} + 2\sqrt{90}$

29. $2\sqrt{x^3} - 4\sqrt{9x^3}$

Simplify.

30. $-\sqrt{\dfrac{81}{100}}$

31. $\sqrt{\dfrac{72}{2}}$

Rationalize the denominator. Assume that the variables are positive.

32. $\sqrt{\dfrac{7}{3}}$

33. $\dfrac{\sqrt{5}}{\sqrt{y}}$

Find the length of the third side of the right triangle (c is the length of the hypotenuse).

34. $a = 5, \quad b = 8$

35. $a = 4, \quad c = 12$

Solve.

36. $\sqrt{x - 2} = 6$

37. $\sqrt{2y + 3} = \sqrt{4y - 3}$

11

Quadratic Equations

11.1 QUADRATIC EQUATIONS AND SOLVING BY SQUARE ROOTS

1 Quadratic Equations

Recall that the degree of a polynomial in one variable is the largest exponent on the variable in any of the terms.

> Quadratic equations are second-degree equations of the form $ax^2 + bx + c = 0$, where a, b, and c are real numbers and $a \neq 0$.

EXAMPLE 1 Are the following quadratic equations?

a. $3x^2 + 2x + 4 = 0$ Yes; the largest exponent on the variable is 2 (second degree)

b. $x + 3 = 8$ No; this is an equation of the first degree

c. $x^2 - x = 0$ Yes

d. $x^4 - x^2 + 2 = 0$ No; this is an equation of the fourth degree ∎

☐ **DO EXERCISE 1.**

> Quadratic equations have exactly two solutions. Sometimes they are identical.

2 Solving by Taking the Square Root

The easiest way to solve equations of the type

$$x^2 = c \quad \text{or} \quad (x + a)^2 = c$$

is to take the square roots of both sides of the equation.

EXAMPLE 2 Solve.

a. $x^2 = 16$

$\quad x = \pm\sqrt{16}$ Since $(\sqrt{16})^2 = 16$ and $(-\sqrt{16})^2 = 16$

$\quad x = \pm 4$

☐ **Exercise 1** Are the following quadratic equations?

a. $x^2 + x = 0$

b. $x + y = 4$

c. $2x + x^2 = 8$

d. $x^3 - x^2 = 0$

Quadratic equations second degree equations of the form $ax^2 + bx + c = 0$ (or $ax^2 + bx + c = y$) where a, b and c are real numbers and $a \neq 0$.

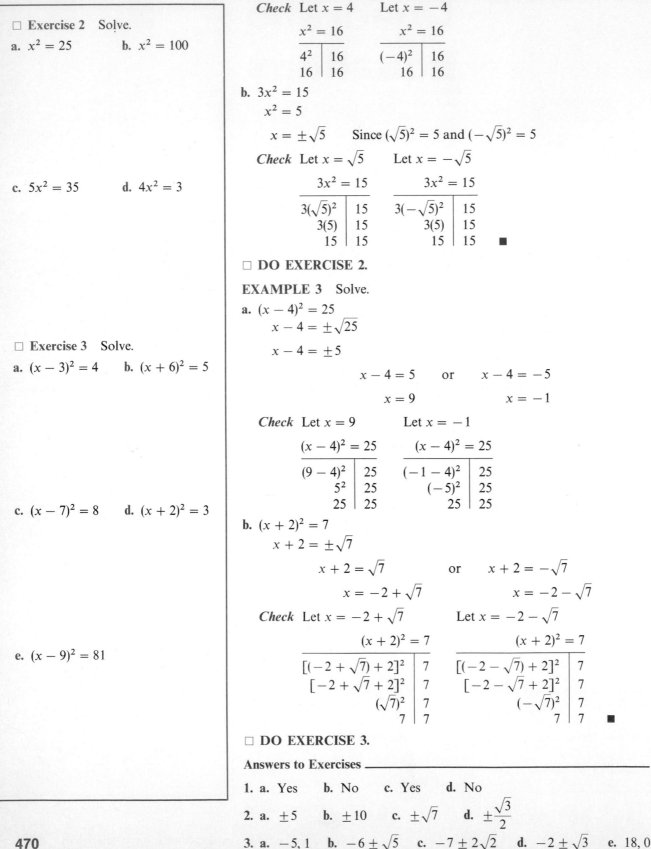

□ **Exercise 2** Solve.

a. $x^2 = 25$ **b.** $x^2 = 100$

c. $5x^2 = 35$ **d.** $4x^2 = 3$

□ **Exercise 3** Solve.

a. $(x - 3)^2 = 4$ **b.** $(x + 6)^2 = 5$

c. $(x - 7)^2 = 8$ **d.** $(x + 2)^2 = 3$

e. $(x - 9)^2 = 81$

Check Let $x = 4$ Let $x = -4$

$$x^2 = 16 \qquad\qquad x^2 = 16$$

4^2	16
16	16

$(-4)^2$	16
16	16

b. $3x^2 = 15$

$x^2 = 5$

$x = \pm\sqrt{5}$ Since $(\sqrt{5})^2 = 5$ and $(-\sqrt{5})^2 = 5$

Check Let $x = \sqrt{5}$ Let $x = -\sqrt{5}$

$$3x^2 = 15 \qquad\qquad 3x^2 = 15$$

$3(\sqrt{5})^2$	15
$3(5)$	15
15	15

$3(-\sqrt{5})^2$	15
$3(5)$	15
15	15	■

□ **DO EXERCISE 2.**

EXAMPLE 3 Solve.

a. $(x - 4)^2 = 25$

$x - 4 = \pm\sqrt{25}$

$x - 4 = \pm 5$

$x - 4 = 5$ or $x - 4 = -5$

$x = 9$ $x = -1$

Check Let $x = 9$ Let $x = -1$

$$(x - 4)^2 = 25 \qquad\qquad (x - 4)^2 = 25$$

$(9 - 4)^2$	25
5^2	25
25	25

$(-1 - 4)^2$	25
$(-5)^2$	25
25	25

b. $(x + 2)^2 = 7$

$x + 2 = \pm\sqrt{7}$

$x + 2 = \sqrt{7}$ or $x + 2 = -\sqrt{7}$

$x = -2 + \sqrt{7}$ $x = -2 - \sqrt{7}$

Check Let $x = -2 + \sqrt{7}$ Let $x = -2 - \sqrt{7}$

$$(x + 2)^2 = 7 \qquad\qquad (x + 2)^2 = 7$$

$[(-2 + \sqrt{7}) + 2]^2$	7
$[-2 + \sqrt{7} + 2]^2$	7
$(\sqrt{7})^2$	7
7	7

$[(-2 - \sqrt{7}) + 2]^2$	7
$[-2 - \sqrt{7} + 2]^2$	7
$(-\sqrt{7})^2$	7
7	7	■

□ **DO EXERCISE 3.**

Answers to Exercises

1. a. Yes   **b.** No   **c.** Yes   **d.** No

2. a. ± 5 **b.** ± 10 **c.** $\pm\sqrt{7}$ **d.** $\pm\dfrac{\sqrt{3}}{2}$

3. a. $-5, 1$ **b.** $-6 \pm \sqrt{5}$ **c.** $-7 \pm 2\sqrt{2}$ **d.** $-2 \pm \sqrt{3}$ **e.** $18, 0$

PROBLEM SET 11.1

Solve.

1. $x^2 = 9$

2. $x^2 = 1$

3. $3x^2 = 48$

4. $2x^2 = 98$

5. $2x^2 = 32$

6. $4x^2 = 100$

7. $x^2 = 3$

8. $x^2 = 11$

9. $x^2 = 8$

10. $x^2 = 12$

11. $3x^2 = 36$

12. $4x^2 = 20$

13. $x^2 - 15 = 0$

14. $x^2 - 7 = 0$

15. $3x^2 - 48 = 0$

16. $2x^2 - 50 = 0$

17. $(y + 3)^2 = 36$

18. $(x - 2)^2 = 25$

19. $(y - 1)^2 = 6$

$y - 1 = \pm \sqrt{6}$

$y - 1 = 1 \pm \sqrt{6}$

20. $(x + 4)^2 = 15$

21. $(x - 7)^2 - 8 = 0$

$(x-7)^2 = 8$ $x = 7 \pm 2\sqrt{2}$

$x - 7 = \pm\sqrt{8}$

$x - 7 = -2\sqrt{2}$

22. $(y + 9)^2 - 24 = 0$

23. Four times the square of a number is 144. What is the number?

24. Twice the square of a number is 34. What is the number?

25. Equations of the form $ax^2 + bx + c = 0$, where a, b, and c are real numbers and $a \neq 0$, are a special type of second-degree equation called _____ equations.

26. Quadratic equations have exactly _____ solutions.

Checkup

The following problems provide a review of some of Sections 5.7 and 10.1. These problems will help you with the next section.
Factor.

27. $x^2 - 9x + 8$

28. $x^2 + 8x + 12$

29. $6x^2 + 11x - 10$

30. $8x^2 - 14x + 3$

31. $8x^2 + 22x + 15$

32. $12x^2 - x - 20$

Find the square roots.

33. $\sqrt{36x^2}$

34. $\sqrt{81y^2}$

35. $\sqrt{x^2 + 10x + 25}$

36. $\sqrt{y^2 - 14x + 49}$

Zero product rule   if the product of two real numbers is zero then at least one of the numbers must be zero.

11.2   SOLVING BY FACTORING

OBJECTIVES

When we multiply two numbers and one of them is zero, the result is always zero.

1. $(4)(0) = 0$
2. $(0)(7) = 0$
3. $(x + 3)(x - 2) = 0$ means that

$$x + 3 = 0 \quad \text{or} \quad x - 2 = 0$$

since $x + 3$ is a number and $x - 2$ is a number.
4. $x(x - 6) = 0$ means that

$$x = 0 \quad \text{or} \quad x - 6 = 0$$

1 Solve quadratic equations by factoring

2 Solve rational equations that when simplified result in quadratic equations

3 Solve equations containing radicals if the result of squaring both sides of the equation is a quadratic equation

> **Zero Product Rule**
>
> If the product of two real numbers is zero, at least one of the numbers must be zero.

1 Factoring to Find Solutions

The solution of a quadratic equation by factoring is based on the zero product rule.

EXAMPLE 1   Solve $(x + 3)(x - 2) = 0$.

The equation is already factored. $(x + 3)(x - 2) = 0$ means that

$$x + 3 = 0 \quad \text{or} \quad x - 2 = 0$$
$$x = -3 \qquad\qquad x = 2$$

Check   Let $x = -3$ Let $x = 2$

$(x + 3)(x - 2) = 0$		$(x + 3)(x - 2) = 0$	
$(-3 + 3)(-3 - 2)$	0	$(2 + 3)(2 - 2)$	0
$(0)(-5)$	0	$(5)(0)$	0
0	0	0	0 ∎

☐ DO EXERCISE 1.

EXAMPLE 2   Solve $x(x - 6) = 0$.

$$x = 0 \quad \text{or} \quad x - 6 = 0$$
$$x = 6$$

Check   Let $x = 0$ Let $x = 6$

$x(x - 6) = 0$		$x(x - 6) = 0$	
$0(0 - 6)$	0	$6(6 - 6)$	0
$0(-6)$	0	$6(0)$	0
0	0	0	0 ∎

☐ DO EXERCISE 2.

☐ **Exercise 1**   Solve.

a. $(x - 1)(x + 4) = 0$

b. $(x + 7)(x + 4) = 0$

c. $(x - 8)(x - 5) = 0$

d. $(x + 3)(x - 6) = 0$

☐ **Exercise 2**   Solve.

a. $x(x - 3) = 0$

b. $x(x + 8) = 0$

c. $x(x - 1) = 0$

d. $x(x + 10) = 0$

If a quadratic equation can be factored, solution by factoring is the easiest way to solve the equation.

EXAMPLE 3 Solve $x^2 + 5x + 6 = 0$.

$$x^2 + 5x + 6 = 0$$
$$(x + 2)(x + 3) = 0$$

$$x + 2 = 0 \qquad \text{or} \qquad x + 3 = 0$$
$$x = -2 \qquad\qquad\qquad x = -3$$

Check Let $x = -2$ $\qquad$ Let $x = -3$

$x^2 + 5x + 6 = 0$		$x^2 + 5x + 6 = 0$	
$(-2)^2 + 5(-2) + 6$	0	$(-3)^2 + 5(-3) + 6$	0
$4 - 10 + 6$	0	$9 - 15 + 6$	0
	$0 \mid 0$		$0 \mid 0$ ∎

□ **DO EXERCISE 3.**

EXAMPLE 4 Solve.

a. $2x^2 - x = 3$

We must set one side of the equation equal to zero to use the zero product rule.

$$2x^2 - x - 3 = 0$$
$$(2x - 3)(x + 1) = 0$$

$$2x - 3 = 0 \qquad \text{or} \qquad x + 1 = 0$$
$$2x = 3 \qquad\qquad\qquad x = -1$$
$$x = \frac{3}{2}$$

When we solve by factoring we do not need to check except to find errors in solving.

b. $3x^2 = 9x$

$3x^2 - 9x = 0$ $\qquad$ Setting the right side of the equation to zero

$3x(x - 3) = 0$

$$3x = 0 \qquad \text{or} \qquad x - 3 = 0$$
$$x = 0 \qquad\qquad\qquad x = 3$$

Notice that there must be two solutions to a quadratic equation and that we may not divide both sides of the equation by x or we will lose a solution.

c. $(y - 2)(y + 3) = 14$

We must set the right-hand side of the equation equal to zero. (If the product of two numbers is 14, it is not necessary for one of them to be 14.)

$$(y - 2)(y + 3) - 14 = 0$$

$$y^2 + y - 6 - 14 = 0 \qquad \text{Multiplying}$$

$$y^2 + y - 20 = 0 \qquad \text{Combining like terms}$$

$$(y + 5)(y - 4) = 0$$

$$y + 5 = 0 \qquad \text{or} \qquad y - 4 = 0$$

$$y = -5 \qquad\qquad y = 4 \quad \blacksquare$$

□ **DO EXERCISE 4.**

2 **Solving Rational Equations**

Sometimes when we multiply both sides of the equation by the LCD to solve a rational equation, we get a quadratic equation. We solve it to find the solution(s). Remember that when we solve a rational equation we must check the answer. If the answer makes the denominator of the original problem zero, it is not a solution.

EXAMPLE 5 Solve $\dfrac{1}{3} = \dfrac{1}{x - 2} - \dfrac{4}{x^2 - 4}$.

The LCD is $3(x - 2)(x + 2)$. Multiply both sides of the equation by the LCD.

$$3(x-2)(x+2)\left(\frac{1}{3}\right) = 3(x - 2)(x + 2)\left(\frac{1}{x - 2} - \frac{4}{x^2 - 4}\right)$$

$$3(x-2)(x+2)\left(\frac{1}{3}\right) = 3(x-2)(x+2)\left(\frac{1}{x-2}\right) - 3(x-2)(x+2)\left(\frac{4}{x^2-4}\right)$$

$$(x-2)(x+2) = 3(x+2) - 3(4)$$

$$x^2 - 4 = 3x + 6 - 12$$

$$x^2 - 4 = 3x - 6$$

$$x^2 - 3x + 2 = 0$$

$$(x-2)(x-1) = 0$$

$$x - 2 = 0 \qquad \text{or} \qquad x - 1 = 0$$

$$x = 2 \qquad\qquad x = 1$$

Check Let $x = 2$ Let $x = 1$

$$\dfrac{1}{3} = \dfrac{1}{x - 2} - \dfrac{4}{x^2 - 4} \qquad \dfrac{1}{3} = \dfrac{1}{x - 2} - \dfrac{4}{x^2 - 4}$$

$$\begin{array}{c|c} \dfrac{1}{3} & \dfrac{1}{2 - 2} - \dfrac{4}{2^2 - 4} \\[2mm] \dfrac{1}{3} & \dfrac{1}{0} - \dfrac{4}{0} \end{array} \qquad \begin{array}{c|c} \dfrac{1}{3} & \dfrac{1}{1 - 2} - \dfrac{4}{1^2 - 4} \\[2mm] \dfrac{1}{3} & \dfrac{1}{-1} - \dfrac{4}{-3} \\[2mm] \dfrac{1}{3} & \dfrac{1}{3} \end{array}$$

The answer 2 makes the denominator zero, so the only solution is 1.

□ **DO EXERCISE 5.** $\blacksquare$

□ **Exercise 4** Solve.

a. $y^2 + 2y - 3 = 12$

b. $5x^2 = 5x$

c. $(x - 8)(x + 1) = -20$

d. $(x - 3)(x + 4) = 8$

□ **Exercise 5** Solve.

a. $\dfrac{2}{x} = \dfrac{x}{5x - 12}$

b. $\dfrac{1}{x - 1} + \dfrac{1}{2} = \dfrac{2}{x^2 - 1}$

□ **Exercise 6** Solve.

a. $x - 9 = \sqrt{x - 3}$

b. $x = 1 + 2\sqrt{x - 1}$

3 **Solving Equations with Radicals**

When we square both sides of a radical equation to solve the equation, we may also get a quadratic equation.

EXAMPLE 6 Solve $\sqrt{2x - 3} = x - 3$.

Square both sides of the equation.

$$(\sqrt{2x - 3})^2 = (x - 3)^2$$
$$2x - 3 = x^2 - 6x + 9 \qquad \text{Recall that } (x - 3)^2 = (x - 3)(x - 3)$$
$$0 = x^2 - 8x + 12$$
$$0 = (x - 2)(x - 6)$$
$$x - 2 = 0 \quad \text{or} \quad x - 6 = 0$$
$$x = 2 \qquad\qquad x = 6$$

Check Let $x = 2$ Let $x = 6$

$\sqrt{2x - 3} = x - 3$		$\sqrt{2x - 3} = x - 3$	
$\sqrt{2(2) - 3}$	$2 - 3$	$\sqrt{2(6) - 3}$	$6 - 3$
$\sqrt{4 - 3}$	-1	$\sqrt{12 - 3}$	3
1	-1	$\sqrt{9}$	3
		3	3

The answer 2 does not check. The solution is 6.   ■

□ **DO EXERCISE 6.**

DID YOU KNOW?

Thomas Harriot's book on algebra, *Artis analyticae praxis*, showed how solutions could be used to form equations. This book was not published until 10 years after Harriot's death.

Answers to Exercises

1. a. $1, -4$ **b.** $-7, -4$ **c.** $8, 5$ **d.** $-3, 6$

2. a. $0, 3$ **b.** $0, -8$ **c.** $0, 1$ **d.** $0, -10$

3. a. $-4, 3$ **b.** $\dfrac{1}{4}, -2$

4. a. $-5, 3$ **b.** $0, 1$ **c.** $3, 4$ **d.** $-5, 4$

5. a. $6, 4$ **b.** -3

6. a. 12 **b.** $5, 1$

PROBLEM SET 11.2

Solve.

1. $(x + 6)(x + 2) = 0$ **2.** $(x + 4)(x - 3) = 0$ **3.** $(x - 7)(x + 2) = 0$ **4.** $(x - 1)(x + 6) = 0$

5. $(2x - 1)(x - 8) = 0$ **6.** $(3x + 4)(x - 7) = 0$ **7.** $(5x + 4)(x - 2) = 0$ **8.** $(x + 5)(2x - 3) = 0$

9. $x^2 - 7x + 10 = 0$ **10.** $x^2 + 2x - 15 = 0$ **11.** $t^2 - 8t + 15 = 0$ **12.** $t^2 - 4t - 5 = 0$

13. $x(x + 4) = 0$ **14.** $y(2y - 3) = 0$ **15.** $(x - 4)x = 0$ **16.** $(y + 8)y = 0$

17. $(x - 7)x = 0$ **18.** $x(2x + 1) = 0$ **19.** $3x^2 - 24x = 0$ **20.** $8t^2 + 12t = 0$

21. $x^2 + 5x = 0$ **22.** $y^2 + 8y = 0$ **23.** $9x^2 - 27x = 0$ **24.** $20y^2 - 5y = 0$

25. $x^2 + 5 = 6x$ **26.** $12x^2 - 2 = -5x$ **27.** $6x^2 + x = 5$ **28.** $3y^2 + y = 2$

29. $6y^2 - 4y = 10$ **30.** $2y^2 + 11y = -12$ **31.** $14 = x^2 - 5x$ **32.** $2 = -3x^2 + 7x$

33. $(x + 1)(x + 3) = -1$ **34.** $(x + 1)(x - 5) = 7$ **35.** $(y + 5)(y + 1) = 21$ **36.** $(s + 3)(s + 5) = 3$

37. $(3t - 1)(t + 2) = -4$ **38.** $(2x + 3)(x - 1) = -2$ **39.** $\dfrac{8x + 3}{x} = 3x$ **40.** $\dfrac{x + 3}{x} = \dfrac{12}{x + 3}$

41. $\dfrac{5}{x-3} - \dfrac{30}{x^2-9} = 1$ **42.** $\dfrac{1}{y} + \dfrac{1}{y+1} = \dfrac{5}{6}$ **43.** $\dfrac{8}{y+2} + \dfrac{8}{y-2} = 3$ **44.** $\dfrac{x}{2x+4} = \dfrac{3}{x+2}$

45. $\dfrac{1}{y} = \dfrac{3}{5} - \dfrac{4}{y+5}$ **46.** $\dfrac{5}{x-1} = \dfrac{x+3}{3-x}$ **47.** $\sqrt{5x+1} = x+1$ **48.** $\sqrt{3y+3} + 5 = y$

49. $x + 4 = 4\sqrt{x+1}$ **50.** $\sqrt{2y-1} + 2 = y$ **51.** $\sqrt{30-3x} = x-4$ **52.** $\sqrt{5y+21} - y = 3$

53. $x - 1 = 6\sqrt{x-9}$ **54.** $x = 4\sqrt{x+1} - 4$

55. If the product of two real numbers is zero, at least one of the numbers must be _____.

56. Usually, the easiest method of solving a quadratic equation is by _____, if possible.

57. All possible solutions of rational equations with variables in the denominator and radical equations should be _____.

Checkup

The following problems provide a review of some of Sections 5.4 and 10.2. These problems will help you with the next section.
Multiply. Do not use FOIL.

58. $(x+3)^2$ **59.** $(x-4)^2$ **60.** $(x-7)^2$ **61.** $(x+2)^2$

Simplify.

62. $\sqrt{48}$ **63.** $\sqrt{72}$ **64.** $\sqrt{98x^2}$ **65.** $\sqrt{150y^2}$

66. $\sqrt{x^{10}}$ **67.** $\sqrt{x^{16}}$ **68.** $\sqrt{y^9}$ **69.** $\sqrt{y^{15}}$

11.3 COMPLETING THE SQUARE

OBJECTIVE

1 Many quadratic equations cannot be factored. To solve them we can use a method called *completing the square*. This method is not as easy to use as the quadratic formula that we will study in the next section. However, the method of completing the square is used to derive the quadratic formula and for other purposes in mathematics, so it is necessary for us to be familiar with it.

1 *Solve equations by completing the square*

Recall the method for squaring a binomial:

$$(x + 3)^2 = x^2 + 2(3)(x) + 3^2 = x^2 + 6x + 9$$

The middle term in the result is twice the product of the two original terms, x and 3, which equals $6x$. Notice that the last term, 9, is then the square of one-half the coefficient, 6, of x.

$$\frac{1}{2}(6) = 3$$

$$3^2 - 9$$

We use this fact to find the last term (or complete the square) of an expression.

EXAMPLE 1 Complete the square.

a. $x^2 + 4x$

$$\frac{1}{2}(4) = 2 \qquad \text{One-half the coefficient of } x$$

$$2^2 = 4 \qquad \text{Square one-half the coefficient}$$

Now add this quantity, 4, to the given expression. We have

$$x^2 + 4x + 4 = (x + 2)^2$$

The square has been completed.

b. $x^2 - 6x$

$$\frac{1}{2}(-6) = -3$$

$$(-3)^2 = 9$$

Add 9 to the original expression.

$$x^2 - 6x + 9 = (x - 3)^2$$

To solve an equation, we want to complete the square on one side of the equation.

c. $x^2 + 2x = 5$

$$\frac{1}{2}(2) = 1 \qquad \text{One-half the coefficient of } x$$

$$1^2 \qquad \text{Square one-half the coefficient}$$

Now add this quantity, 1^2, to both sides of the original equation. Remember that an equation is a balance and if we add a quantity to one side, we must also add the same quantity to the other side.

□ **Exercise 1** Complete the square.

a. $x^2 + 6x = 4$

b. $x^2 + 4x = -3$

c. $x^2 + 8x = 2$

d. $x^2 + 12x = 2$

□ **Exercise 2** Solve by completing the square.

a. $x^2 + 2x = 8$

b. $x^2 + 14x = -2$

$$x^2 + 2x = 5 \qquad \text{Original equation}$$
$$x^2 + 2x + 1^2 = 5 + 1^2$$
$$x^2 + 2x + 1^2 = 5 + 1$$
$$(x + 1)^2 = 6$$

The square has been completed. Now the equation can be solved using methods from Section 11.1.

Notice that to begin the process of completing the square, all terms with variables must be on one side of the equation and the constant term must be on the other side.

d. $x^2 + 10x - 4 = 0$

$$x^2 + 10x = 4 \qquad \begin{array}{l}\text{Adding 4 to both sides to get}\\\text{the variables on one side}\end{array}$$

$$\frac{1}{2}(10) = 5$$

Add 5^2 to both sides of the equation.

$$x^2 + 10x + 5^2 = 4 + 5^2$$
$$x^2 + 10x + 5^2 = 4 + 25$$
$$(x + 5)^2 = 29$$

Notice that the numbers connected by the arrows are always the same, so the factorization is easy. ■

□ **DO EXERCISE 1.**

Once we have completed the square we can solve the equation by taking square roots.

EXAMPLE 2 Solve $x^2 + 8x - 9 = 0$.

$$x^2 + 8x = 9 \qquad \text{Adding 9 to both sides}$$

$$\frac{1}{2}(8) = 4$$

Add 4^2 to both sides of the equation.

$$x^2 + 8x + 4^2 = 9 + 4^2$$
$$x^2 + 8x + 4^2 = 9 + 16$$
$$(x + 4)^2 = 25$$
$$x + 4 = \pm 5$$
$$x + 4 = 5 \qquad \text{or} \qquad x + 4 = -5$$
$$x = 1 \qquad\qquad\qquad x = -9 \quad ■$$

□ **DO EXERCISE 2.**

EXAMPLE 3 Solve $x^2 + 3x = 10$.

$$\frac{1}{2}(3) = \frac{3}{2}$$

480 Chapter 11 Quadratic Equations

© 1989 by Prentice Hall

Add $\left(\frac{3}{2}\right)^2$ to both sides of the equation.

$$x^2 + 3x + \left(\frac{3}{2}\right)^2 = 10 + \left(\frac{3}{2}\right)^2$$

$$x^2 + 3x + \left(\frac{3}{2}\right)^2 = 10 + \frac{9}{4}$$

$$x^2 + 3x + \left(\frac{3}{2}\right)^2 = \frac{40 + 9}{4}$$

$$\left(x + \frac{3}{2}\right)^2 = \frac{49}{4}$$

$$x + \frac{3}{2} = \pm\sqrt{\frac{49}{4}}$$

$$x + \frac{3}{2} = \pm\frac{7}{2}$$

$$x + \frac{3}{2} = \frac{7}{2} \quad \text{or} \quad x + \frac{3}{2} = -\frac{7}{2}$$

$$x = \frac{4}{2} \qquad\qquad x = -\frac{10}{2} \qquad \text{Subtracting } -\tfrac{3}{2} \text{ from both sides of the equation}$$

$$x = 2 \qquad\qquad x = -5 \quad \blacksquare$$

☐ **DO EXERCISE 3.**

The preceding method of completing the square should not be used unless the coefficient of x^2 is 1. If the coefficient of x^2 is not 1, we can divide each side of the equation by this coefficient.

EXAMPLE 4 Solve $2x^2 - 8x + 3 = 0$.

$$x^2 - 4x + \frac{3}{2} = 0 \qquad \text{Dividing each side of the equation by 2}$$

$$x^2 - 4x = -\frac{3}{2} \qquad \text{Adding } -\tfrac{3}{2} \text{ to both sides of the equation}$$

$$\frac{1}{2}(-4) = -2$$

Add $(-2)^2$ to both sides of the equation.

$$x^2 - 4x + (-2)^2 = -\frac{3}{2} + (-2)^2$$

$$x^2 - 4x + (-2)^2 = -\frac{3}{2} + 4$$

$$x^2 - 4x + (-2)^2 = \frac{-3 + 8}{2}$$

$$(x - 2)^2 = \frac{5}{2}$$

$$x - 2 = \pm\sqrt{\frac{5}{2}}$$

☐ **Exercise 3** Solve.

a. $x^2 - 5x = 6$

b. $x^2 + x - 1 = 0$
Add 1 to both sides of the equation

□ **Exercise 4** Solve by completing the square.

a. $4x^2 + 12x - 5 = 0$

b. $2x^2 - 3x - 8 = 0$

$$x - 2 = \sqrt{\frac{5}{2}} \qquad \text{or} \qquad x - 2 = -\sqrt{\frac{5}{2}}$$

$$x = 2 + \sqrt{\frac{5}{2}} \qquad\qquad x = 2 - \sqrt{\frac{5}{2}}$$

$$x = 2 + \frac{\sqrt{5}}{\sqrt{2}} \qquad\qquad x = 2 - \frac{\sqrt{5}}{\sqrt{2}}$$

$$x = 2 + \frac{\sqrt{5}}{\sqrt{2}} \cdot \frac{\sqrt{2}}{\sqrt{2}} \qquad x = 2 - \frac{\sqrt{5}}{\sqrt{2}} \cdot \frac{\sqrt{2}}{\sqrt{2}} \qquad \text{Rationalizing the denominator}$$

$$x = 2 + \frac{\sqrt{10}}{2} \qquad\qquad x = 2 - \frac{\sqrt{10}}{2} \qquad \text{Recall that } \sqrt{2}\sqrt{2} = 2$$

$$x = \frac{4 + \sqrt{10}}{2} \qquad\qquad x = \frac{4 - \sqrt{10}}{2} \qquad \text{Finding a common denominator}$$

■

□ **DO EXERCISE 4.**

Solving a Quadratic Equation by Completing the Square

1. If the coefficient of x^2 is 1, go to step 2. If the coefficient of x^2 is not 1, divide each side of the equation by this coefficient.
2. Be sure that terms containing variables are on one side of the equation and constant terms are on the other side.
3. Add the square of half the coefficient of the x term to both sides of the equation. The side containing the variables is now a perfect square trinomial.
4. Factor the perfect square trinomial.
5. Solve the equation by taking the square root of both sides and solving the resulting equations.

Answers to Exercises

1. a. $(x + 3)^2 = 13$ **b.** $(x + 2)^2 = 1$ **c.** $(x + 4)^2 = 18$
d. $(x + 6)^2 = 38$
2. a. $2, -4$ **b.** $-7 \pm \sqrt{47}$
3. a. $6, -1$ **b.** $\dfrac{-1 \pm \sqrt{5}}{2}$
4. a. $\dfrac{-3 \pm \sqrt{14}}{2}$ **b.** $\dfrac{3 \pm \sqrt{73}}{4}$

PROBLEM SET 11.3

Solve by completing the square.

1. $x^2 + 4x = 21$

2. $x^2 + 4x = 12$

3. $x^2 - 6x = 3$

4. $x^2 - 10x = 3$

5. $x^2 + 2x - 8 = 0$

6. $x^2 - 18x + 17 = 0$

7. $x^2 + 14x - 3 = 0$

8. $x^2 - 22x + 20 = 0$

9. $x^2 + 3x - 10 = 0$

10. $x^2 + 5x - 4 = 0$

11. $x^2 - 7x - 2 = 0$

12. $x^2 + 5x - 6 = 0$

13. $2x^2 + 7x = 4$

14. $2y^2 + y = 6$

15. $3x^2 - 9x = -6$

16. $2x^2 - 9x = 5$

17. $3x^2 + 2x - 14 = 0$

18. $2x^2 + 3x - 15 = 0$

19. $4x^2 + 12x - 3 = 0$

20. $9x^2 - 15x - 2 = 0$

21. The first step in solving an equation by completing the square is to divide both sides of the equation by the _____ of x^2 if it is not 1.

22. The second step in solving an equation by completing the square is to get terms containing variables on one side of the equation and _____ on the other side.

23. In the method of solving equations by completing the square, taking the square root of both sides of the equation gives _____ equations to solve.

Checkup

The following problems provide a review of some of Section 10.3. These problems will help you with the next section. Assume that all variables under the radical sign are positive.
Multiply and simplify.

24. $\sqrt{3}\sqrt{24}$

25. $\sqrt{32}\sqrt{2}$

26. $\sqrt{3x}\sqrt{3x}$

27. $\sqrt{7y}\sqrt{7y}$

28. $\sqrt{5x^2}\sqrt{15x^3}$

29. $\sqrt{5y^4}\sqrt{12y^3}$

Add or subtract.

30. $7\sqrt{3} + 9\sqrt{3}$

31. $4\sqrt{15} - 6\sqrt{15}$

32. $5\sqrt{3} - 6\sqrt{12}$

33. $8\sqrt{5} + 3\sqrt{20}$

34. $3\sqrt{27} + \sqrt{48}$

35. $2\sqrt{90} - 6\sqrt{40}$

11.4 QUADRATIC FORMULA

1 Solving Quadratic Equations

We may use the method of completing the square to solve the general quadratic equation.

$$ax^2 + bx + c = 0 \qquad a \neq 0$$

This gives us a formula, called the **quadratic formula**, which may be used to solve any quadratic equation. The solution of the general quadratic equation to find the quadratic formula is shown below.

$$ax^2 + bx + c = 0$$

$$ax^2 + bx = -c$$

$$x^2 + \frac{bx}{a} = -\frac{c}{a} \qquad \text{Dividing each term by } a \text{ to make the coefficient of } x^2 \text{ equal to 1}$$

$$\frac{1}{2}\left(\frac{b}{a}\right) = \frac{b}{2a}$$

Add $(b/2a)^2$ to both sides of the equation.

$$x^2 + \frac{bx}{a} + \left(\frac{b}{2a}\right)^2 = -\frac{c}{a} + \left(\frac{b}{2a}\right)^2$$

$$\left(x + \frac{b}{2a}\right)^2 = -\frac{c}{a} + \frac{b^2}{4a^2}$$

$$\left(x + \frac{b}{2a}\right)^2 = \frac{-c(4a) + b^2}{4a^2} \qquad \text{Finding a common denominator}$$

$$\left(x + \frac{b}{2a}\right)^2 = \frac{b^2 - 4ac}{4a^2} \qquad \text{Rearranging terms in the numerator}$$

$$x + \frac{b}{2a} = \pm\sqrt{\frac{b^2 - 4ac}{4a^2}}$$

$$x + \frac{b}{2a} = \pm\frac{\sqrt{b^2 - 4ac}}{2a}$$

$$x = \frac{-b}{2a} \pm \frac{\sqrt{b^2 - 4ac}}{2a}$$

$$x = \frac{-b \pm \sqrt{b^2 - 4ac}}{2a} \qquad \text{Quadratic formula}$$

The quadratic formula is

$$x = \frac{-b \pm \sqrt{b^2 - 4ac}}{2a}$$

□ **Exercise 1** Solve using the quadratic formula.

$$x^2 + x - 1 = 0$$

b. $3x^2 + 2x - 4 = 0$

c. $x^2 + 4x = -2$

d. $-2x^2 + x = -3$

EXAMPLE 1 Solve.

a. $2x^2 + 2x - 1 = 0$

$ax^2 + bx + c = 0$ General quadratic equation

$a = 2, \quad b = 2, \quad c = -1$

$$x = \frac{-b \pm \sqrt{b^2 - 4ac}}{2a}$$ Quadratic formula

$$x = \frac{-2 \pm \sqrt{(2)^2 - 4(2)(-1)}}{2(2)}$$

$$x = \frac{-2 \pm \sqrt{4 + 8}}{4}$$

$$x = \frac{-2 \pm \sqrt{12}}{4}$$

$$x = \frac{-2 \pm \sqrt{4(3)}}{4}$$

$$x = \frac{-2 \pm 2\sqrt{3}}{4}$$

$$x = \frac{2(-1 \pm \sqrt{3})}{4}$$ Factoring

$$x = \frac{-1 \pm \sqrt{3}}{2}$$

b. $x^2 - 3x = -1$

$x^2 - 3x + 1 = 0$ Setting the right side of the equation equal to zero

$ax^2 + bx + c = 0$ General quadratic equation

$a = 1, \quad b = -3, \quad c = 1$

$$x = \frac{-b \pm \sqrt{b^2 - 4ac}}{2a}$$ Quadratic formula

$$x = \frac{-(-3) \pm \sqrt{(-3)^2 - 4(1)(1)}}{2(1)}$$

$$x = \frac{3 \pm \sqrt{9 - 4}}{2}$$

$$x = \frac{3 \pm \sqrt{5}}{2}$$ ■

□ **DO EXERCISE 1.**

EXAMPLE 2   Solve $3x^2 - 2x - 1 = 0$ using the quadratic formula.

$$a = 3, \quad b = -2, \quad c = -1$$

$$x = \frac{-b \pm \sqrt{b^2 - 4ac}}{2a} \qquad \text{Quadratic formula}$$

$$x = \frac{-(-2) \pm \sqrt{(-2)^2 - 4(3)(-1)}}{2(3)}$$

$$x = \frac{2 \pm \sqrt{4 + 12}}{6}$$

$$x = \frac{2 \pm \sqrt{16}}{6}$$

$$x = \frac{2 + 4}{6} \qquad \text{or} \qquad x = \frac{2 - 4}{6}$$

$$x = \frac{6}{6} \qquad\qquad\qquad x = -\frac{2}{6}$$

$$x = 1 \qquad\qquad\qquad x = -\frac{1}{3} \quad \blacksquare$$

It is easier to solve the equation in Example 2 by factoring as follows.

$$3x^2 - 2x - 1 = 0$$

$$(x - 1)(3x + 1) = 0$$

$$x - 1 = 0 \qquad \text{or} \qquad 3x + 1 = 0$$

$$x = 1 \qquad\qquad\qquad 3x = -1$$

$$x = -\frac{1}{3}$$

To solve a quadratic equation, first try factoring. If you cannot factor, use the quadratic formula.

□ **DO EXERCISE 2.**

2 Approximating Solutions

Solutions to quadratic equations may be approximated using a calculator or a square root table.

□ **Exercise 2**   Solve, using either factoring or the quadratic formula.

a. $2x^2 + x - 1 = 0$

b. $x^2 + 5x - 2 = 0$

c. $3x^2 - 5x = 0$

□ **Exercise 3** Solve the following quadratic equations and approximate the solution to three decimal places.

a. $x^2 + 4x - 2 = 0$

b. $2x^2 + 5x - 1 = 0$

EXAMPLE 3 Solve $x^2 + 3x - 1 = 0$.

$$a = 1, \quad b = 3, \quad c = -1$$

$$x = \frac{-b \pm \sqrt{b^2 - 4ac}}{2a} \qquad \text{Quadratic formula}$$

$$x = \frac{-3 \pm \sqrt{(3)^2 - 4(1)(-1)}}{2}$$

$$x = \frac{-3 \pm \sqrt{9 + 4}}{2}$$

$$x = \frac{-3 \pm \sqrt{13}}{2}$$

$$x = \frac{-3 + \sqrt{13}}{2} \qquad \text{or} \qquad x = \frac{-3 - \sqrt{13}}{2}$$

From the square root table, we find that $\sqrt{13} \approx 3.606$.

$$x \approx \frac{-3 + 3.606}{2} \qquad \text{or} \qquad x \approx \frac{-3 - 3.606}{2}$$

$$x \approx \frac{0.606}{2} \qquad\qquad\qquad x \approx \frac{-6.606}{2}$$

$$x \approx 0.303 \qquad\qquad\qquad x \approx -3.303 \quad ■$$

□ **DO EXERCISE 3.**

Sometimes when we try to solve a quadratic equation, we get a negative number under the radical sign. When this happens, the quadratic equation has no real number solutions.

DID YOU KNOW?

Our knowledge of Babylonian mathematics comes from the thousands of clay tablets unearthed by archaeologists in Mesopotamia. Some tablets from about 1600 B.C. show tables of squares and square roots. Other tablets show numbers and their reciprocals.

Answers to Exercises

1. **a.** $\frac{-1 \pm \sqrt{5}}{2}$ **b.** $\frac{-1 \pm \sqrt{13}}{3}$ **c.** $-2 \pm \sqrt{2}$ **d.** $-1, \frac{3}{2}$

2. **a.** $\frac{1}{2}, -1$, by factoring **b.** $\frac{-5 \pm \sqrt{33}}{2}$ **c.** $0, \frac{5}{3}$, by factoring

3. **a.** $0.449, -4.449$ **b.** $0.186, -2.686$

PROBLEM SET 11.4

A. *Solve using the quadratic formula.*

1. $x^2 + 5x - 1 = 0$ **2.** $y^2 - 3y - 2 = 0$ **3.** $2x^2 + 7x - 2 = 0$ **4.** $3x^2 + x - 3 = 0$ **5.** $x^2 - 3x = 3$

6. $t^2 + 6t = 4$ **7.** $4y^2 + 5y = 2$ **8.** $5x^2 - 2x = 1$ **9.** $3x^2 - 2x = 4$ **10.** $3t^2 + 2t = 7$

B. *Solve. Use either factoring or the quadratic formula.*

11. $x^2 + 4x = -4$ **12.** $x^2 + 3x = 3$ **13.** $y^2 - 4y = 21$ **14.** $t^2 + 5t = 2$

15. $4x^2 + 11x - 3 = 0$ **16.** $6y^2 + 17y - 3 = 0$ **17.** $x^2 = -5x$ **18.** $3t^2 = 6t$

19. $5x^2 - 2x = 2$ **20.** $3z^2 - 2z = 2$ **21.** $(x + 3)(x - 1) = 7$ **22.** $(y - 3)(y - 4) = 16$

23. $t(t - 2) = 9$ **24.** $z(z + 8) = 1$

C. *Use the square root table to approximate the solutions to three places.*

25. $x^2 - 4x - 1 = 0$ **26.** $x^2 - 5x - 4 = 0$ **27.** $3x^2 + x - 1 = 0$ **28.** $4y^2 + y = 4$

29. If the quadratic equation $ax^2 + bx + c = 0$ is solved for x by completing the square, we get the _____ _____.

30. Before solving a quadratic equation by the quadratic formula, try solving by _____.

31. A quadratic equation has no _____ number solutions if its solutions contain a radicand that is negative.

Checkup

The following problems provide a review of some of Sections 6.2 and 10.4. Some of these problems will help you with the next section.
Graph using intercepts.

32. $y = 2x + 4$

33. $y = 3x - 6$

34. $5x + 2y = 10$

35. $2x - 3y = 6$

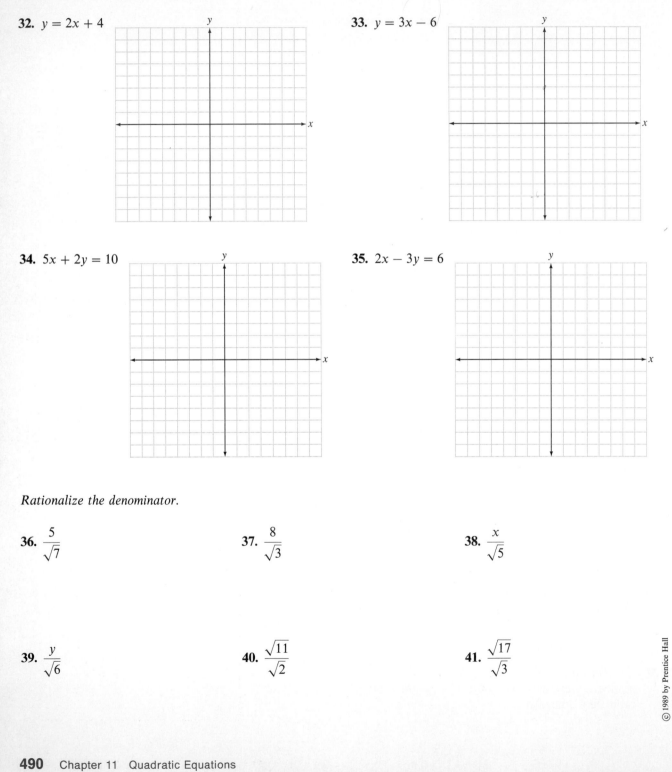

Rationalize the denominator.

36. $\dfrac{5}{\sqrt{7}}$

37. $\dfrac{8}{\sqrt{3}}$

38. $\dfrac{x}{\sqrt{5}}$

39. $\dfrac{y}{\sqrt{6}}$

40. $\dfrac{\sqrt{11}}{\sqrt{2}}$

41. $\dfrac{\sqrt{17}}{\sqrt{3}}$

Vertex the highest point on the graph of $y = ax^2 + bx + c$ if it opens downward or the lowest point if it opens upward.

11.5 GRAPHS OF QUADRATIC EQUATIONS

OBJECTIVE

1 Quadratic equations in two variables of the form

$$y = ax^2 + bx + c \qquad a \neq 0$$

1 *Graph quadratic equations using the vertex and the x-intercepts*

may be graphed. If we set $y = 0$, we find the x-intercepts. The y-intercept of the graph is found by letting $x = 0$. The graph also has a high point or low point. This is called the **vertex**. It may be shown that for any graph of $y = ax^2 + bx + c$, the x-coordinate of the vertex is $x = -(b/2a)$. The y-coordinate of the vertex may be obtained by substituting this x value into the given equation.

EXAMPLE 1

a. Sketch the graph of $y = x^2 - x - 6$.

1. Find the coordinates of the vertex.
Compare the equation $y = x^2 - x - 6$ with the general equation $y = ax^2 + bx + c$. Notice that for this equation $a = 1$ and $b = -1$. The x-coordinate of the vertex is

$$x = -\frac{b}{2a} = \frac{-(-1)}{2(1)} = \frac{1}{2}$$

To find the y-coordinate, substitute the value of the x-coordinate of the vertex into the original equation.

$$y = x^2 - x - 6$$

$$y = \left(\frac{1}{2}\right)^2 - \frac{1}{2} - 6$$

$$y = \frac{1}{4} - \frac{1}{2} - 6 = -6\frac{1}{4}$$

The coordinates of the vertex are $(\frac{1}{2}, -6\frac{1}{4})$.

2. To find the x-intercepts, let $y = 0$.

$$0 = x^2 - x - 6$$

$$0 = (x - 3)(x + 2)$$

$$x - 3 = 0 \qquad \text{or} \qquad x + 2 = 0$$

$$x = 3 \qquad\qquad\qquad x = -2$$

The x-intercepts of the graph are $(3, 0)$ and $(-2, 0)$.

3. To find the y-intercept, let $x = 0$.

$$y = x^2 - x - 6$$

$$y = 0^2 - 0 - 6$$

$$y = -6$$

The y-intercept is $(0, -6)$.

4. Plot the *x*-intercepts, the *y*-intercept, and the vertex. Draw a smooth curve through these points.

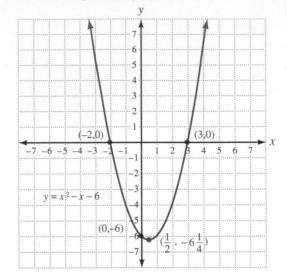

When we connect the points, we get a U-shaped curve. Graphs of quadratic equations of the form $y = ax^2 + bx + c$ are always U-shaped. Sometimes the U is inverted. These curves are called *parabolas*.

b. Sketch the graph of $y = x^2 - 3x - 4$.

  1. Find the coordinates of the vertex.
  Compare the equation $y = x^2 - 3x - 4$ with the general equation $y = ax^2 + bx + c$. Notice that $a = 1$ and $b = -3$. The *x*-coordinate of the vertex is

$$x = -\frac{b}{2a} = -\frac{(-3)}{2(1)} = \frac{3}{2} = 1\frac{1}{2}$$

The *y*-coordinate of the vertex is

$$y = \left(\frac{3}{2}\right)^2 - 3\left(\frac{3}{2}\right) - 4 \qquad \text{Substituting } \tfrac{3}{2} \text{ into the original equation}$$

$$y = \frac{9}{4} - \frac{9}{2} - 4 = \frac{9}{4} - \frac{18}{4} - \frac{16}{4}$$

$$y = -\frac{25}{4} = -6\frac{1}{4}$$

The coordinates of the vertex are $(1\frac{1}{2}, -6\frac{1}{4})$.

  2. To find the *x*-intercepts, let $y = 0$.

$$0 = (x - 4)(x + 1)$$

$$x - 4 = 0 \qquad \text{or} \qquad x + 1 = 0$$

$$x = 4 \qquad\qquad\qquad x = -1$$

The *x*-intercepts are $(4, 0)$ and $(-1, 0)$.

3. To find the y-intercept, let $x = 0$.

$$y = x^2 - 3x - 4$$

$$y = 0^2 - 3(0) - 4$$

$$y = -4$$

4. The y-intercept is $(0, -4)$.

Plot the vertex, the y-intercept and the x-intercepts and draw a smooth curve through these points. ■

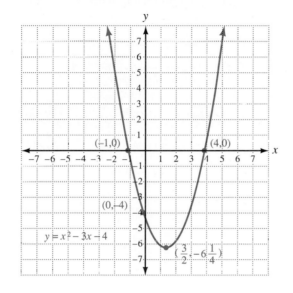

□ **DO EXERCISE 1.**

If in the equation $y = ax^2 + bx + c$, b equals zero, so that the equation does not have an x term, the y-intercept and the vertex are the same point.

EXAMPLE 2 Sketch the graph of $y = -2x^2 + 4$.

1. Find the coordinates of the vertex.
Compare the general equation $y = ax^2 + bx + c$ with the given equation $y = -2x^2 + 4$. For this equation $a = -2$ and $b = 0$. The x-coordinate of the vertex is

$$x = -\frac{b}{2a} = \frac{0}{2(-2)} = 0$$

The y-coordinate of the vertex is

$$y = -2(0)^2 + 4 = 4$$

The coordinates of the vertex are $(0, 4)$.

□ **Exercise 1** Sketch the graph of the following equations.

a. $y = x^2 + 4x - 5$

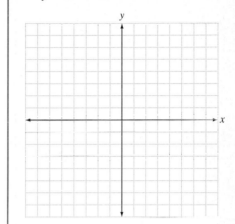

b. $y = 2x^2 - 5x - 3$

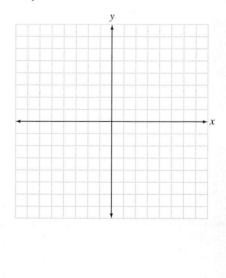

□ **Exercise 2**   Sketch the graph of the equations.

a. $y = -x^2 + 9$

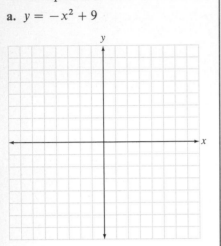

b. $y = -2x^2 + x + 1$

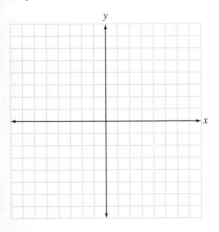

2. Find the x-intercepts. Let $y = 0$.

$$0 = -2x^2 + 4$$
$$2x^2 = 4$$
$$x^2 = 2$$
$$x = \pm\sqrt{2}$$
$$x \approx 1.4 \quad \text{or} \quad x \approx -1.4$$

The x-intercepts are about $(1.4, 0)$ and $(-1.4, 0)$.

3. Find the y-intercept. Let $x = 0$.

$$y = -2x^2 + 4$$
$$y = -2(0)^2 + 4$$
$$y = 4$$

The y-intercept is $(0, 4)$.

4. Plot the x-intercepts and the vertex. Sketch a smooth curve through these points.

Notice that since this equation does not have an x term, the y-intercept has the same coordinates as the vertex.

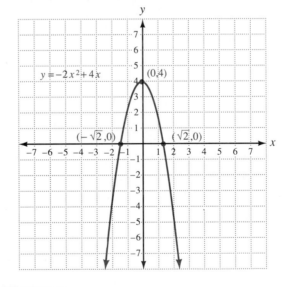

□ **DO EXERCISE 2.**

Sometimes the x-intercept, the y-intercept, and the vertex are the same point. Then we choose at least two other values for x (one negative and one positive), find the corresponding y values, and plot these points and the vertex.

EXAMPLE 3   Sketch the graph of $y = x^2$.

1. Find the coordinates of the vertex. For the given equation $a = 1$ and $b = 0$. The x-coordinate of the vertex is

$$x = \frac{b}{2a} = \frac{0}{2(1)} = \frac{0}{2} = 0$$

The y-coordinate of the vertex is

$$y = (0)^2 = 0$$

The coordinates of the vertex are $(0, 0)$.

2. Find the x-intercepts. Let $y = 0$.

$$0 = x^2$$

$$0 = x^2$$

$$x = 0$$

The x-intercept and the vertex are the same point whose coordinates are $(0, 0)$. Since this equation does not have an x term, the coordinates of the y-intercept are also $(0, 0)$.

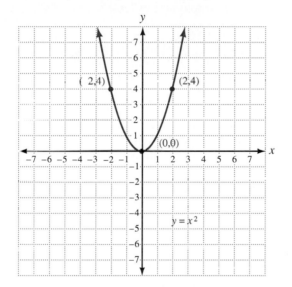

3. Choose at least two other values for x. We chose 2 and -2. Find the corresponding y values.

Let $x = 2$ Let $x = -2$

$y = x^2$ $y = x^2$

$y = (2)^2$ $y = (-2)^2$

$y = 4$ $y = 4$

x	y
2	4
-2	4

4. Plot the vertex and the points from the table. Connect the points with a smooth curve.

Note It may be easier to plot points than to find the vertex and x-intercepts when graphing equations that do not have an x term. There are also other methods available for graphing parabolas, but we leave those to more advanced courses. ∎

☐ **DO EXERCISE 3.**

☐ **Exercise 3**   Sketch the graph of $y = -x^2$.

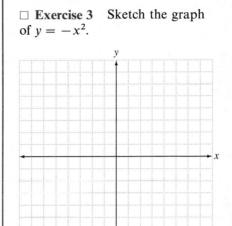

Graphing Equations of the Form $y = ax^2 + bx + c$.

1. Find the coordinates of the vertex. The x-coordinate is $-(b/2a)$. Use this value to find the y-coordinate in the given equation.
2. Find the x-intercepts by letting $y = 0$.
3. Find the y-intercept by letting $x = 0$. If the y-intercept and a point from steps 1 or 2 are the same, find additional points by choosing values for x to plot at least three points. Be sure to plot both negative and positive values for x.
4. Plot all of the points above and connect them with a smooth curve.

For a more accurate graph, plot at least five points.

Answers to Exercises

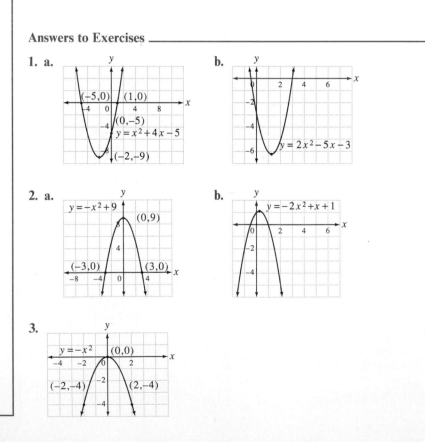

1. a. (−5,0) (1,0) (0,−5) $y = x^2 + 4x - 5$ (−2,−9)

   b. $y = 2x^2 - 5x - 3$

2. a. $y = -x^2 + 9$ (0,9) (−3,0) (3,0)

   b. $y = -2x^2 + x + 1$

3. $y = -x^2$ (0,0) (−2,−4) (2,−4)

PROBLEM SET 11.5

Sketch the graph of each equation.

1. $y = x^2 - 4$

2. $y = -x^2 + 4$

3. $y = -3x^2$

4. $y = 4x^2$

5. $y = x^2 - 4x - 5$

6. $y = x^2 - x - 2$

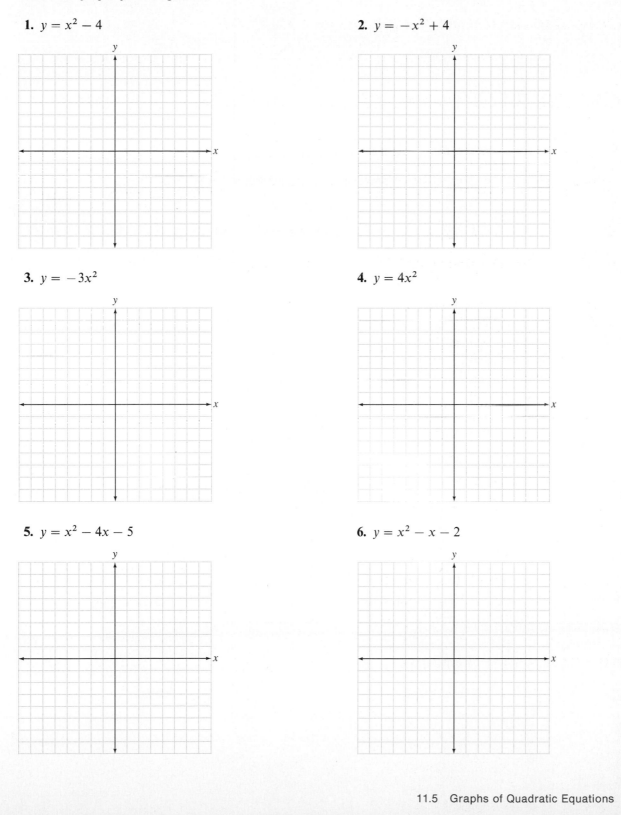

7. $y = -x^2 - 4x - 4$

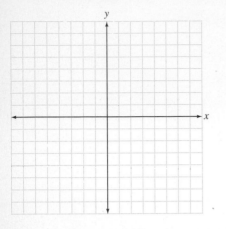

8. $y = x^2 + 2x + 1$

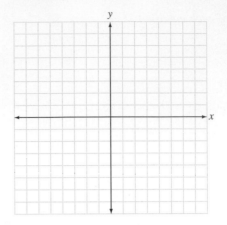

9. $y = x^2 + x - 1$

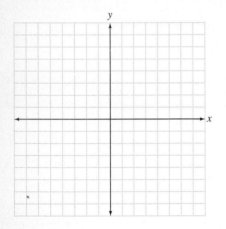

10. $y = -x^2 + 2x + 1$

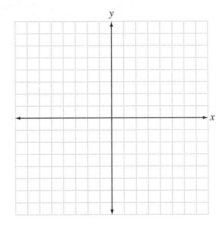

11. $y = 2x^2 + 5x - 12$

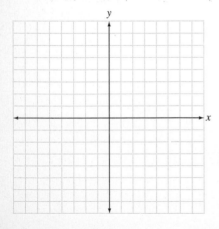

12. $y = -2x^2 - 3x + 2$

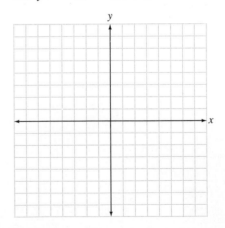

13. $y = x^2 + 2x - 3$

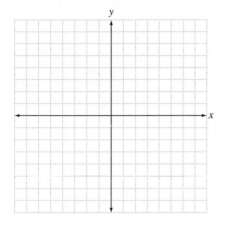

14. $y = x^2 - 4x - 5$

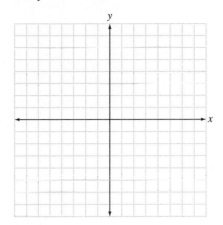

15. $y = -2x^2 - 4x$

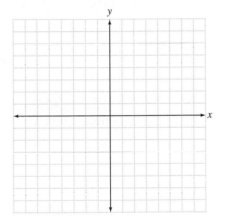

16. $y = 3x^2 - 6x$

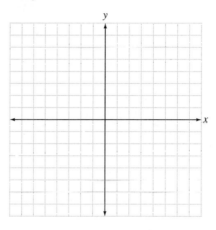

17. To find the y-intercept of the graph of $y = ax^2 + bx + c$, we set x equal to _____.

18. The highest or lowest point on the graph of $y = ax^2 + bx + c$ is called the _____.

19. The _____ of the vertex may be found by substituting $-b/2a$ for x in the given equation.

20. Graphs of quadratic equations in two variables are called _____.

21. For the graph of $y = ax^2 + c$, the y-intercept and the _____ are the same point.

22. If an equation is of the form $y = ax^2$, the graph of the equation has the same point as y-intercept, vertex and _____.

Checkup

The following problems provide a review of some of Section 10.5. These problems will help you with the next section. Find the length of the third side of each right triangle if c is the hypotenuse.

23. $a = 8$, $b = 4$

24. $a = 7$, $b = 2$

25. $a = 4$, $c = 12$

26. $b = 5$, $c = 10$

27. A ladder 16 feet long is leaning against a wall. If the bottom of the ladder is 3 feet from the wall, how far is the top of the ladder from the bottom of the wall?

28. A carpenter built a house that is 90 feet by 40 feet. What is the length of the diagonal across the house?

11.6 APPLIED PROBLEMS

OBJECTIVE

1 It may be necessary to use quadratic equations to solve word problems.

1 *Solve applied problems using quadratic equations*

EXAMPLE 1 John wants to fence his rectangular garden so that it has an area of 112 square yards. He wants the length to be 9 yards greater than the width. Find the length and width of the garden.

Variable Let x = width of the garden

$x + 9$ = length of the garden

Drawing

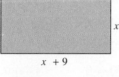

x

$x + 9$

□ **Exercise 1** The length of a rectangle is 5 meters greater than the width. If the area is 84 square meters, find the width and the length.

Equation Area = width · length

$$112 = x(x + 9)$$

$$112 = x^2 + 9x$$

This is a quadratic equation, so set one side equal to zero to solve it.

$$0 = x^2 + 9x - 112$$

$$0 = (x + 16)(x - 7)$$

$$x + 16 = 0 \qquad \text{or} \qquad x - 7 = 0$$

$$x = -16 \qquad\qquad x = 7$$

It does not make sense for the width of the garden to be negative, so we discard this value. Hence $x = 7$ and $x + 9 = 16$. The width is 7 yards and the length is 16 yards. ■

□ **DO EXERCISE 1.**

Quadratic equations may be used to solve right triangles.

EXAMPLE 2   If one leg of a right triangle is 2 centimeters longer than the other leg and the length of the hypotenuse is 12 centimeters, what are the lengths of the legs?

Variable   Let x = length of one leg

$x + 2$ = length of the other leg

Drawing

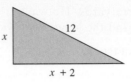

Equation $x^2 + (x + 2)^2 = (12)^2$ Pythagorean theorem

$$x^2 + x^2 + 4x + 4 = 144$$

$$2x^2 + 4x - 140 = 0$$

$$x^2 + 2x - 70 = 0 \qquad \text{Dividing by 2}$$

The equation does not factor, so we use the quadratic formula.

$$a = 1, \quad b = 2, \quad c = -70$$

$$x = \frac{-b \pm \sqrt{b^2 - 4ac}}{2a} \qquad \text{Quadratic formula}$$

$$x = \frac{-2 \pm \sqrt{2^2 - 4(1)(-70)}}{2}$$

$$x = \frac{-2 \pm \sqrt{4 + 280}}{2}$$

$$x = \frac{-2 \pm \sqrt{284}}{2} = \frac{-2 \pm \sqrt{4 \cdot 71}}{2}$$

$$x = \frac{-2 \pm 2\sqrt{71}}{2}$$

From the square root table, $\sqrt{71} \approx 8.426 \approx 8.4$ (rounding to tenths).

$$x \approx \frac{-2 \pm 2(8.4)}{2}$$

$$x \approx \frac{-2 + 16.8}{2} \qquad \text{or} \qquad x \approx \frac{-2 - 16.8}{2}$$

Since we cannot have a negative length,

$$x \approx \frac{-2 + 16.8}{2} \approx \frac{14.8}{2} \approx 7.4 \qquad \text{and} \qquad x + 2 \approx 9.4$$

The lengths of the legs are approximately 7.4 centimeters and 9.4 centimeters.   ■

□ **DO EXERCISE 2.**

Occasionally, when we solve right triangles, the x^2 term is eliminated and we can solve the remaining linear equation for x.

EXAMPLE 3 The top is broken out of a 25-foot tree. It is still attached to the tree and touches the ground 10 feet from the base of the tree. How far up the tree did the top break off?

Variable $25 = $ original height of the tree

$x = $ distance from the ground to the point at which the tree broke off

$25 - x = $ length of the broken part of the tree

Drawing

Equation $x^2 + (10)^2 = (25 - x)^2$

Solve $x^2 + 100 = 625 - 50x + x^2$

$x^2 - x^2 + 100 - 625 = -50x$

$-525 = -50x$

$10.5 = x$

The tree broke off 10.5 feet from the ground. ■

□ **DO EXERCISE 3.**

□ **Exercise 3** A pole breaks over and touches the ground 8 feet from its base. If the pole was originally 20 feet high, how far up did it break over?

Answers to Exercises ──────────────────────────────────

1. Let x = width
   $x + 5$ = length
   $A = 1w$
   $84 = (x + 5)x$
   $0 = x^2 + 5x - 84$
   $0 = (x - 7)(x + 12)$
   $x - 7 = 0$ oɪ $x + 12 = 0$
   $x = 7$ $x = -12$ not allowed
   $x + 5 = 12$
   The width is 7 meters and the length is 12 meters.

2. Let x = length of one leg
   $x + 1$ = length of the other leg
   $x^2 + (x + 1)^2 = 3^2$
   $x^2 + x^2 + 2x + 1 = 9$
   $2x^2 + 2x - 8 = 0$
   $x^2 + x - 4 = 0$ Dividing by 2

   $x = \dfrac{-1 \pm \sqrt{1 + 4}}{2}$ (1)(4)   Using the quadratic formula

   $x = \dfrac{-1 \pm \sqrt{17}}{2} \approx \dfrac{-1 \pm 4.1}{2}$

   $x \approx \dfrac{-1 + 4.1}{2}$ or $x \approx \dfrac{-1 - 4.1}{2}$ not allowed

   $x \approx \dfrac{3.1}{2} \approx 1.6,$ $x + 1 \approx 2.6$

   The lengths of the legs are approximately 2.6 feet and 1.6 feet.

3.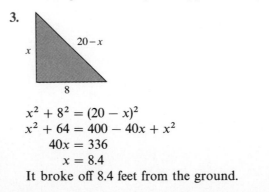

   $x^2 + 8^2 = (20 - x)^2$
   $x^2 + 64 = 400 - 40x + x^2$
   $40x = 336$
   $x = 8.4$
   It broke off 8.4 feet from the ground.

PROBLEM SET 11.6

Solve.

1. The length of a rectangle is 3 centimeters greater than the width. The area is 40 square centimeters. Find the length and width.

2. The length of a rectangle is 4 meters greater than the width. The area is 96 square meters. Find the length and width.

3. The width of a rectangle is 2 centimeters less than the length. The area is 21 square centimeters. Find the length and width.

4. The width of a rectangle is 1 centimeter less than the length. The area is 22 square centimeters. Find the length and width.

5. Susan wants her rectangular vegetable garden to have an area of 98 square meters. Find the dimensions of the garden if the length is to be twice the width.

6. Tracy wants to fence her rectangular garden so that the length is twice the width. What should be the dimensions of the garden so that the area is 128 square meters?

7. The hypotenuse of a right triangle is 5 feet long. One leg is 2 feet longer than the other. Find the lengths of the legs.

8. The hypotenuse of a right triangle is 8 yards long. One leg is 2 yards longer than the other. Find the lengths of the legs.

9. The hypotenuse of a right triangle is 9 meters long. One leg is 3 meters longer than the other. Find the lengths of the legs.

10. The hypotenuse of a right triangle is 6 centimeters long. One leg is 1 centimeter longer than the other. Find the lengths of the legs.

11. The hypotenuse of a right triangle is 10 yards long. Find the lengths of the legs if one leg is 2 yards longer than the other.

12. The hypotenuse of a right triangle is 5 inches long. Find the lengths of the legs if one leg is 1 inch longer than the other.

13. The sum of two numbers is 8. The sum of their squares is 40. Find the numbers.

14. The sum of the squares of two numbers is 130. If the larger number minus the smaller number is 2, find the numbers.

15. The area of a rectangle is 180 square centimeters. If the width is 3 inches less than the length, find the dimensions of the rectangle.

16. The area of a rectangle is 143 square feet. If the width is 2 feet less than the length, what are the dimensions of the rectangle?

17. A 15-foot tree breaks over and touches the ground 7 feet from its base. How far up the tree did it break over?

18. At Pizza Heaven, Kevin and Keith decide to buy one large pizza instead of two small ones. What should be the diameter, d, of a pizza so that it has the same area as two 8-inch pizzas? *Hint:* For a circle, area $= \pi(d/2)^2$, where d is the diameter.

Checkup

The following problems provide a review of some of Section 10.6. Solve.

19. $\sqrt{x - 3} = 3$

20. $\sqrt{y + 5} = 4$

21. $\sqrt{3y + 4} = -2$

22. $\sqrt{7x - 8} = -1$

23. $\sqrt{2x + 8} = 2$

24. $\sqrt{3y + 4} = 5$

25. $\sqrt{3x - 2} = \sqrt{x + 10}$

26. $\sqrt{x - 3} = \sqrt{4x - 18}$

27. $3\sqrt{y} = \sqrt{16y - 7}$

28. $\sqrt{5x - 4} = \sqrt{x + 12}$

CHAPTER 11   ADDITIONAL EXERCISES (OPTIONAL)

Section 11.1

Are the following quadratic equations?

1. $4x - 2x + 3 = 0$

2. $5x^3 + 7x - 25 = 0$

3. $4x^2 - 5x + 7 - 0$

Solve.

4. $x^2 = 32$

5. $5x^2 = 30$

6. $ax^2 = b, \quad a > 0, \quad b > 0$

7. $x^2 = d^2, \quad d > 0$

8. $(x - 18)^2 = 450$

9. $(x + 14)^2 = 288$

Section 11.2

Solve.

10. $(2x - 3)(x - 5) = 0$

11. $6x^2 + 7x = 3$

12. $5x^2 + 22x = -8$

13. $x^2 + 8x - 105$

14. $x^2 - 90 = 9x$

15. $\dfrac{4}{(x - 3)^2} + \dfrac{7}{x - 3} + 3 = 0$

16. $\dfrac{x - 4}{x^2 - 5x} = \dfrac{2}{x^2 - 25}$

17. $\sqrt{x + 2} = x + 2$

18. $y + 5 = \sqrt{2y + 9}$

Section 11.3

Solve by completing the square.

19. $x^2 - 2x = 7$

20. $x^2 + 12x = -2$

21. $x^2 + 5x - 3 = 0$

22. $x^2 - 3x - 4 = 0$

23. $3t^2 + 1 = 8t$

24. $4y^2 = 3y + 5$

Section 11.4

Solve using factoring or the quadratic formula.

25. $6x^2 - 19x + 3 = 0$

26. $2x^2 + 5x - 4 = 0$

27. $3x^2 - 2x = 5$

28. $4x^2 + x = 10$

29. $\dfrac{t^2}{3} - \dfrac{5}{6}t = \dfrac{1}{2}$

30. $\dfrac{x^2}{2} - \dfrac{2}{3} = -\dfrac{2}{3}x$

31. $\dfrac{2}{3}x^2 - x = -\dfrac{1}{6}$

32. $\dfrac{y^2}{6} - \dfrac{5}{6} = -\dfrac{y}{3}$

Section 11.5

Graph.

33. $y = x^2 - 3$

34. $y = -x^2 + 2$

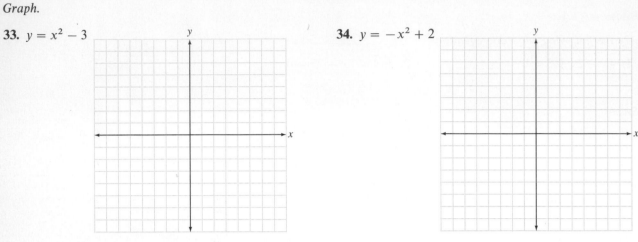

Graph. Use the quadratic formula to find the x-intercepts.

35. $y = x^2 + 4x - 3$

36. $y = x^2 + 2x - 1$

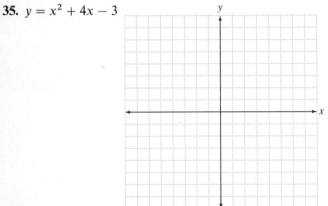

Section 11.6

37. Laura wants to add a rectangular room with an area of 285 square feet to her house. Find the dimensions of the room if the length is to be 4 feet greater than the width.

38. David has a rectangular flower bed with an area of 176 square feet. The width is 5 feet less than the length. What is the width of the flower bed?

39. The length of the hypotenuse of a right triangle is 5 feet. Find the lengths of the legs if one leg is 2 feet longer than the other.

40. The hypotenuse of a right triangle is 7 meters long. Find the lengths of the legs if one leg is 3 meters shorter than the other.

41. The sum of two numbers is 12. The sum of their squares is 234. Find the numbers.

42. The sum of the squares of two numbers is 65. If the larger minus the smaller is 11, find the numbers.

CHAPTER 11   PRACTICE TEST

Solve.

1. $x^2 = 24$

2. $(x - 2)^2 = 36$

3. $x^2 + 2x - 8 = 0$

4. $y^2 + 8y = 0$

5. $x^2 + 8x - 1 = 0$

6. Solve $x^2 + 6x + 2 = 0$ by completing the square.

1. _____

2. _____

3. _____

4. _____

5. _____

6. _____

7. _____

7. Solve $2x^2 + 4x = 4$ by completing the square.

8. _____

8. Sketch the graph of $y = x^2 + 3x - 4$.

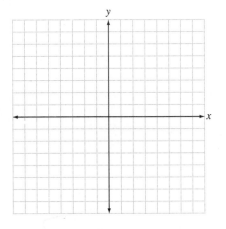

9. _____

9. The length of the hypotenuse of a right triangle is $\sqrt{45}$ meters. One leg is 3 meters longer than the other. Find the length of each leg.

10. _____

10. The area of a rectangular garden is 96 square meters. If the width is 4 meters less than the length, find the dimensions of the garden.

FINAL EXAMINATION

Chapter 1

1. Expand 7^4.

2. Change to percent: $\dfrac{7}{8}$.

3. Find the area.

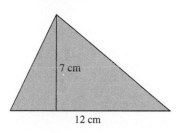

7 cm

12 cm

4. Find the volume. Use 3.14 for π.

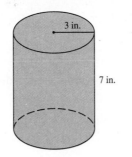

3 in.

7 in.

Chapter 2

5. Simplify: $-|-4|$.

6. Divide: $-\dfrac{3}{4} \div -\dfrac{5}{8}$.

7. Simplify: $7 - (5 - 2x)$.

8. Divide and write with a positive exponent:

$$\dfrac{x}{x^4}, \quad x \neq 0.$$

9. Evaluate $x^2 + 3x - 4$ for $x = -1$.

Chapter 3

10. Solve: $x - 6 = 7$.

11. Divide: $\dfrac{0}{5}$.

12. Solve: $5x + 2 = 3x - 6$.

1. _____

2. _____

3. _____

4. _____

5. _____

6. _____

7. _____

8. _____

9. _____

10. _____

11. _____

12. _____

13. _____

14. _____

15. _____

16. _____

17. _____

18. _____

19. _____

20. _____

21. _____

22. _____

23. _____

24. _____

25. _____

13. Solve $I = \dfrac{E}{R}$ for R.

14. Solve and *graph*: $4y + 2 < 5y + 6$.

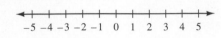

Chapter 4

15. A landscaper has 108 feet of fencing. He wants to build a fence with length 3 feet greater than the width. What should be the dimensions of the fence?

16. Julie invested $5000, part at 11% and the remainder at 9%. How much did she invest at 9% if her yearly interest was $510?

17. How many liters of 45% acid solution must be added to 20 liters of a 25% acid solution to make a 30% solution?

18. Jeff drove to San Francisco. He averaged 50 miles per hour going to San Francisco and 55 miles per hour returning. If the entire trip took 6 hours, how long did it take him to get to San Francisco?

Chapter 5

19. Combine like terms and arrange in descending order: $3x - 4x^2 + 2x^4 - x^2 - 5x^4 - x + 3$.

20. Subtract $(4x^3 - 5x^2 + 3x) - (-6x^3 + 2x^2 + x)$.

Multiply.

21. $(4y - 3)(2y - 5)$

22. $(x - 3)(x + 3)$

23. Divide:

$$\dfrac{x^2 - 3x - 10}{x + 2}$$

Factor completely.

24. $8x^2 - 20x - 12$

25. $x^4 - 1$

Chapter 6

26. Graph: $3x - 4y = 12$.

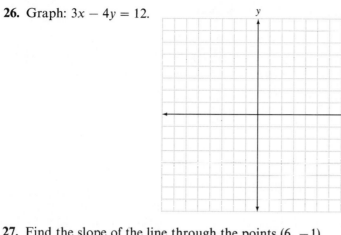

27. Find the slope of the line through the points $(6, -1)$ and $(-2, -3)$.

Graph.

28. $y = \dfrac{1}{2}x - 1$

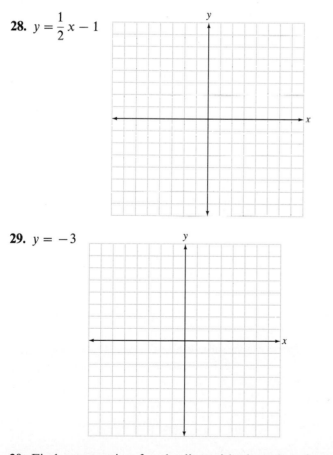

29. $y = -3$

30. Find an equation for the line with slope 2 and y-intercept $(0, 5)$.

Chapter 7

31. _____

31. Solve using the addition method:

$$2x + 5y = 19$$
$$3x - 2y = 0$$

32. _____

32. Solve using the substitution method:

$$4y - x = 7$$
$$x = 2y - 1$$

33. _____

33. Solve by any method:

$$x - 4y = 1$$
$$x + 4y = 17$$

34. _____

34. Mark's wages were indicated by 5% to $4.73 per hour. What were his former wages?

35. _____

35. Graph: $2x + 5y \leq 10$

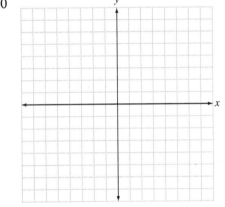

Chapter 8

36. _____

36. Simplify:

$$\frac{4x - 8}{3x^2 + x - 14}$$

37. _____

37. Divide:

$$\frac{x^2 - y^2}{x^5} \div \frac{x + y}{x^2}$$

38. Add:

$$\frac{3}{7-x} + \frac{-4}{x-7}$$

39. Subtract:

$$\frac{4x}{x^2 - 4x - 12} - \frac{5}{x^2 - 4}$$

Chapter 9

Solve.

40. $\dfrac{6}{5-x} = \dfrac{18}{x+7}$

41. $\dfrac{x+2}{x-1} = \dfrac{3}{x-1} + 2$

42. Assume that y varies inversely as x. If $y = 7$ when $x = 4$, find y when $x = 2$.

43. Simplify:

$$\frac{\dfrac{4}{6x}}{\dfrac{3}{2x^2}}$$

Chapter 10

Find the square root.

44. $-\sqrt{100}$

45. $\sqrt{25x^2}$

46. Multiply and simplify, if possible: $\sqrt{3}\sqrt{8}$.

47. Subtract: $2\sqrt{32} - 4\sqrt{2}$.

48. Rationalize the denominator: $\sqrt{\dfrac{2}{5}}$.

38. _____

39. _____

40. _____

41. _____

42. _____

43. _____

44. _____

45. _____

46. _____

47. _____

48. _____

49. _____

49. Divide and simplify: $\dfrac{\sqrt{72}}{\sqrt{2}}$.

50. _____

50. Find the third side of the right triangle: $a = 5$, $b = 7$, $c = ?$

Chapter 11

Solve.

51. _____

51. $x^2 = 48$

52. _____

52. $x^2 - 7x + 10 = 0$

53. _____

53. $x^2 + 9x = 0$

54. _____

54. $x^2 + 3x - 2 = 0$

55. _____

55. Sketch the graph of the equation: $y = x^2 - 3$.

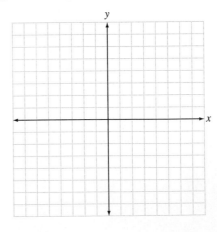

56. _____

56. The length of the hypotenuse of a right triangle is $4\sqrt{5}$ centimeters. One leg is 4 centimeters longer than the other. Find the length of each leg.

Appendices

APPENDIX A: SCIENTIFIC NOTATION

Very large and very small numbers are often used in the sciences. One application of the theory of exponents is in **scientific notation**. The following chart helps us write numbers in scientific notation. The chart may be extended.

$1,000,000 = 10^6$ Since $1,000,000 = 10 \cdot 10 \cdot 10 \cdot 10 \cdot 10 \cdot 10$

$100,000 = 10^5$ Since $100,000 = 10 \cdot 10 \cdot 10 \cdot 10 \cdot 10$

$10,000 = 10^4$

$1000 = 10^3$

$100 = 10^2$

$10 = 10^1$

$1 = 10^0$ By definition

$\dfrac{1}{10} = 10^{-1}$

$\dfrac{1}{100} = 10^{-2}$ Since $1/100 = 1/10^2$

$\dfrac{1}{1000} = 10^{-3}$ Since $1/1000 = 1/10^3$

$\dfrac{1}{10,000} = 10^{-4}$

$\dfrac{1}{100,000} = 10^{-5}$

$\dfrac{1}{1,000,000} = 10^{-6}$

OBJECTIVES

1. Change from decimal notation to scientific notation

2. Change from scientific notation to decimal notation

3. Use scientific notation to multiply or divide

517

a. 9300

b. 850

c. 0.0004

d. 0.02

◼ Scientific Notation

A number is in scientific notation if it is a number between 1 and 10 times an integer power of 10. Powers of 10 are also in scientific notation.

The following numbers are in scientific notation.

$$3.2 \times 10^4 \qquad 5.13 \times 10^{-6} \qquad 10^{-3}$$

EXAMPLE 1

a. Write 8500 in scientific notation.
Notice that we must have a number between 1 and 10, so we write

$$8.5$$

But now we have moved the decimal point three places to the left, which is the same as dividing by 1000, so we must multiply by $1000 = 10^3$ to give an equivalent number.

$$8500 = 8.5 \times 10^3$$

b. Write 0.00036 in scientific notation.
We write 3.6 to get a number between 1 and 10, but we have moved the decimal point four places to the right. This is the same as multiplying by 10,000, so we must divide by 10,000 or multiply by $\frac{1}{10,000} = 10^{-4}$. Hence

$$0.00036 = 3.6 \times 10^{-4} \qquad ◼$$

☐ **DO EXERCISE 1.**

2 Decimal Notation

We may also change from scientific notation to decimal notation.

EXAMPLE 2 Change to decimal notation.

a. 6.24×10^3

$$6.24 \times 10^3 = 6.24 \times 1000 = 6240$$

b. 3.2×10^{-3}

$$3.2 \times 10^{-3} = 3.2 \times \frac{1}{1000}$$

$$= \frac{3.2}{1000} = 0.0032 \qquad \text{Dividing by 1000 is the same as moving the decimal point three places to the left} \quad \blacksquare$$

☐ DO EXERCISE 2.

☐ **Exercise 2** Change to decimal notation.

a. 3.25×10^4

b. 2.8×10^6

c. 4.1×10^{-6}

d. 7.34×10^{-2}

□ **Exercise 3** Calculate. Write answers in scientific notation.

a. $(4.5 \times 10^5) \times (3 \times 10^3)$

b. $(6.2 \times 10^{-6}) \times (7.1 \times 10^3)$

c. $\dfrac{9.3 \times 10^{-4}}{3 \times 10^{-6}}$

d. $\dfrac{10.8 \times 10^{-3}}{2 \times 10^2}$

■ Calculations

When we calculate with very large numbers we sometimes exceed the capacity of calculators. It is necessary to use scientific notation.

EXAMPLE 3 Calculate. Write answers in scientific notation.

a. $(3.2 \times 10^4) \times (2.1 \times 10^2)$

$$(3.2 \times 10^4) \times (2.1 \times 10^2) = (3.2)(2.1) \times 10^4 \times 10^2$$
$$= 6.72 \times 10^6$$

b. $(5.1 \times 10^2) \times (3 \times 10^{-8})$

$$(5.1 \times 10^2) \times (3 \times 10^{-8}) = (5.1)(3) \times 10^2 \times 10^{-8}$$
$$= 15.3 \times 10^{-6}$$
$$= 1.53 \times 10 \times 10^{-6}$$
$$= 1.53 \times 10^{-5} \qquad \text{Recall that } 10 = 10^1$$

c. $\dfrac{8.4 \times 10^8}{2 \times 10^2}$

$$\frac{8.4 \times 10^8}{2 \times 10^2} = \frac{8.4}{2} \times \frac{10^8}{10^2}$$
$$= 4.2 \times 10^{8-2}$$
$$= 4.2 \times 10^6$$

d. $\dfrac{6.3 \times 10^{-6}}{3 \times 10^{-2}}$

$$\frac{6.3 \times 10^{-6}}{3 \times 10^{-2}} = \frac{6.3}{3} \times \frac{10^{-6}}{10^{-2}}$$
$$= 2.1 \times 10^{-6-(-2)}$$
$$= 2.1 \times 10^{-4} \qquad ■$$

□ **DO EXERCISE 3.**

Answers to Exercises

1. a. 9.3×10^3　**b.** 8.5×10^2　**c.** 4×10^{-4}　**d.** 2×10^{-2}

2. a. 32,500　**b.** 2,800,000　**c.** 0.0000041　**d.** 0.0734

3. a. 1.35×10^9　**b.** 4.402×10^{-2}　**c.** 3.1×10^2　**d.** 5.4×10^{-5}

PROBLEM SET: APPENDIX A

A. *Change to scientific notation.*

1. 85,000

2. 2300

3. 805,000,000,000

4. 721,000,000,000

5. 0.00000345

6. 0.00000000284

7. 0.000000016

8. 0.0000000003

9. 10,000

10. 100,000,000

11. 0.000001

12. 0.00001

B. *Convert to decimal notation.*

13. 6.25×10^8

14. 3.74×10^6

15. 5.862×10^{-5}

16. 7.034×10^{-3}

17. 10^6

18. 10^2

19. 10^{-3}

20. 10^{-6}

C. *Multiply or divide and write the result in scientific notation.*

21. $(5 \times 10^4) \times (1.4 \times 10^5)$

22. $(1.8 \times 10^8) \times (2.3 \times 10^{-3})$

23. $(4.2 \times 10^5) \times (5.1 \times 10^{-2})$

24. $(6.5 \times 10^{-7}) \times (2.8 \times 10^{-5})$

25. $(9.1 \times 10^{-6}) \times (8.2 \times 10^{-8})$

26. $(1.1 \times 10^4) \times (10^{-9})$

27. $\dfrac{8.6 \times 10^8}{2 \times 10^{-5}}$

28. $\dfrac{4.8 \times 10^{-2}}{4 \times 10^5}$

29. $(4 \times 10^6) \div (2 \times 10^9)$

30. $(8 \times 10^{-3}) \div (4 \times 10^{-6})$

31. $\dfrac{1.25 \times 10^{-9}}{0.25 \times 10^{11}}$

32. $\dfrac{6 \times 10^{-3}}{3 \times 10^{18}}$

APPENDIX B: ALTERNATIVE METHOD OF FACTORING TRINOMIALS WHEN THE COEFFICIENT OF x^2 IS OTHER THAN 1

Recall that we multiply two binomials as follows:

$$\overset{\text{F \quad O \quad I \quad L}}{(3x + 2)(2x + 1)} = 6x^2 + 3x + 4x + 2$$
$$= 6x^2 + 7x + 2$$

1 In general,

$$(ax + b)(cx + d) = acx^2 + adx + bcx + bd$$
$$= acx^2 + (ad + bc)x + bd$$

We factor by reversing the procedure.

$$acx^2 + (ad + bc)x + bd = (ax + b)(cx + d)$$

In order to do this, we multiply $(ac)(bd)$ and factor the product in such a way that we can write the middle term as a sum. Then we factor the resulting polynomial.

EXAMPLE   Factor.

a. $2x^2 + 5x - 12$

1. Multiply the coefficient of x^2, 2, by the constant term, -12.

$$2(-12) = -24$$

2. Factor -24 so that the sum of the factors is 5 (the coefficient of the middle term).

Factors of -24	Sum of Factors of -24
1, -24	-23
-1, 24	23
2, -12	-10
-2, 12	10
3, -8	-5
-3, 8	5
4, -6	-2
-4, 6	2

The factors are -3 and 8.

3. Write the middle term, $5x$, as a sum using the factors -3 and 8.

$$5x = -3x + 8x$$

a. $3x^2 - 14x + 16$

4. Rewrite the expression to be factored, writing the middle term as a sum.

$$2x^2 + 5x - 12 = 2x^2 - 3x + 8x - 12$$

5. Factor by separating the expression into groups of two terms and complete the factorization.

$$2x^2 - 3x + 8x - 12 = (2x^2 - 3x) + (8x - 12)$$
$$= x(2x - 3) + 4(2x - 3)$$
$$= (x + 4)(2x - 3) \quad \text{Factoring out } 2x - 3$$

b. $3x^2 - 13x - 10$

b. $8x^2 - 10x - 3$

    1. Multiply the coefficient of x^2 by the constant term.

$$3(-10) = -30$$

    2. Factor -30 so that the sum of the factors is -13.

Factors of -30	Sum of Factors of -30
$1, -30$	-29
$-1, \quad 30$	29
$2, -15$	-13

       We may stop at this point since the sum of the factors is -13. The factors are 2 and -15.

    3. Rewrite the expression, writing the middle term as a sum, using the factors 2 and -15. Then factor.

c. $7x^2 - 27x - 4$

$$3x^2 - 13x - 10 = 3x^2 + 2x - 15x - 10$$
$$= x(3x + 2) - 5(3x + 2) \quad \begin{array}{l}\text{Factoring } -5 \\ \text{out of the} \\ \text{second group}\end{array}$$
$$= (x - 5)(3x + 2) \quad \blacksquare$$

☐ **DO THE EXERCISE at the left.**
  Do Problem Set for Appendix B on p. 527.

d. $5x^2 + 21x + 4$

Answers to the Exercise _____

a. $(3x - 8)(x - 2)$ **b.** $(4x + 1)(2x - 3)$ **c.** $(7x + 1)(x - 4)$
d. $(5x + 1)(x + 4)$

APPENDIX C: FACTORING THE SUM OR DIFFERENCE OF TWO CUBES

1 **The Sum of Two Cubes**

Notice the result of the following multiplication.

$(A + B)(A^2 - AB + B^2)$
$$= A(A^2 - AB + B^2) + B(A^2 - AB + B^2)$$
$$= A^3 - A^2B + AB^2 + A^2B - AB^2 + B^3$$
$$= A^3 - A^2B + A^2B + AB^2 - AB^2 + B^3 \quad \text{Rearranging terms}$$
$$= A^3 + B^3$$

Recall that factoring reverses multiplication. Therefore, $A^3 + B^3$ may be factored as follows:

$$A^3 + B^3 = (A + B)(A^2 - AB + B^2)$$

The following list of the cubes of 1 through 10 helps us factor.

Number	Cube
1	1
2	8
3	27
4	64
5	125
6	216
7	343
8	512
9	729
10	1000

EXAMPLE 1 Factor.

a. $x^3 + 27$

Recall the rule.

$$A^3 + B^3 = (A + B)(A^2 - AB + B^2)$$

Then

$$x^3 + 27 = x^3 + 3^3 = (x + 3)[x^2 - x(3) + 3^2]$$
$$= (x + 3)(x^2 - 3x + 9)$$

b. $8x^3 + 125$

The rule is

$$A^3 + B^3 = (A + B)(A^2 - AB + B^2).$$
$$8x^3 + 125 = (2x)^3 + 5^3 = (2x + 5)[(2x)^2 - (2x)5 + 5^2]$$
$$= (2x + 5)(4x^2 - 10x + 25) \quad \blacksquare$$

□ **DO EXERCISE 1.**

□ **Exercise 1** Factor.

a. $x^3 + 8$

b. $y^3 + 1$

c. $125y^3 + 64$

d. $27x^3 + 1$

a. $x^3 - 27$

b. $8y^3 - 125$

c. $128x^3 - 2$

d. $27x^4 - 8x$

2 The Difference of Two Cubes

$$(A - B)(A^2 + AB + B^2) = A(A^2 + AB + B^2) - B(A^2 + AB + B^2)$$
$$= A^3 + A^2B + AB^2 - A^2B - AB^2 - B^3$$
$$= A^3 + A^2B - A^2B + AB^2 - AB^2 - B^3$$
$$= A^3 - B^3$$

Therefore, $A^3 - B^3$ may be factored as follows.

$$A^3 - B^3 = (A - B)(A^2 + AB + B^2)$$

EXAMPLE 2   Factor.

a. $x^3 - 8$

$$x^3 - 8 = x^3 - 2^3 = (x - 2)[x^2 + x(2) + 2^2]$$
$$= (x - 2)(x^2 + 2x + 4)$$

b. $64y^3 - 27$

$$64y^3 - 27 = (4y)^3 - 3^3 = (4y - 3)[(4y)^2 + (4y)3 + 3^2]$$
$$= (4y - 3)(16y^2 + 12y + 9)$$

c. $32x^3 - 4$

$$32x^3 - 4 = 4(8x^3 - 1)$$
$$= 4[(2x)^3 - 1^3]$$
$$= 4(2x - 1)[(2x)^2 + 2x(1) + 1^2]$$
$$= 4(2x - 1)(4x^2 + 2x + 1)   ∎$$

□ **DO EXERCISE 2.**
Problem Set for Appendix C is on page 528.

Answers to Exercises _____

1. a. $(x + 2)(x^2 - 2x + 4)$ **b.** $(y + 1)(y^2 - y + 1)$
c. $(5y + 4)(25y^2 - 20y + 16)$ **d.** $(3x + 1)(9x^2 - 3x + 1)$
2. a. $(x - 3)(x^2 + 3x + 9)$ **b.** $(2y - 5)(4y^2 + 10y + 25)$
c. $2(4x - 1)(16x^2 + 4x + 1)$ **d.** $x(3x - 2)(9x^2 + 6x + 4)$

PROBLEM SET: APPENDIX B

Factor.

1. $2x^2 + 11x + 5$

2. $12x^2 + 19x + 4$

3. $6x^2 - 13x + 6$

4. $3x^2 - 7x + 2$

5. $12x^2 - 31x + 20$

6. $10x^2 + x - 3$

$(3x - 4)(4x - 5)$

See Problem Set 5.7.

PROBLEM SET: APPENDIX C

Factor

1. $x^3 + 1$ **2.** $y^3 - 1$ **3.** $125a^3 + 1$

4. $125a^3 - 1$ **5.** $27x^3 - 125$ **6.** $125x^3 + 8$

7. $8x^3 + 1$ **8.** $27y^3 - 8$ **9.** $27a^3 - 64$

10. $125t^3 + 216$ **11.** $64x^3 + 125$ **12.** $1000z^3 - 1$

13. $125x^3 - 8$ **14.** $343y^3 + 1000$

15. $1000z^3 + 216$ **16.** $64y^3 - 125$

17. $24x^3 - 3$ **18.** $54x^3 + 2$

19. $250y^3 + 128$ **20.** $40x^3 - 135$

21. $y^6 + 1$ **22.** $x^6 - 1$

APPENDIX D: SETS

OBJECTIVES

1 *List the elements of a set and determine if an object is an element of a set*

2 *Determine whether a set is a subset of another set*

3 *Find the union and intersection of two sets*

1 **Elements of a Set**

A **set** is a collection of objects. The objects in the set are called **elements** of the set. We use braces, { }, to enclose the elements of the set. The elements of the set may be listed in any order.

EXAMPLE 1 List the elements of the following sets.

a. The set of all days of the week

{Sunday, Monday, Tuesday, Wednesday, Thursday, Friday, Saturday}

b. The set of all counting numbers less than five

$$\{1, 2, 3, 4\} \quad \blacksquare$$

□ **DO EXERCISE 1.**

Sets in which the number of elements can be counted are called **finite** sets. We can not determine the number of elements in some sets. These sets are called **infinite** sets. We list some elements of the set and indicate the rest of the elements with three dots. For example, the set of counting or natural numbers is

$$\{1, 2, 3, 4, \ldots\}.$$

Capital letters are usually used to name sets and elements of the set are denoted by lower case letters.

> The symbol ∈ is used to indicate membership in a set.

EXAMPLE 2 Determine whether the following are true or false.

a. $5 \in \{1, 3, 5, 7, \ldots\}$
    True

b. $u \in \{t, v, w\}$
    False ■

□ **DO EXERCISE 2.**

2 **Subsets**

When all the elements of a set A are also elements of a set B, we say that A is a **subset** of B, written $A \subset B$.

> $A \subset B$ means that A is a subset of B.

EXAMPLE 3 Decide if the following are true or false.

a. $\{2, 3, 4\} \subset \{1, 2, 3, 4, 5\}$
    True

b. $\{a, b, c\} \subset \{b, c, d, e\}$
    False

c. The set of all Siamese cats is a subset of all cats

True ■

□ **DO EXERCISE 3.**

□ **Exercise 1** List the elements of the following sets.

a. The set of all months that begin with the letter J

b. The set of counting numbers between 12 and 18

□ **Exercise 2** Determine whether the following are true or false for the set.

$$R = \{u, v, w, x, z\}.$$

a. $z \in R$ **b.** $s \in R$

c. $y \in R$ **d.** $w \in R$

□ **Exercise 3** Decide if the following are true or false.

a. $\{5, 7, 9\} \subset \{1, 3, 5, 7\}$

b. $\{p, q, r\} \subset \{p, q, r, s, t\}$

c. The set of all airplanes is a subset of the set of all Boeing airplanes

□ **Exercise 4** Find the following if $A = \{d, e, f\}$, $B = \{g, h, i\}$ and $C = \{h, i, j\}$.

a. $A \cup B$ b. $B \cup C$

□ **Exercise 5** Find the following if $A = \{u, v, w, x, y\}$, $B = \{t, u, v\}$ and $C = \{t\}$.

a. $A \cap B$

b. $A \cap C$

c. $B \cap C$

3   Unions and Intersections

The **union** of two sets, written $A \cup B$, is the set of all elements in A *or* in B.

> $A \cup B$ is the set of all elements in A *or* in B.

We do not repeat elements in a set.

EXAMPLE 4   Find $A \cup B$ if $A = \{1, 2, 3, 4\}$ and $B = \{3, 5, 7\}$

$$A \cup B = \{1, 2, 3, 4, 5, 7\} \quad \blacksquare$$

□ **DO EXERCISE 4.**

A set with no elements is called the **empty set**, denoted by $\varnothing$.

> The empty set, $\varnothing$, has no elements.

The **intersection** of two sets, denoted $A \cap B$, is the set of all elements that are in both A and B. We say that the elements are common to A and B.

> $A \cap B$ is the set of all elements in both A *and* B.

The intersection of two sets may be the empty set.

EXAMPLE 5   Find the following if $R = \{2, 3, 17, 18\}$, $S = \{3, 11, 17\}$ and $T = \{11, 25\}$.

a. $R \cap S = \{3, 17\}$

b. $S \cap T = \{11\}$

c. $R \cap T = \varnothing \quad \blacksquare$

□ **DO EXERCISE 5.**

Answers to Exercises _____

1. a. $\{$January, June, July$\}$ b. $\{13, 14, 15, 16, 17\}$
2. a. True   b. False   c. False   d. True
3. a. False   b. True   c. False
4. a. $\{d, e, f, g, h, i\}$ b. $\{g, h, i, j\}$
5. a. $\{u, v\}$ b. $\varnothing$ c. $\{t\}$

PROBLEM SET: APPENDIX D

A. *List the elements of the following sets.*

1. The set of all coins currently used in the United States

2. The set of all days of the week that begin with the letter T

3. The set of all counting numbers less than 7

4. The set of all counting numbers between 32 and 35

5. The set of letters of the alphabet between h and m

6. The set of the first three letters of the alphabet

B. *Determine if the following are true or false.*

7. $5 \in \{3, 5, 7\}$

8. $0 \in \{1, 3, 7, 9\}$

9. $d \in \{a, b, c\}$

10. $6 \in \{2, 4, 6, 8\}$

11. $\{2, 5, 9\} \subset \{2, 3, 5, 7, 9\}$

12. $\{4, 7\} \subset \{3, 7, 12\}$

13. $\{e, f, x\} \subset \{a, e, f, r\}$

14. $\{c, g\} \subset \{c, d, g, i\}$

15. $\{7, 11, 16\} \subset \{5, 7, 13, 16\}$

16. $\{5, 8\} \subset \{8, 27\}$

C. *Let* $A = \{1, 2, 3, 4, 5\}$
$B = \{2, 4, 6\}$
$C = \{6, 7\}$
$D = \{2, 5, 9\}$
$E = \{6, 7, 9\}$
Find the following.

17. $A \cup B$

18. $A \cup D$

19. $C \cup D$

20. $A \cup E$

21. $A \cap B$

22. $A \cap C$

23. $C \cap D$

24. $D \cap E$

25. $D \cup E$

26. $B \cap D$

27. $A \cap D$

28. $B \cup E$

Bibliography

BELL, E. T., *The Development of Mathematics*. New York: McGraw-Hill, 1945.

——, *Men of Mathematics*. New York: Simon & Schuster, 1986.

BOYER, CARL B., *A History of Mathematics*. New York: John Wiley, 1968.

BURTON, DAVID M., *The History of Mathematics: An Introduction*. Newton, Mass.: Allyn and Bacon, 1985.

DANZIG, TOBIAS, *Number, The Language of Science*. New York: Free Press, 1967.

EVES, HOWARD, *In Mathematical Circles*. Boston: Prindle, Weber & Schmidt, 1969.

——, *An Introduction to the History of Mathematics*, 5th ed. New York: Saunders College Publishing, 1983.

——, *Mathematical Circles Squared*. Boston: Prindle, Weber & Schmidt, 1972.

——, *Mathematical Circles Revisited*. Boston: Prindle, Weber & Schmidt, Inc., 1971.

——, and CARROLL V. NEWSOM, *An Introduction to the Foundations and Fundamental Concepts of Mathematics*. New York: Holt, Rinehart and Winston, 1958.

GAMOW, GEORGE, *One Two Three . . . Infinity*. New York: Bantam Books, 1971.

Historical Topics for the Mathematics Classroom (Thirty-first Yearbook). Washington, D.C.: National Council of Teachers of Mathematics, 1969.

HOGBEN, LANCELOT, *Mathematics in the Making*. New York: Doubleday & Company, 1960.

HOOPER, ALFRED, *Makers of Mathematics*. London: Faber & Faber, 1948.

LARRIVEE, JULES A., "A History of Computers I," *Mathematics Teacher* LI, October, 1958, 469–473.

KLINE, MORRIS, *Mathematics and the Physical World*. New York: Thomas Y. Crowell, 1959.

——, *Mathematics in Western Culture*. New York: Oxford University Press, 1953.

——, *Mathematics for the Liberal Arts*. Reading, Mass.: Addison-Wesley, 1967.

SAWYER, W. W., *The Search for Pattern*. New York: Penguin Books, 1970.

Scientific American, June, 1975, p. 49.

SCOTT, J. F., *A History of Mathematics*. London: Taylor & Francis, 1960.

SMITH, DAVID E., *History of Mathematics* (2 vols.). New York: Dover, 1923.

SMITH, KARL J., *The Nature of Modern Mathematics*, 3rd ed. Belmont, Calif.: Wadsworth, 1980.

WAMPLER, J. F., "The Concept of Function," *Mathematics Teacher* LIII, November, 1960, 581–583.

WEST, BEVERLY HENDERSON, et al., *The Prentice-Hall Encyclopedia of Mathematics*, Englewood Cliffs, N.J.: Prentice Hall, 1982.

Answers

CHAPTER 1

Problem Set 1.1

1. $2 \cdot 2 \cdot 7$ **3.** $3 \cdot 5$ **5.** $2 \cdot 2 \cdot 2 \cdot 2 \cdot 2$ **7.** $2 \cdot 11$ **9.** $\frac{1}{3}$ **11.** $\frac{3}{2}$ **13.** 3 **15.** $\frac{3}{2}$ **17.** $\frac{3}{28}$ **19.** $\frac{3}{16}$
21. $\frac{6}{5}$ **23.** $\frac{3}{8}$ **25.** $\frac{1}{5}$ **27.** 3 **29.** $\frac{15}{56}$ **31.** $\frac{11}{6}$ **33.** $\frac{5}{3}$ **35.** $\frac{18}{5}$ **37.** $\frac{35}{32}$ **39.** $\frac{5}{12}$ **41.** $\frac{1}{25}$ **43.** 8
45. $\frac{7}{3}$ **47.** $\frac{96}{5}$ **49.** 4 **51.** $\frac{8}{15}$ **53.** 6 **55.** $\frac{9}{2}$ **57.** 7 **59.** $2\frac{1}{4}$ cups **61.** 2 **63.** Proper **65.** Greater
67. Numerators, denominators

Problem Set 1.2

1. $\frac{4}{3}$ **3.** 3 **5.** $\frac{30}{10}$ **7.** $\frac{81}{72}$ **9.** 14 **11.** 15 **13.** 72 **15.** 160 **17.** $\frac{5}{12}$ **19.** $\frac{43}{24}$ **21.** $\frac{19}{24}$ **23.** $\frac{47}{24}$
25. $\frac{4}{7}$ **27.** $\frac{11}{6}$ **29.** $\frac{2}{9}$ **31.** $\frac{5}{6}$ **33.** $\frac{2}{15}$ **35.** $\frac{1}{4}$ **37.** $\frac{17}{4}$ **39.** $\frac{7}{8}$ **41.** $\frac{137}{15}$ **43.** $\frac{53}{96}$ **45.** $\frac{33}{28}$ miles
47. $3\frac{3}{8}$ yards **49.** Numerators **51.** Lowest common denominator

Problem Set 1.3

1. $\frac{17}{25}$ **3.** $\frac{7}{200}$ **5.** $\frac{317}{1000}$ **7.** $\frac{3}{100}$ **9.** $15\frac{2}{25}$ **11.** $121\frac{1}{5}$ **13.** 0.5 **15.** 0.875 **17.** 0.4 **19.** 0.364
21. 6.4 **23.** 7.714 **25.** 3^2 **27.** 2^5 **29.** x^4 **31.** $5 \cdot 5 \cdot 5 \cdot 5$ **33.** $7 \cdot 7$ **35.** $x \cdot x \cdot x$ **37.** 1 **39.** 6
41. 1 **43.** b **45.** Appoximately **47.** Cubed

Problem Set 1.4

1. 0.6 **3.** 0.07 **5.** 0.323 **7.** 1 **9.** 0.0025 **11.** 0.00667 **13.** 26% **15.** 6% **17.** 100% **19.** 4.5%
21. 340% **23.** 245% **25.** $\frac{7}{25}$ **27.** $\frac{39}{100}$ **29.** $\frac{77}{200}$ **31.** $\frac{3}{125}$ **33.** $\frac{21}{5000}$ **35.** $\frac{2}{25}$ **37.** 50% **39.** 55.6%
41. 70% **43.** 33.3% **45.** 220% **47.** 114.3% **49.** Hundred **51.** $\frac{1}{100}$

Problem Set 1.5

1. (a) 12; (b) 12 **3.** (a) 72; (b) 72 **5.** 18 **7.** 60 **9.** (a) 14; (b) 14 **11.** (a) 240; (b) 240 **13.** (a) 30; (b) 30
15. (a) 60; (b) 60 **17.** Associative law of addition **19.** Distributive law **21.** Commutative law of addition
23. Commutative law of addition **25.** Distributive law **27.** Associative law of multiplication
29. Commutative law of multiplication **31.** Natural **33.** Rational **35.** Addition

Problem Set 1.6

1. 21 **3.** 32 **5.** 30 **7.** 43 **9.** 52 cm **11.** 68 ft **13.** 46.2 m **15.** 192 cm **17.** 75.36 in.
19. 59.66 cm **21.** 24 ft **23.** 92 ft **25.** 12.56 in. **27.** Variable **29.** Relationship **31.** Circumference
33. Approximately

Problem Set 1.7

1. 260 ft^2 **3.** 14.44 in.^{2} **5.** 216 cm^2 **7.** 35 in.^{2} **9.** 28.26 cm^2 **11.** 113.04 ft^2 **13.** 1130.4 in.^{2}
15. 60.288 cm^2 **17.** 4860 m^3 **19.** 2210.56 m^3 **21.** 9074.6 cm^3 **23.** 336 m^2 **25.** 51 cm^2 **27.** 452.16 in.^{2}
29. 1440 cm^3 **31.** Area **33.** Triangle **35.** Unit

Chapter 1 Additional Exercises

1. $3 \cdot 3 \cdot 5$ **3.** $2 \cdot 2 \cdot 2 \cdot 2 \cdot 2 \cdot 2$ **5.** $\frac{7}{2}$ **7.** $\frac{36}{5}$ **9.** $\frac{32}{21}$ **11.** $\frac{1}{12}$ **13.** $\frac{9}{2}$ or $4\frac{1}{2}$ **15.** $\frac{31}{24}$ **17.** $\frac{31}{18}$ **19.** $\frac{63}{20}$
21. $\frac{65}{28}$ **23.** $\frac{3}{2}$ **25.** $\frac{7}{4}$ **27.** Thirty-seven hundredths **29.** 2.429 **31.** $\frac{17}{250}$ **33.** $2^2 \cdot 3^3$ **35.** $3 \cdot 3 \cdot 3 \cdot 3$
37. 1 **39.** 0.548 **41.** 0.005 **43.** $\frac{129}{500}$ **45.** 70% **47.** 500% **49.** Commutative law of addition
51. Associative law of multiplication **53.** Commutative law of addition **55.** No **57.** 8 **59.** 40 **61.** 66 ft
63. 78.5 cm **65.** 102 in.^{2} **67.** 212.4 in.^{2} **69.** 200.96 in.^{2} **71.** 100.48 in.^{2} **73.** 33.48 cm^3

1. $\frac{2}{3}$ **2.** $\frac{15}{16}$ **3.** $\frac{7}{12}$ **4.** $\frac{8}{21}$ **5.** 0.778   **6.** $4\frac{1}{20}$ **7.** $8 \cdot 8 \cdot 8$ **8.** 0.392   **9.** 62.5%   **10.** 27
11. Distributive law   **12.** 10   **13.** 15 m   **14.** 37.68 in.   **15.** 5.76 cm^2 **16.** 18 ft^2 **17.** 200.96 m^2
18. 1256 in.^{2} **19.** 24 m^3 **20.** 282.6 cm^3

CHAPTER 2

Problem Set 2.1

1. -5 **3.** 10   **5.** $\frac{3}{5}$ **7.** -2.4 **9.** $-\frac{11}{7}$ **11.** 10.4   **13.** -3 **15.** 4   **17.** -2 **19.** 99   **21.** $-\frac{2}{5}$
23. $\frac{12}{7}$ **25.** 7.8   **27.** -71.4 **29.** $9 > 5$ **31.** $-4 < -3$ **33.** $0 > -3$ **35.** $-7 < 3$ **37.** $1.5 > 0$
39. $-1.5 < 1$ **41.** 5   **43.** 3   **45.** 0   **47.** 15   **49.** $\frac{1}{3}$ **51.** 3.4   **53.** 10.1   **55.** $\frac{15}{8}$ **57.** -8.6
59. $-\frac{11}{5}$ **61.** Negative   **63.** Simplify   **65.** Absolute value   **67.** 1.4   **69.** 1.286   **71.** $\frac{9}{25}$ **73.** $\frac{21}{250}$

Problem Set 2.2

1. 6   **3.** 6   **5.** -7 **7.** -15 **9.** -11 **11.** -8 **13.** -3 **15.** -3 **17.** 1   **19.** 2   **21.** 1
23. -6 **25.** 0   **27.** 0   **29.** 5   **31.** 4   **33.** -2 **35.** -10 **37.** -6 **39.** -8 **41.** -11 **43.** 12
45. -13 **47.** 4   **49.** 7   **51.** -15 **53.** -1 **55.** 0   **57.** 5   **59.** 1   **61.** Common   **63.** Zero
65. 2.4   **67.** 0.09   **69.** 30%   **71.** 56%   **73.** $\frac{2}{25}$ **75.** $\frac{97}{1000}$ **77.** 50%   **79.** 128.6%

Problem Set 2.3

1. -2 **3.** 2   **5.** -11 **7.** -7 **9.** -3 **11.** 4   **13.** 2   **15.** -1 **17.** -6 **19.** -17 **21.** 6
23. 1   **25.** 14   **27.** -1 **29.** -3 **31.** -12 **33.** -5 **35.** -9 **37.** 0   **39.** 4   **41.** 10   **43.** -7
45. 4   **47.** 2   **49.** -18 **51.** 10   **53.** -8 **55.** -16 **57.** $+2$ **59.** Distributive law
61. Associative law of multiplication   **63.** Commutative law of addition   **65.** Commutative law of multiplication

Problem Set 2.4

1. 56   **3.** 8   **5.** -21 **7.** -63 **9.** 32   **11.** 20   **13.** 40   **15.** -54 **17.** 30   **19.** 400   **21.** 0
23. 0   **25.** 4   **27.** 3   **29.** 2   **31.** -3 **33.** -6 **35.** 0   **37.** 5   **39.** -10 **41.** 0   **43.** Not allowed
45. -7 **47.** 10   **49.** 3   **51.** -54 **53.** 2   **55.** -48 **57.** 28   **59.** -16 **61.** 12   **63.** -4 **65.** 2
67. -1 **69.** 11   **71.** $-\frac{9}{4}$ **73.** $\frac{1}{2}$ **75.** 33   **77.** Negative   **79.** Positive   **81.** Multiplication, division
83. $\frac{10}{3}$ **85.** $\frac{45}{49}$ **87.** $\frac{29}{28}$ **89.** $\frac{1}{3}$

Problem Set 2.5

1. $\frac{2}{5}$ **3.** $-\frac{1}{2}$ **5.** $-\frac{8}{15}$ **7.** $-\frac{37}{30}$ **9.** 3.7   **11.** -10.3 **13.** -6.7 **15.** $-\frac{3}{25}$ **17.** -0.9 **19.** -2.1
21. 4.3   **23.** -5.7 **25.** $-\frac{3}{4}$ **27.** $-\frac{1}{5}$ **29.** $-\frac{29}{24}$ **31.** $\frac{29}{18}$ **33.** -4.8 **35.** 0.06   **37.** -15.2 **39.** 4.51
41. $-\frac{5}{6}$ **43.** $\frac{3}{49}$ **45.** -3 **47.** 4   **49.** -15 **51.** 15   **53.** $-\frac{5}{21}$ **55.** $\frac{2}{3}$ **57.** -2 **59.** $\frac{5}{2}$ **61.** -5
63. 2.1   **65.** $\frac{4}{3}$ **67.** $\frac{5}{-11}$ **69.** $\frac{1}{3}$ **71.** $\frac{1}{-10}$ **73.** $+\frac{-1}{-8}, -\frac{-1}{+8}, -\frac{+1}{-8}$ **75.** $+\frac{+2}{-7}, -\frac{+2}{+7}, -\frac{-2}{-7}$
77. $+\frac{+11}{-5}, -\frac{+11}{+5}, -\frac{-11}{-5}$ **79.** Integers   **81.** Integers   **83.** 1   **85.** -72 **87.** -26

Problem Set 2.6

1. $7x, -y$ **3.** $-3x, 2y, 2$ **5.** $x, 4$ **7.** $\frac{x}{2}, 3$ **9.** $x, -y$ **11.** 6, -1, 3, respectively   **13.** $\frac{1}{4}, -7$, respectively
15. $8a, 4a$ **17.** 3, 9   **19.** $10x - 2$ **21.** $8a + 3$ **23.** $4 - 2s$ **25.** $3b + 10$ **27.** $a^2 + a$ **29.** $5x - 9$
31. $1.5x - 1.5y$ **33.** $6x - 6$ **35.** $-3y + 3$ **37.** $-12 + 2t$ **39.** $-3a - 12b$ **41.** $-6r - 18s + 12$
43. $-3a - 4$ **45.** $-3x + 3y - 6z$ **47.** $-7r - 6s - 4$ **49.** $-3a - 4b + 7$ **51.** $2x + 4y + 17$ **53.** $5x - 2$
55. $7y - 7$ **57.** $7x - 6$ **59.** $9y - 14x$ **61.** $-y - w$ **63.** $-5 - 3x$ **65.** $18 - 4a$ **67.** $5x - 7$
69. $22 - 2x$ **71.** $3x + 42$ **73.** Number   **75.** 1   **77.** Innermost   **79.** 5^4 **81.** $b \cdot b \cdot b$ **83.** 1   **85.** a

Problem Set 2.7

1. $\frac{1}{3^2}$ **3.** $\frac{1}{a^7}$ **5.** $\frac{1}{x}$ **7.** $\frac{1}{r^8}$ **9.** 3^{-2} **11.** a^{-5} **13.** x^{-4} **15.** u^{-1} **17.** 2^7 **19.** $\frac{1}{7^2}$ **21.** $\frac{1}{x^{10}}$
23. a^2 **25.** z^7 **27.** 1   **29.** 6   **31.** $\frac{1}{x^8}$ **33.** a^4 **35.** $\frac{1}{z}$ **37.** 1   **39.** $\frac{1}{x}$ **41.** 3^6 **43.** $\frac{1}{8^6}$ **45.** a^8

47. 3^6 **49.** 9   **51.** -9 **53.** 25   **55.** x^9y^6 **57.** $\dfrac{a^6}{b^{12}}$ **59.** $4x^4$ **61.** $16y^8$ **63.** $-8x^6$

65. The exponent and the number of zeros are the same.   **67.** Exponent   **69.** Multiply   **71.** 127   **73.** 28.6 m
75. 115.552 in.

Problem Set 2.8

1. -5 **3.** -17 **5.** 9   **7.** -2 **9.** -8 **11.** 3   **13.** 39   **15.** 1   **17.** -3 **19.** 3   **21.** $-\frac{5}{2}$
23. -4 **25.** 3   **27.** -14 **29.** \$140   **31.** $\frac{11}{2}$ amperes   **33.** 200 ft   **35.** 20.25 in.^{2} **37.** 81.6 ft^2
39. 226.865 in.^{2} **41.** 113.4 in.^{3}

Chapter 2   Additional Exercises

1. $-\frac{3}{4}$ **3.** 9.3   **5.** $<$ **7.** $<$ **9.** $\frac{11}{5}$ **11.** -4.2 **13.** 4   **15.** -9 **17.** -3 **19.** 1   **21.** -2
23. -15 **25.** 7   **27.** 85   **29.** -56 **31.** 56   **33.** $-\frac{1}{2}$ **35.** 2   **37.** $-\frac{19}{8}$ **39.** -3.1 **41.** $-\frac{3}{40}$
43. -30.0 **45.** 17.76   **47.** $-\frac{34}{5}$ **49.** $\frac{1}{12}$ **51.** $\dfrac{-3}{-4}, -\dfrac{3}{4}, \dfrac{3}{-4}$ **53.** $\frac{2}{3}x^2, x, -4$ **55.** $1, -\frac{1}{4}$
57. $5, -2, 3$ **59.** $5x - 4$ **61.** 3^6 **63.** $\dfrac{1}{x^n}$ **65.** x^6 **67.** $\dfrac{1}{y^{4n}}$ **69.** x^{8n} **71.** -13 **73.** -7.

Chapter 2   Practice Test

1. $-5 < -3$ **2.** 8.5   **3.** -3 **4.** -3 **5.** -5.5 **6.** $\dfrac{-7}{12}$ **7.** -6 **8.** -10.2 **9.** $-\frac{13}{8}$ **10.** 28
11. $-\frac{1}{3}$ **12.** -4 **13.** $\frac{3}{14}$ **14.** $-\frac{1}{3}$ **15.** $x^2, -6x, 4$ **16.** $a - 11$ **17.** $-6x + 6$ **18.** $10a - 3$
19. $9x - 4$ **20.** 3^{-4} **21.** $\dfrac{1}{5^2}$ **22.** y^2 **23.** b^3 **24.** $\dfrac{1}{x^{15}}$ **25.** -9 **26.** 300 miles

Cumulative Review   Chapters 1 and 2

1. 3.667   **2.** $\frac{1}{250}$ **3.** 16   **4.** 0.005   **5.** 30%   **6.** 10.08 m^2 **7.** 6 in.^{2} **8.** 200.96 cm^2 **9.** 15.7 ft
10. 23 in.^{3} **11.** 576.975 cm^3 **12.** $>$ **13.** -8 **14.** 0   **15.** 6   **16.** $\dfrac{9}{-7}$ **17.** $-4a - 7b$
18. $-9x + 12y$ **19.** $9x - 11y$ **20.** $15a - 21$ **21.** $\dfrac{18}{x}$ **22.** $\dfrac{1}{a^3}$ **23.** $9x^8$ **24.** $\dfrac{1}{16b^{12}}$ **25.** -44
26. \$135

CHAPTER 3

Problem Set 3.1

1. True   **3.** Conditional   **5.** False   **7.** True   **9.** Conditional   **11.** True   **13.** Linear   **15.** Linear
17. Not linear   **19.** Linear   **21.** Linear   **23.** Not linear   **25.** Not linear   **27.** Yes   **29.** No   **31.** Yes
33. Yes   **35.** Yes   **37.** No   **39.** Conditional   **41.** Solution, root   **43.** $>$ **45.** $<$ **47.** -8 **49.** -3.6
51. $\frac{3}{7}$

Problem Set 3.2

1. 11   **3.** -2 **5.** 5   **7.** -2 **9.** 1   **11.** $\frac{7}{12}$ **13.** 5.5   **15.** -3.8 **17.** 3   **19.** -14 **21.** -3
23. $\frac{1}{4}$ **25.** 1.4   **27.** 5.9   **29.** 0   **31.** Same   **33.** $\frac{8}{7}$ **35.** $\frac{1}{7}$ **37.** $\dfrac{1}{-2}$ **39.** $-\dfrac{1}{4}, -\dfrac{1}{-4}, \dfrac{-1}{-4}$
41. $-\dfrac{3}{5}, -\dfrac{3}{-5}, \dfrac{3}{-5}$ **43.** $\dfrac{-5}{2}, \dfrac{5}{-2}, -\dfrac{5}{-2}$

Problem Set 3.3

1. 3   **3.** -4 **5.** 4   **7.** $\frac{27}{2}$ **9.** 50   **11.** $-\frac{1}{7}$ **13.** 3   **15.** -4 **17.** 7   **19.** $-\frac{32}{25}$ **21.** $\frac{1}{2}$ **23.** 6.8
25. Not possible   **27.** 0   **29.** Not possible   **31.** 0   **33.** Nonzero   **35.** Same   **37.** Not possible   **39.** 1
41. -8 **43.** 4

Problem Set 3.4

1. 4   **3.** 4   **5.** -4 **7.** 7   **9.** 2   **11.** -5 **13.** 0.9   **15.** 0.1   **17.** -7 **19.** 5   **21.** 5   **23.** 8
25. 7   **27.** 9   **29.** 4   **31.** -5 **33.** $-\frac{1}{5}$ **35.** 8   **37.** -4 **39.** Addition   **41.** Variables   **43.** $-2x+3$
45. $x+12$ **47.** $10x-8$ **49.** $-5x+10$

Problem Set 3.5

1. $4x-12$ **3.** $-3x+2$ **5.** $-3x-12$ **7.** -2 **9.** 4   **11.** 2   **13.** 2   **15.** 8   **17.** $\frac{9}{13}$ **19.** -1
21. 17   **23.** 8   **25.** -3 **27.** 0   **29.** 2   **31.** $\frac{58}{7}$ **33.** Distributive   **35.** 4   **37.** 3   **39.** $\frac{2}{5}$ **41.** -21

Problem Set 3.6

1. $x=y-3$ **3.** $x=y-z+6$ **5.** $x=-z+3$ **7.** $x=y+8$ **9.** $H=\dfrac{V}{LW}$ **11.** $r=\dfrac{C}{2\pi}$ **13.** $y=\dfrac{z}{x}$

15. $r^2=\dfrac{A}{4\pi}$ **17.** $t=\dfrac{A-p}{pr}$ **19.** $h=\dfrac{S-2\pi r^2}{2\pi r}$ **21.** $x=\dfrac{y+2}{6}$ **23.** $x=\dfrac{10-z}{3}$ **25.** $x=\dfrac{b-y}{a}$

27. $x=4y-3$ **29.** $x=\dfrac{ay-c}{b}$ **31.** $h=\dfrac{3A}{B}$ **33.** $x=\dfrac{4y}{z}$ **35.** $C=\dfrac{5F-160}{9}$ **37.** $z=3A-x-y$

39. $g=\dfrac{mv^2}{2k}$ **41.** $a=s-sr$ **43.** Formula   **45.** $\dfrac{1}{3^2}$ **47.** $\dfrac{1}{a^4}$ **49.** 7^{-3} **51.** x^{-4} **53.** $\frac{4}{9}$

Problem Set 3.7

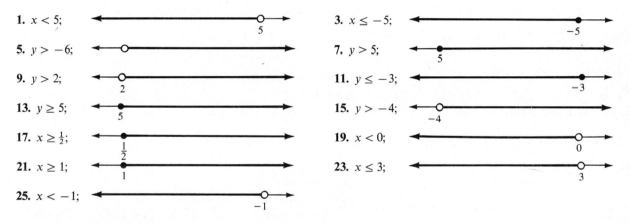

1. $x<5$;
3. $x\le-5$;
5. $y>-6$;
7. $y>5$;
9. $y>2$;
11. $y\le-3$;
13. $y\ge5$;
15. $y>-4$;
17. $x\ge\frac{1}{2}$;
19. $x<0$;
21. $x\ge1$;
23. $x\le3$;
25. $x<-1$;

27. Left   **29.** Same   **31.** Four   **33.** \$3480   **35.** 35   **37.** $\frac{14}{9}$ **39.** 0   **41.** Not possible

Problem Set 3.8

1. $x<28$ **3.** $y<-6$ **5.** $x\le-21$ **7.** $x\le15$ **9.** $x<4$ **11.** $x\le-2$ **13.** $x\ge3$ **15.** $x>8$
17. $y\le-9$ **19.** $y\ge\frac{10}{3}$ **21.** $x>1$ **23.** $x\ge-2$ **25.** $y\le6$ **27.** $y<2$ **29.** $x<\frac{3}{4}$ **31.** $y>0$
33. Positive   **35.** Addition   **37.** Not linear   **39.** Linear   **41.** Not linear   **43.** No

Chapter 3   Additional Exercises

1. False   **3.** Conditional   **5.** True   **7.** Yes   **9.** No   **11.** 23   **13.** -1 **15.** $\frac{59}{40}$ **17.** 2.6   **19.** $\frac{50}{3}$
21. -9 **23.** -5 **25.** 0   **27.** Not possible   **29.** 6   **31.** 0   **33.** 20   **35.** $\frac{25}{9}$ **37.** $-\frac{5}{6}$ **39.** -7
41. 1   **43.** 5   **45.** $x=\dfrac{V}{yz}$ **47.** $b=\dfrac{2A-ch}{h}$ **49.** $y=-\frac{1}{3}x+\frac{4}{3}$

51. $x<-2$;
53. $x\ge-1$;
55. $x>-3$;
57. $x\le2$;

1. (b) **2.** (b) **3.** 4 **4.** 5.1 **5.** 4 **6.** $-\frac{49}{15}$ **7.** 0 **8.** Not possible **9.** $-\frac{1}{4}$ **10.** 4 **11.** 4

12. $-\frac{7}{2}$ **13.** $x = \dfrac{4-y}{5}$ **14.** $I = \dfrac{E}{R}$ **15.** $x \geq -3$;

16. $x > -3$;   **17.** $y \geq 2$ **18.** $y > -\frac{7}{4}$

CHAPTER 4

Problem Set 4.1

1. $x + 6$ **3.** $x + 4$ **5.** $x + 20$ **7.** $x + 4$ **9.** $x - (-2)$ **11.** $x - 7$ **13.** $15 - x$ **15.** $4x$ **17.** $\frac{3}{7}x$

19. $\dfrac{x}{9}$ **21.** $\dfrac{x}{10}$ **23.** $x + 5 = 24$ **25.** $\dfrac{36}{x} = -3$ **27.** $x + 4 = \dfrac{x}{6}$ **29.** $5 - 2x = 4x + 7$ **31.** $3 - 2x = 5x + 6$

33. $2(3 - x) = 5x + 6$ **35.** Two, a number **37.** Equals **39.** -4 **41.** 3 **43.** -3 **45.** 1

Problem Set 4.2

1.
$$4x - 8 = 32$$
$$4x = 40$$
$$x = 10$$
The number is 10.

3.
$$5x + 4 = 3x - 6$$
$$2x = -10$$
$$x = -5$$
The number is -5.

5.
$$\frac{x}{2} + \frac{3}{4} = 2x$$
$$4\left(\frac{x}{2} + \frac{3}{4}\right) = 4(2x)$$
$$2x + 3 = 8x$$
$$3 = 6x$$
$$\frac{1}{2} = x$$
The number is $\frac{1}{2}$.

7.
$$3x + 5 = 4(x - 4)$$
$$3x + 5 = 4x - 16$$
$$21 = x$$
The number is 21.

9.
$$x + 2x + x + 20 = 180$$
$$4x + 20 = 180$$
$$4x = 160$$
$$x = 40$$
$$2x = 80$$
$$x + 20 = 60$$
The angles are 40°, 80°, and 60°.

11.
$$x + (x - 4) + 2(x - 4) = 24$$
$$x + x - 4 + 2x - 8 = 24$$
$$4x - 12 = 24$$
$$4x = 36$$
$$x = 9$$
$$x - 4 = 5$$
$$2(x - 4) = 10$$
Joyce buys 9 hot dogs, Karen buys 5, and Linda buys 10.

13.
$$2W + 2(3W) = 32$$
$$2W + 6W = 32$$
$$8W = 32$$
$$W = 4$$
$$3W = 12$$
The length is 12 in. and the width is 4 in.

15.
$$P = 2L + 2W$$
$$4L - 12 = 2L + 2(\tfrac{1}{3}L)$$
$$4L - 12 = 2L + \tfrac{2}{3}L$$
$$3(4L - 12) = 3(2L + \tfrac{2}{3}L)$$
$$12L - 36 = 6L + 2L$$
$$12L - 36 = 8L$$
$$4L = 36$$
$$L = 9$$
$$\tfrac{1}{3}L = 3$$
The dimensions are 3 ft by 9 ft.

17.
$$x + (x + 1) + (x + 2) = 108$$
$$x + x + 1 + x + 2 = 108$$
$$3x + 3 = 108$$
$$3x = 105$$
$$x = 35$$
$$x + 1 = 36$$
$$x + 2 = 37$$
The integers are 35, 36, and 37.

19.
$$x + (x + 2) = 54$$
$$x + x + 2 = 54$$
$$2x + 2 = 54$$
$$2x = 52$$
$$x = 26$$
$$x + 2 = 28$$
The integers are 26 and 28.

21.
$$x + (x + 2) = 112$$
$$x + x + 2 = 112$$
$$2x + 2 = 112$$
$$2x = 110$$
$$x = 55$$
$$x + 2 = 57$$
The integers are 55 and 57.

23. Read **25.** Formula **27.** Consecutive

29. $r^3 = \dfrac{V}{k\pi}$ **31.** $r = \dfrac{A - p}{pt}$ **33.** $h = \dfrac{V}{\pi r^2}$ **35.** $y = 2x - 3$

Problem Set 4.3

1. $I = Prt$
$I = 2000(0.09)(2)$
$I = 360$
The interest earned is $360.

3. $I = Prt$
$1200 = P(0.08)(3)$
$1200 = 0.24P$
$5000 = P$
Kevin invested $5000.

5. $I = Prt$
$I = 800(0.11)(5) = 440$
$A = 800 + 440 = 1240$
$1240 must be returned.

7. $I = Prt$
$63 = (1000 - x)(0.07)(1)$
$63 = 70 - 0.07x$
$-7 = -0.07x$
$x = 100$
Joan withdrew $100.

9. $I = Prt + Prt$
$460 = x(0.07)(1) + 2x(0.08)(1)$
$460 = 0.07x + 0.16x$
$460 = 0.23x$
$x = 2000$
$2x = 4000$
$2000 is invested at 7%,
$4000 is invested at 8%.

11. $Prt = Prt$
$(x - 400)(0.09)(1) = x(0.07)(1)$
$0.09x - 36 = 0.07x$
$0.02x = 36$
$x = 1800$
$1800 is invested at 7%.

13. $Prt + Prt = I$
$x(0.08)(1) + (6000 - x)(0.09)(1) = 525$
$0.08x + 540 - 0.09x = 525$
$-0.01x = -15$
$x = 1500$
$6000 - x = 4500$
Michael invested $1500 at 8% and $4500 at 9%.

15. Principal

17. $x > -2$; (number line, open circle at -2)

19. $x \le -2$; (number line, closed circle at -2)

21. $x > -2$ (number line, open circle at -2)

23. $x \le 4$; (number line, closed circle at 4)

Problem Set 4.4

1. $0.10x + 0.05(21 - x) = 1.40$
$10x + 5(21 - x) = 140$
$10x + 105 - 5x = 140$
$5x = 35$
$x = 7$
$21 - x = 14$
There are 7 dimes and 14 nickels in the collection.

3. $0.10x + 0.25(30 - x) = 4.20$
$10x + 25(30 - x) = 420$
$10x + 750 - 25x = 420$
$-15x = -330$
$x = 22$
$30 - x = 8$
Susan has 22 dimes and 8 quarters.

5. $0.10x + 0.05(3x) = 2.75$
$10x + 5(3x) = 275$
$10x + 15x = 275$
$25x = 275$
$x = 11$
$3x = 33$
Karen has 11 dimes and 33 nickels.

7. 4.5   **9.** $0.45x$

11. $0.40x + 0.15(50) = 0.20(x + 50)$
$40x + 15(50) = 20(x + 50)$
$40x + 750 = 20x + 1000$
$20x = 250$
$x = 12.5$
Charlene must add 12.5 liters of 40% solution.

13. $0.55(30) + 0.90x = 0.65(30 + x)$
$55(30) + 90x = 65(30 + x)$
$1650 + 90x = 1950 + 65x$
$25x = 300$
$x = 12$
Diane must use 12 gallons of 90% solution.

15. $0.35x + 0.65(20 - x) = 0.40(20)$
$35x + 65(20 - x) = 40(20)$
$35x + 1300 - 65x = 800$
$-30x = -500$
$x = 16.667$
$20 - x = 3.333$
16.667 kg of 35% alloy and 3.333 kg of 65% alloy should be used.

17. $1.75   **19.** $x > -2$ **21.** $x > 4$ **23.** $x \quad \frac{9}{2}$ **25.** $x \le \frac{1}{4}$

Problem Set 4.5

1. $420t + 510t = 2790$
$930t = 2790$
$t = 3$
They will be 2790 miles apart in 3 hours.

3. $10 \cdot \frac{1}{2} + 8 \cdot \frac{1}{2} = 5 + 4 = 9$ miles

5. $48t + 42 = 54t$
$-6t = -42$
$t = 7$
They will be 42 miles apart in 7 hours.

7. $25(t + \frac{1}{3}) = 30t$
$25t + \frac{25}{3} = 30t$
$-5t = -\frac{25}{3}$
$t = 1\frac{2}{3}$
Debra catches up with Jean in $1\frac{2}{3}$ hours.

9. $900t = 800(5 - t)$
$900t = 4000 - 800t$
$1700t = 4000$
$t = \frac{40}{17} = 2\frac{6}{17}$
It flew north $2\frac{6}{17}$ hours.

11. $10t = 12(\frac{22}{5} - t)$
$10t = \frac{264}{5} - 12t$
$22t = \frac{264}{5}$
$t = \frac{264}{5} \cdot \frac{1}{22} = \frac{12}{5} = 2\frac{2}{5}$
It took $2\frac{2}{5}$ hours to get to Jones Island.

13. Time   **15.** $3y, 5y$ **17.** $-8, 4$ **19.** $4x - 4$ **21.** $-7x + 17$

23. $-8a + 3b$ **25.** $-5y + 3$

Chapter 4   Additional Exercises

1. $x + 5$ **3.** $x + 8$ **5.** $\frac{3}{5}x$ **7.** $x - 4$ **9.** $8 - x$ **11.** $\frac{2}{3}x$ **13.** $-\frac{3}{2}$ **15.** 44 in., 40 in.   **17.** 30°, 90°, 60°
19. 61, 62, 63   **21.** 33, 35, 37   **23.** $3800   **25.** $16,650   **27.** $4200 at 8%, $3300 at 6.5%   **29.** $2890
31. 24   **33.** 18 nickels, 24 dimes, 15 quarters   **35.** 1.667 gal   **37.** 4.545 oz   **39.** 2.769 gal   **41.** 0.75 hour
43. 12.5 hours   **45.** 6.4 hours   **47.** 1820 ft

Chapter 4   Practice Test

1. $8 - x$ **2.** $\frac{x}{4} - 7x$ **3.** -17 **4.** 20 ft, 28 ft   **5.** $1296   **6.** $2000   **7.** 25 liters   **8.** 5 dimes, 17 quarters

9. 4 hours   **10.** $4\frac{8}{9}$ hours

Cumulative Review   Chapters 3 and 4

1. (b) and (c)   **2.** (a)   **3.** 0   **4.** Not possible   **5.** -8 **6.** -38 **7.** 11   **8.** 4.24   **9.** $y = -\frac{5}{2}x + 5$
10. $y = \frac{3}{4}x - \frac{7}{4}$ **11.** $r^3 = \frac{3V}{4\pi}$ **12.** $R = \frac{S - \pi hr}{\pi h}$ **13.** $x \geq -4$;
14. $y < -2$;   **15.** 8   **16.** 35°, 55°, 90°   **17.** 5 by 9   **18.** 27, 29, 31
19. $5760   **20.** $1750 at 6%, $3500 at 8%   **21.** 15 dimes, 60 quarters   **22.** 8.571 liters   **23.** $\frac{12}{5}$ or 2.4 hours
24. 2.813 hours   **25.** 2.455 hours

CHAPTER 5

Problem Set 5.1
1. Yes   **3.** Yes   **5.** No   **7.** No   **9.** Binomial   **11.** Trinomial   **13.** Monomial   **15.** None of these
17. Monomial   **19.** $-3x + 4$ **21.** $-10x^2 + 3x$ **23.** x **25.** $-4x^2y + xy^2$ **27.** $\frac{5}{4}x^2 + x$
29. $-1.5x - 2x^3 - 4$ **31.** 2, 1, 0, respectively; 2   **33.** 1, 0 respectively; 1   **35.** $-x^2 + x + 7$
37. $9x^7 - 3x^5 + x^2 + 2x$ **39.** $6x^7 + 4x^2 + 8x - 9$ **41.** 0 feet   **43.** Whole   **45.** Exponent   **47.** $x - 4$
49. $6 - x$ **51.** $5x - 6 = 4$ **53.** $2(4 - x) = 9$

Problem Set 5.2
1. $4x^2 - 3x - 7$ **3.** $x^4 - x^2 + 3$ **5.** $-8x^3 - 4x^2 + 9x - 3$ **7.** $10x^2 + 6x + 6$ **9.** $x^2 + xy + y^2$
11. $4x^2y^2 - 4xy$ **13.** $\frac{13}{15}x^4 - \frac{3}{2}x^2 + \frac{7}{16}$ **15.** $2.2x^3 - 7.0x^2 + 0.2$ **17.** $6x^2 + x + 6$ **19.** $2x^4 + x^3 + 3x^2 - x$
21. $-6x^2 - x - 10$ **23.** $17x^3 - 5x^2 + 5$ **25.** $15x^5 - 8y^4 - 4y^2 + 5$ **27.** $3x^2 - 9xy + 2y^2$ **29.** $\frac{6}{5}x^3 - \frac{1}{4}x^2 + \frac{7}{24}$
31. $-1.2x^3 - 0.72x + 0.36$ **33.** $-2z^2$ **35.** $2x^2 + 13x$ **37.** x **39.** $7x^3 + 6x^2 + 6$ **41.** -8 **43.** 0
45. $7x^2 - 6x - 3$ **47.** $8x^5 - 7x^4 + 3x^2 + 2x$ **49.** $3x^4 - 5x^2 - 10$ **51.** Like   **53.** -14 **55.** 56   **57.** 17
59. $10,000

Problem Set 5.3
1. $-12x$ **3.** $20x$ **5.** $-8x^3$ **7.** $7x^3$ **9.** x^5 **11.** $-12x^6y^3$ **13.** $42x^4y^9$ **15.** $3x^6y^4$ **17.** $3y^5 - 6y$
19. $4y^3 - 8y^2 + 16$ **21.** $-x^5 + 2x^4$ **23.** $-5y^3 - 15y^2 + 20y$ **25.** $x^2 - 8x + 7$ **27.** $2y^2 - y - 6$
29. $x^3 - x^2 - 7x + 3$ **31.** $2x^3 + 5x^2 - 11x + 4$ **33.** $3z^5 + 8z^4 - 3z^2 + 7z + 40$ **35.** $x^4 - x^2 + 2x - 1$
37. $y^4 - 3y^3 - 2y^2 + 10y - 12$ **39.** $y^2 + 3y - 4$ **41.** $y^2 - 9y + 18$ **43.** $6x^2 - 8x + 2$ **45.** $7y^2 - 22y + 16$

47. $6x^2 - 18x + 12$ **49.** $5x^2 - xy - 6y^2$ **51.** $10y^2 + 23y + 12$ **53.** $x^2 - 81$ **55.** $x^2 + 16x + 64$
57. $2x^5 - 3x^4 + 4x - 6$ **59.** $7x^4 - x^3 - 21x + 3$ **61.** $2x^2 - 0.08$ **63.** Distributive   **65.** Binomial   **67.** $120
69. $2800

Problem Set 5.4
1. $x^2 - 16$ **3.** $x^2 - 4$ **5.** $25 - x^2$ **7.** $1 - y^2$ **9.** $4x^2 - 9$ **11.** $49x^2 - 4$ **13.** $25x^2 - 4y^2$
15. $4x^2 - 9y^2$ **17.** $x^4 - 4$ **19.** $x^2 + 4x + 4$ **21.** $y^2 - 8y + 16$ **23.** $y^2 - 14y + 49$ **25.** $4x^2 + 4x + 1$
27. $25x^2 - 10x + 1$ **29.** $16y^2 - 16y + 4$ **31.** $x^2 + 2xy + y^2$ **33.** $25x^2 - 10xy + y^2$ **35.** $64x^2 - 80xy + 25y^2$
37. $x^2 - 4$ **39.** $x^2 + 4x + 4$ **41.** $25x^2 - 80x + 64$ **43.** $9x^2 - 4$ **45.** $36x^2 - 1$ **47.** $36x^2 - 12x + 1$
49. $9x^2 - y^2$ **51.** $9x^2 - 6xy + y^2$ **53.** $4x^2 - 12xy + 9y^2$ **55.** $x^2 - \frac{1}{16}$ **57.** $0.16x^2 - 2.4x + 9$ **59.** Subtract
61. x^2 **63.** x^3 **65.** 46   **67.** 6 kg

Problem Set 5.5
1. $x^2 + 2x + 3$ **3.** $2x^2 + 4x - 1$ **5.** $x^2 - \frac{3}{5}x + \frac{4}{5}$ **7.** $2y - 1$ **9.** $x + 2$ **11.** $y - 4$ **13.** $x - 3$
15. $x + 1 + \dfrac{6}{x + 2}$ **17.** $x - 6 + \dfrac{10}{x + 1}$ **19.** $x^2 - x + 1 + \dfrac{-2}{x + 1}$ **21.** $2y + 3$ **23.** $3x + 8 + \dfrac{66}{5x - 7}$ **25.** Term
27. Polynomial   **29.** $\frac{2}{5}$ hour   **31.** $2\frac{23}{31}$ hours   **33.** $5x - 15$ **35.** $x^4 - 2x^3$ **37.** $15x^5 + 10x^3 - 20x^2$

Problem Set 5.6
1. $4(x + 2)$ **3.** $x^2(x + 4)$ **5.** $6x^3(x + 3)$ **7.** $2x(2x^2 + x + 1)$ **9.** $5(2x^3 - x^2 + 1)$ **11.** $2x(x - 2y)$
13. $4xy(2x^2y^2 + 3xy + 1)$ **15.** $7x(x - y)$ **17.** $x = \dfrac{7}{1 + y}$ **19.** $t = \dfrac{A}{x + g}$ **21.** $x = \dfrac{r}{1 - r}$ **23.** $(y + 5)(y + 2)$
25. $(x + 5)(x - 2)$ **27.** $(2x - 3)(3x - 1)$ **29.** $(a - b)(x + 1)$ **31.** $(y + a)(y + x)$ **33.** $(3y + x)(1 - 2y)$
35. $(x^2 + 2y^2)(x - 2y)$ **37.** Exponent   **39.** Common   **41.** $x^2 + 6x + 8$ **43.** $2x^2 + x - 1$ **45.** $8x^2 - 6x + 1$
47. $6x^2 - 3x - 18$ **49.** $3x^4 + 14x^3 - 5x^2$ **51.** $3x^4 - 12x^3 + 9x^2$

Problem Set 5.7
1. $(x + 5)(x + 1)$ **3.** $(x + 4)(x + 5)$ **5.** $(y - 12)(y - 2)$ **7.** $(y - 6)(y + 2)$ **9.** $(x + 7)(x - 4)$ **11.** $(x + 1)^2$
13. $(x^2 - 3)(x^2 - 1)$ **15.** $(y + 1)(y + 9)$ **17.** $(x - 3)(x - 5)$ **19.** $(y^2 - 7)(y^2 + 2)$ **21.** $(y + 6)(y + 7)$
23. $(3 - x)(2 + x)$ **25.** $(y - 5)(y - 2)$ **27.** $(x - \frac{1}{2})(x - \frac{1}{2})$ **29.** $(x + 9)(x - 12)$ **31.** $3(x + 1)(x + 5)$
33. $3(y + 8)(y - 1)$ **35.** $(3x + 2)(3x + 4)$ **37.** $(y - 7)^2$ **39.** $3(y + 2)(y + 1)$ **41.** $2(3x + 1)(x + 1)$
43. $7(2y + 1)(y + 2)$ **45.** $4(x + 6)(x - 1)$ **47.** $3(x + 4)(x + 5)$ **49.** $3x(x + 1)(x + 3)$ **51.** $3x^2(x - 6)(x + 5)$
53. $(7y - 1)(2y + 3)$ **55.** $3x(2x - 3)(3x + 1)$ **57.** $(9x - 4)(2x + 1)$ **59.** $3(2x - 7)(x + 3)$ **61.** $x(x - 5)(2x + 3)$
63. $2x(x - 8)(x - 4)$ **65.** First   **67.** $y^2 - 6y + 9$ **69.** $4x^2 + 12x + 9$ **71.** $x^2 - 25$ **73.** $36x^2 - 49$
75. $49x^2 - 14x + 1$

Problem Set 5.8
1. $(x + 2)^2$ **3.** $(x - 8)^2$ **5.** $(2x + 3)^2$ **7.** $(4x - 1)^2$ **9.** $(x - 2y)^2$ **11.** $(3x + y)^2$ **13.** $(x^2 - 2)^2$
15. $(y^3 + 5)^2$ **17.** $(6 - x)^2$ **19.** $(x + 5)(x - 5)$ **21.** $(x - 1)(x + 1)$ **23.** $(2y - 3)(2y + 3)$ **25.** $(6x - 5)(6x + 5)$
27. $(3x + y)(3x - y)$ **29.** $(6x - 7y)(6x + 7y)$ **31.** $(x^2 - 2)(x^2 + 2)$ **33.** $(x^3 - 5)(x^3 + 5)$ **35.** $(3 - x)(3 + x)$
37. $x(x + 3)(x - 3)$ **39.** $5(x + 1)^2$ **41.** $4(x + 2)(x - 1)$ **43.** $x(x + 4)(x - 2)$ **45.** $x(x + 5)^2$
47. $8x^2(x + 3)(x - 3)$ **49.** $(3x - 1)(x + 2)$ **51.** $(2x + y)(b + 1)$ **53.** $(x + 1)(x - 1)(x^2 + 1)$ **55.** $(2x - y)^2$
57. $(x + 3y)(a + 4)$ **59.** $2x^2(x - 9)^2$ **61.** $10x(1 - x)(1 + x)$ **63.** $(5 - x^2)^2$ **65.** Binomial
67. Greatest common factor   **69.** Yes   **71.** No   **73.** 5   **75.** 2

Chapter 5   Additional Exercises
1. Yes   **3.** No   **5.** Binomial   **7.** Trinomial   **9.** $-\frac{4}{15}x^2 - 13x - \frac{1}{16}$ **11.** 2   **13.** $9x^3 + 4x^2 - 6$
15. $7.4x^3 - 7.8x^2 - 3.1x - 1.7$ **17.** $-x^4 + 9x^2 - 7$ **19.** $-8x^3 + 6x^2 - 4x - 6$ **21.** $3x^3 - 6x^2 + 12x$
23. $-72x^{6n}y^{8n}$ **25.** $12x^{6n} - 32x^{3n}$ **27.** $15x^2 + 2xy - 8y^2$ **29.** $8x^{3n} - 16x^{2n}y^{2n} - 3x^ny^{3n} + 6y^{5n}$ **31.** $x^2 - 64$
33. $4x^2 - \frac{9}{16}$ **35.** $4x^2 + 20x + 25$ **37.** $x^{4n} - 18x^{2n}y^{3n} + 81y^{6n}$ **39.** $2x^2 + 3x - \dfrac{3}{4}$ **41.** $x - 4 + \dfrac{5}{x + 5}$
43. $4a^2 + 7 + \dfrac{-2}{3a^2 - 2}$ **45.** $27xy^3$ **47.** $5x^{2n}(2 - 5x^{2n})$ **49.** $h = \dfrac{2A}{b + c}$ **51.** $(4x - 3z)(y - w)$
53. $(y + 2)(3x^2 - 1)$ **55.** $(x - 6)(x - 1)$ **57.** $(3x - 5)(2x + 3)$ **59.** $2x(2x - 5)(x - 3)$ **61.** $x^{2n}(3x^n + 4)(2x^n - 1)$
63. $(3x - 4)^2$ **65.** $(y^{3n} - 5)^2$ **67.** $(5x - 2)(5x + 2)$ **69.** $(8x - 3)(12x + 5)$ **71.** $x^2(x^2 + 9)(x - 3)(x + 3)$

Chapter 5   Practice Test

1. 2 **2.** $-y^4 - 11y^2 + 4y$ **3.** $12x^2 - 3xy - 9y^2$ **4.** $-10x$ **5.** $-x^3 - 10x^2 + 2x$ **6.** $-6x^4 + 8x^3 - 16x^2$

7. $12y^2 + 7y - 10$ **8.** $2x^3 - 9x^2 + 13x - 12$ **9.** $4x^2 - 16$ **10.** $4x^2 - 16x + 16$ **11.** $x - 1 + \dfrac{3}{x - 2}$

12. $x = \dfrac{b}{1 + b}$ **13.** $3x(2x - y + 3y^2)$ **14.** $(y + x)(p + q)$ **15.** $(x + 5)(x - 4)$ **16.** $6(3x + 1)(x + 1)$

17. $3x(x - 2y)^2$ **18.** $(x^2 + 4)(x + 2)(x - 2)$

CHAPTER 6

Problem Set 6.1

1–7.

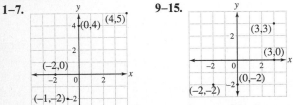

9–15.

17. $(4, 0)$ **19.** $(2, 3)$ **21.** $(-2, 5)$ **23.** $(4, -4)$ **25.** $(0, 0)$ **27.** $(0, 5)$ **29.** $(-4, -2)$ **31.** $(-4, 0)$

33. III **35.** II **37.** II **39.** III **41.** x-axis **43.** Ordinate **45.** Positive **47.** Positive

49. $-5x^2 - 4x - 3$ **51.** $\frac{23}{40}x^2 - 6x + \frac{1}{2}$ **53.** $4y^4 - y^3 - 14y^2 - 4$ **55.** $-2a^2 + 9ab - 4b^2$

Problem Set 6.2

1. Yes

3. No

5. Yes

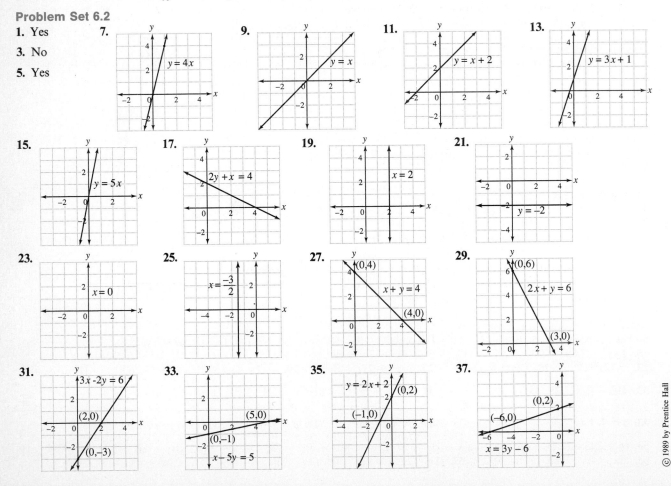

39. Constant   **41.** Ordered pairs   **43.** Straight line   **45.** y-axis   **47.** $3y + 1 + \dfrac{4}{3y}$ **49.** $x - 8 + \dfrac{3}{x + 3}$

51. $x + 4 + \dfrac{3}{3x - 2}$ **53.** $x - 1 + \dfrac{2}{x + 1}$

Problem Set 6.3

1. $\frac{2}{3}$ **3.** 0   **5.** -1 **7.** $\frac{1}{6}$ **9.** $\frac{3}{11}$ **11.** 0   **13.** 1   **15.** $-\frac{7}{13}$ **17.** $-\frac{8}{7}$ **19.** 0   **21.** 0   **23.** No slope
25. 0   **27.** Vertical, horizontal   **29.** Horizontal   **31.** $7x^2(1 + 2x)$ **33.** $6(2x^2 + 4x - 3)$ **35.** $4x(xy^2 + 4y + 2)$
37. $(a + b)(x - 1)$ **39.** $(y - 1)(y - x)$ **41.** $(2x - 3)(x + 4)$

Problem Set 6.4

1. 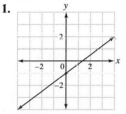   **3.** 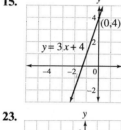   **5.**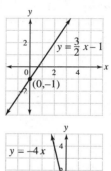

7. Slope: 2; y-intercept: $(0, -3)$ **9.** Slope: -2; y-intercept: $(0, 8)$ **11.** Slope: $-\frac{3}{2}$; y-intercept: $(0, 4)$
13. Slope: $\frac{2}{3}$; y-intercept: $\left(0, -\frac{5}{3}\right)$

15. 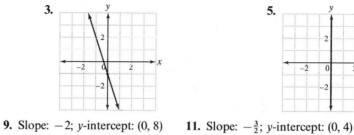   **17.**   **19.**   **21.**

23. 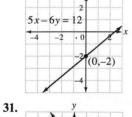   **25.** 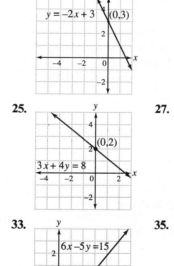   **27.** 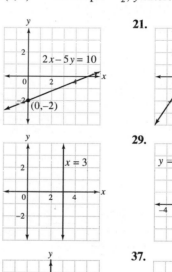   **29.**

31. 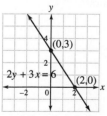   **33.** 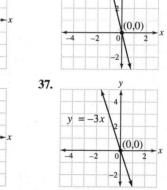   **35.**   **37.**

39. Slope, y-intercept   **41.** $\frac{2}{3}$ **43.** $\dfrac{1}{-3}$ **45.** $\dfrac{8}{-1}$ **47.** $(x - 2)(x + 5)$ **49.** $(2x + 3)(x - 2)$
51. $3(x - 1)(x - 8)$ **53.** $2x(x + 5)(2x - 1)$

Problem Set 6.5

1. $y = 2x - 1$ **3.** $y = -4x$ **5.** $y = -8$ **7.** $y = \frac{1}{2}x + 5$ **9.** $y = 2x + 6$ **11.** $y = -3x + 1$ **13.** $y = 5x - 4$
15. $y = \frac{3}{2}x - 7$ **17.** $y = -2$ **19.** $y = x - 3$ **21.** $y = \frac{1}{2}x + \frac{5}{2}$ **23.** $y = -\frac{1}{4}x + 1$ **25.** $y = 4$
27. $y = -4x - 8$ **29.** Parallel   **31.** Perpendicular   **33.** Neither   **35.** Parallel   **37.** Perpendicular
39. Slope-intercept   **41.** Parallel   **43.** $(3x - 2)^2$ **45.** $(x - 7)(x + 7)$ **47.** $(2x - 3)(3x - 4)$
49. $x^2(x + 3)(x - 3)$ **51.** $(x + 4y)(b + 3)$

Problem Set 6.6

1. $m = \dfrac{y_2 - y_1}{x_2 - x_1} = \dfrac{35 - 10}{95 - 50} = \dfrac{25}{45} = \dfrac{5}{9}$

$y - y_1 = m(x - x_1)$

$y - 10 = \dfrac{5}{9}(x - 50)$

$y = \dfrac{5}{9}x - \dfrac{250}{9} + 10$

$y = \dfrac{5}{9}x - \dfrac{160}{9}$ (equation)

Let $x = 77$.

$y = \dfrac{5}{9}(77) - \dfrac{160}{9}$

$y = \dfrac{385 - 160}{9} = 25$

When the temperature is 77°F, it is 25°C.

3. $V - V_1 = m(t - t_1)$

$V - 60,000 = -3000t$

$V = -3000t + 60,000$ (equation)

Let $t = 11$.

$V = -3000(11) + 60,000 = 27,000$

After 11 years the book value is $27,000.

5. $m = \dfrac{y_2 - y_1}{x_2 - x_1} = \dfrac{26 - 20}{68 - 44} = \dfrac{6}{24} = \dfrac{1}{4}$

$c - c_1 = m(L - L_1)$

$c - 20 = \dfrac{1}{4}(L - 44)$

$c - 20 = \dfrac{1}{4}L - 11$

$c = \dfrac{1}{4}L + 9$ (equation)

Let $L = 88$.

$c = \dfrac{1}{4}(88) + 9 = 31$

When 88 square feet of lumber are used, the cost is $31.

7. $c - c_1 = m(e - e_1)$

$c - 120 = 7(e - 0)$

$c - 120 = 7e$

$c = 7e + 120$ (equation)

When there are 116 passengers,

$e = 125 - 116 = 9$

and the cost is

$c = 7(9) + 120 = \$183$

9. $d - d_1 = m(p - p_1)$

$d - 88 = \frac{11}{5}(p - 55)$

$d - 88 = \frac{11}{5}p - 121$

$d = \frac{11}{5}p - 33$

Let $p = 75$.

$d = \frac{11}{5}(75) - 33 = 132$

When the pressure is 75 psi the depth is 132 feet.

11. $V - V_1 = m(t - t_1)$

$V - 21,000 = -3000t$

$V = -3000t + 21,000$ (equation)

Let $t = 3\frac{1}{2}$.

$V = -3000(\frac{7}{2}) + 21,000 = 10,500$

After $3\frac{1}{2}$ years the book value of the machine is $10,500.

13.

$x - y = 4$

15.

$2x - 3y = 6$

17.

$x + 4y = 4$

19.

$y = 2x$

21.

$y = 3x - 2$

23.

$y = -2x + 1$

Chapter 6 Additional Exercises

1,3. 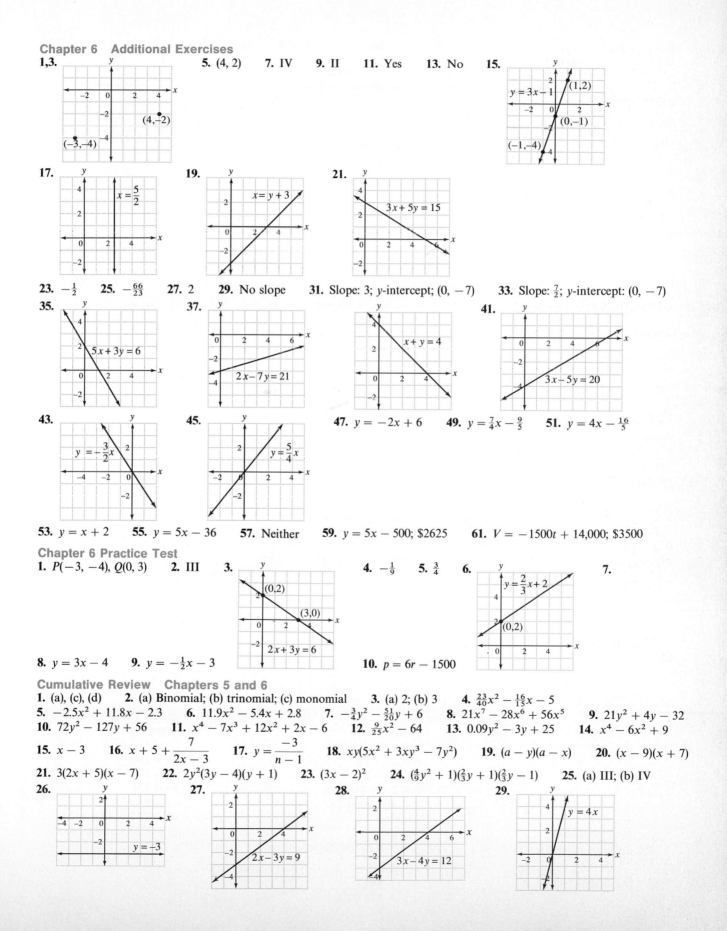   **5.** $(4, 2)$ **7.** IV   **9.** II   **11.** Yes   **13.** No   **15.**

17.   **19.**   **21.**

23. $-\frac{1}{2}$ **25.** $-\frac{66}{23}$ **27.** 2 **29.** No slope   **31.** Slope: 3; y-intercept; $(0, -7)$ **33.** Slope: $\frac{7}{2}$; y-intercept: $(0, -7)$

35.   **37.**   **41.**

43.   **45.**   **47.** $y = -2x + 6$ **49.** $y = \frac{7}{4}x - \frac{9}{5}$ **51.** $y = 4x - \frac{16}{5}$

53. $y = x + 2$ **55.** $y = 5x - 36$ **57.** Neither   **59.** $y = 5x - 500$; \$2625   **61.** $V = -1500t + 14{,}000$; \$3500

Chapter 6 Practice Test

1. $P(-3, -4), Q(0, 3)$ **2.** III   **3.**   **4.** $-\frac{1}{9}$ **5.** $\frac{3}{4}$ **6.**   **7.**

8. $y = 3x - 4$ **9.** $y = -\frac{1}{2}x - 3$ **10.** $p = 6r - 1500$

Cumulative Review   Chapters 5 and 6

1. (a), (c), (d)   **2.** (a) Binomial; (b) trinomial; (c) monomial   **3.** (a) 2; (b) 3   **4.** $\frac{23}{40}x^2 - \frac{16}{15}x - 5$
5. $-2.5x^2 + 11.8x - 2.3$ **6.** $11.9x^2 - 5.4x + 2.8$ **7.** $-\frac{3}{4}y^2 - \frac{51}{20}y + 6$ **8.** $21x^7 - 28x^6 + 56x^5$ **9.** $21y^2 + 4y - 32$
10. $72y^2 - 127y + 56$ **11.** $x^4 - 7x^3 + 12x^2 + 2x - 6$ **12.** $\frac{9}{25}x^2 - 64$ **13.** $0.09y^2 - 3y + 25$ **14.** $x^4 - 6x^2 + 9$
15. $x - 3$ **16.** $x + 5 + \frac{7}{2x - 3}$ **17.** $y = \frac{-3}{n - 1}$ **18.** $xy(5x^2 + 3xy^3 - 7y^2)$ **19.** $(a - y)(a - x)$ **20.** $(x - 9)(x + 7)$
21. $3(2x + 5)(x - 7)$ **22.** $2y^2(3y - 4)(y + 1)$ **23.** $(3x - 2)^2$ **24.** $(\frac{4}{9}y^2 + 1)(\frac{2}{3}y + 1)(\frac{2}{3}y - 1)$ **25.** (a) III; (b) IV
26.   **27.**   **28.**   **29.**

30. $\frac{3}{2}$ **31.** No slope   **32.** $\frac{2}{7}$ **33.** $y = \frac{3}{2}x - 7$ **34.** $-2, -2$; parallel   **35.** $\frac{3}{5}, -\frac{5}{3}$; perpendicular
36. $V = -800t + 6000$; \$3000

CHAPTER 7

Problem Set 7.1
1. Yes   **3.** No   **5.** No   **7.** Yes   **9.**   **11.**   **13.** No solution

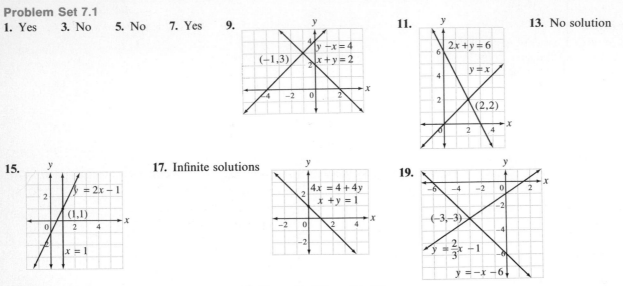

15.   **17.** Infinite solutions   **19.**

21. Points   **23.** Inconsistent   **25.** I   **27.** II   **29.** III   **31.** IV

Problem Set 7.2
1. $(9, 2)$ **3.** $(2, 0)$ **5.** $(1, -3)$ **7.** $(4, 8)$ **9.** No solution   **11.** No solution   **13.** Infinite solutions
15. $(-\frac{13}{4}, -\frac{11}{4})$ **17.** $(2, 3)$ **19.** $(6, -5)$ **21.** $(\frac{1}{8}, \frac{9}{8})$ **23.** $(3, -7)$ **25.** Variable   **27.** Multiplied   **29.** $-\frac{5}{2}$
31. $-\frac{1}{8}$ **33.** $\frac{4}{3}$ **35.** 0   **37.** 0   **39.** No slope

Problem Set 7.3
1. $(2, -1)$ **3.** $(2, 1)$ **5.** $(-1, -3)$ **7.** $(2, 2)$ **9.** $(7, 10)$ **11.** $(2, 1)$ **13.** $(-2, 0)$ **15.** $(0, 0)$ **17.** $(3, -1)$ **19.** $(6, -5)$
21. $(2, -2)$ **23.** $(18, -12)$ **25.** Substitution   **27.** Slope: 3; y-intercept $(0, -4)$ **29.** Slope: $\frac{4}{3}$; y-intercept: $(0, -3)$

31.   **33.**   **35.**   **37.**

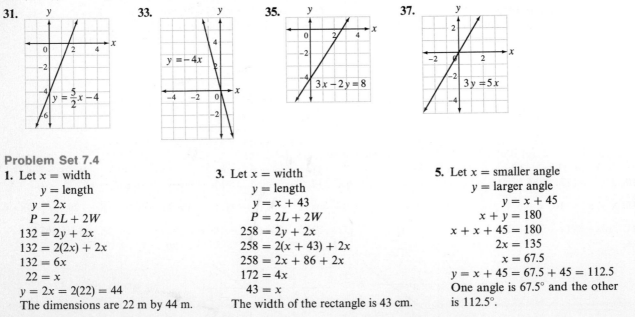

Problem Set 7.4
1. Let x = width
    y = length
    $y = 2x$
    $P = 2L + 2W$
    $132 = 2y + 2x$
    $132 = 2(2x) + 2x$
    $132 = 6x$
    $22 = x$
    $y = 2x = 2(22) = 44$
The dimensions are 22 m by 44 m.

3. Let x = width
    y = length
    $y = x + 43$
    $P = 2L + 2W$
    $258 = 2y + 2x$
    $258 = 2(x + 43) + 2x$
    $258 = 2x + 86 + 2x$
    $172 = 4x$
    $43 = x$
The width of the rectangle is 43 cm.

5. Let x = smaller angle
    y = larger angle
    $y = x + 45$
    $x + y = 180$
    $x + x + 45 = 180$
    $2x = 135$
    $x = 67.5$
    $y = x + 45 = 67.5 + 45 = 112.5$
One angle is 67.5° and the other
is 112.5°.

7. Let x = smaller angle
y = larger angle
$y = 2x - 6$
$x + y = 90$
$x + 2x - 6 = 90$
$3x = 96$
$x = 32$
$y = 2(32) - 6 = 64 - 6 = 58$
One angle is 32° and the other is 58°.

9. Let x = original price
y = amount of increase
$x + y = 78400$
$y = 0.12x$
$x + 0.12x = 78400$
$1.12x = 78400$
$x = 70000$
The original price was $70,000.

11. Let x = original price
y = amount of decrease
$x - y = 8245$
$y = 0.15x$
$x - 0.15x = 8245$
$0.85x = 8245$
$x = 9700$
The original price was $9700.

13. Let x = original amount
y = interest earned
$x + y = 856$
$y = 0.07x$
$x + 0.07x = 856$
$1.07x = 856$
$x = 800$
John put $800 in the bank.

15. Let x = amount invested at 7%
y = amount invested at 10%
$y = x - 2500$
$0.07x + 0.10y = 600$
$0.07x + 0.10(x - 2500) - 600$
$0.07x + 0.10x - 250 = 600$
$0.17x = 850$
$x = 5000$
$y = x - 2500 = 5000 - 2500 = 2500$
$5000 is invested at 7%, $2500 is invested at 10%.

17. Let x = amount invested at 7%
y = amount invested at 9%
$y = 2x$
$0.07x + 0.09y = 600$
$0.07x + 0.09(2x) = 600$
$0.07x + 0.18x = 600$
$0.25x = 600$
$x = 2400$
$y = 2x = 2(2400) = 4800$
$2400 is invested at 7%, $4800 is invested at 9%.

19. Let x = liters of 40% solution
y = liters of 80% solution
$x + y = 12$ (first equation)
$0.40x + 0.80y = 0.50(12)$
$4x + 8y = 5(12)$
$4x + 8y = 60$
$\underline{-4x - 4y = -48}$ (first equation times -4)
$4y = 12$
$y = 3$
$x + y = 12$
$x + 3 = 12$
$x = 9$
She should use 3 liters of 80% solution and 9 liters of 40% solution.

21. Let x = quarts of 55% solution
y = quarts of 10% solution
12 = quarts of 25% solution
$x + y = 12$ (first equation)
$0.55x + 0.10y = 0.25(12)$
$55x + 10y = 25(12)$
$55x + 10y = 300$
$\underline{-10x - 10y = -120}$ (first equation times -10)
$45x = 180$
$x = 4$
Four quarts of 55% antifreeze should be used.

23. 6 = liters of 35% solution
Let x = liters of 50% solution
y = liters of 45% solution
$6 + x = y$
$0.35(6) + 0.50x = 0.45y$
$0.35(6) + 0.50x = 0.45(6 + x)$
$35(6) + 50x = 45(6 + x)$
$210 + 50x = 270 + 45x$
$5x = 60$
$x = 12$
Juan should use 12 liters of 50% solution.

25. Word

27. No

29. Yes

31.

33.

35. $f = -100t + 8000$; 4500 ft   **37.** $80,000

Problem Set 7.5

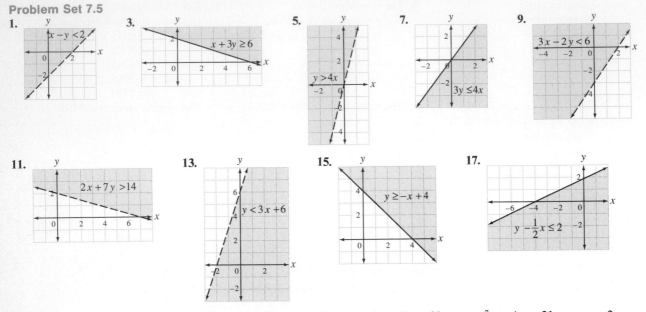

1. $x - y < 2$ **3.** $x + 3y \geq 6$ **5.** $y > 4x$ **7.** $3y \leq 4x$ **9.** $3x - 2y < 6$

11. $2x + 7y > 14$ **13.** $y < 3x + 6$ **15.** $y \geq -x + 4$ **17.** $y - \frac{1}{2}x \leq 2$

19. Ordered pairs **21.** Shade **23.** -7 **25.** 5 **27.** $y = -5x + 3$ **29.** $y = -\frac{7}{3}x - 4$ **31.** $y = x - 2$
33. $y = \frac{3}{4}x - \frac{13}{4}$

Chapter 7 Additional Exercises

1. Yes **3.** No **5.** **7.** No solution **9.** (5, 2) **11.** $(-1, -2)$

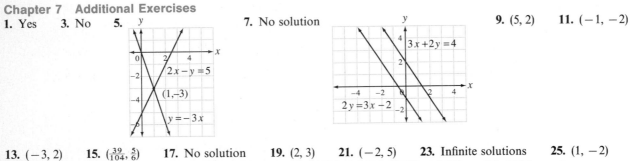

$2x - y = 5$ $(1, -3)$ $y = -3x$ $3x + 2y = 4$ $2y = 3x - 2$

13. $(-3, 2)$ **15.** $(\frac{39}{104}, \frac{5}{6})$ **17.** No solution **19.** (2, 3) **21.** $(-2, 5)$ **23.** Infinite solutions **25.** $(1, -2)$
27. $(-6, 4)$ **29.** $33°, 57°$ **31.** $7.50 per hour **33.** $1800 at 5.5%, $3600 at 8.5% **35.** 15.75 quarts

37. **39.** **41.**

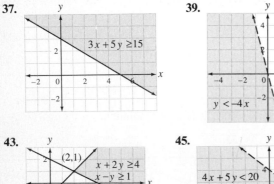

$3x + 5y \geq 15$ $y < -4x$ $x - 3y < 4 + 5y$

43. **45.**

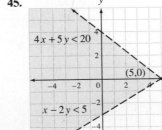

$(2, 1)$ $x + 2y \geq 4$ $x - y \geq 1$ $4x + 5y < 20$ $(5, 0)$ $x - 2y < 5$

Chapter 7 Practice Test

1. No   **2.**   **3.** (2, 2)   **4.** (−2, 1)   **5.** (−2, 5)   **6.** (3, −2)   **7.** (−4, 1)

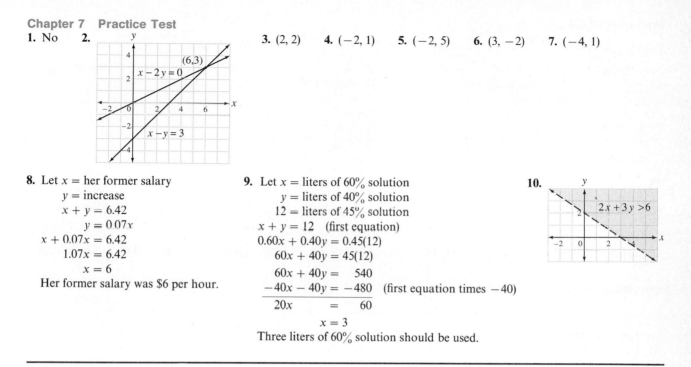

8. Let x = her former salary
   y = increase
   $x + y = 6.42$
   $y = 0.07x$
   $x + 0.07x = 6.42$
   $1.07x = 6.42$
   $x = 6$
Her former salary was $6 per hour.

9. Let x = liters of 60% solution
   y = liters of 40% solution
   12 = liters of 45% solution
   $x + y = 12$ (first equation)
   $0.60x + 0.40y = 0.45(12)$
   $60x + 40y = 45(12)$
   $60x + 40y = 540$
   $\underline{-40x - 40y = -480}$ (first equation times −40)
   $20x \qquad = 60$
   $x = 3$
Three liters of 60% solution should be used.

10.

CHAPTER 8

Problem Set 8.1

1. Yes   **3.** No   **5.** Yes   **7.** Yes   **9.** (a) $\frac{5}{2}$; (b) $-\frac{5}{4}$ **11.** (a) −1; (b) $\frac{1}{5}$ **13.** (a) 0; (b) −3   **15.** (a) $\frac{2}{5}$; (b) $-\frac{2}{7}$

17. (a) $-\frac{1}{8}$; (b) −2   **19.** Whole   **21.** y^3 **23.** $\dfrac{1}{x^5}$ **25.** 1   **27.** $x(x + y)(x - y)$ **29.** $(2x - 3)(3x + 2)$

31. $3(x - 4)^2$ **33.** $y(y - 2)(3y - 1)$

Problem Set 8.2

1. xy **3.** $\dfrac{2y}{x}$ **5.** 5   **7.** $\frac{6}{5}$ **9.** $\frac{4}{3}$ **11.** $\dfrac{x + 2}{2(x - 2)}$ **13.** $\dfrac{3x + 1}{2}$ **15.** $\dfrac{1}{x + y}$ **17.** $x + 4$ **19.** $\dfrac{x + 2}{x + 4}$

21. $\dfrac{3x - 1}{2x - 1}$ **23.** 1   **25.** −5   **27.** $-(x + y)$ **29.** Factors   **31.** Greatest common factor   **33.** No   **35.** Yes

37.   **39.** Infinite solutions

Problem Set 8.3

1. $\dfrac{4}{xy^3}$ **3.** $\dfrac{7}{5x^4}$ **5.** $\dfrac{4(x - 2)}{3(x + 3)}$ **7.** $\dfrac{(2x - 3)(x + 1)}{3(x - 4)}$ **9.** $\dfrac{3(x - 5)}{5}$ **11.** $3(x + 3)$ **13.** $(x + 1)(x + 2)$ **15.** $\dfrac{y - 2}{y + 2}$

17. $-\frac{5}{4}$ **19.** $\dfrac{-x(x + 3)}{2}$ **21.** $\frac{7}{3}$ **23.** $\frac{5}{7}$ **25.** $\dfrac{x^2}{y}$ **27.** $\dfrac{x}{2}$ **29.** $\dfrac{2(x + 3)}{x(x + 4)}$ **31.** $\dfrac{2}{x + y}$ **33.** $\dfrac{(x + 2)^2}{2x}$

35. $\dfrac{(x + 3)^2}{x^2 + 5}$ **37.** $\dfrac{x + 6}{x + 3}$ **39.** $\dfrac{x + 2}{x + 4}$ **41.** $\dfrac{2x + 3}{(x - 3)(x - 6)^2}$ **43.** Denominators   **45.** Negative 1   **47.** $\frac{13}{10}$

49. $\frac{7}{16}$ **51.** (1, −2)   **53.** (−2, −3)

Problem Set 8.4

1. $\frac{11}{5}$ **3.** $\frac{14}{x}$ **5.** $\frac{11}{x+1}$ **7.** $\frac{3x^2-x-7}{3x-4}$ **9.** $\frac{5}{9}$ **11.** $\frac{3-2y}{y^2}$ **13.** -2 **15.** $\frac{-x^2-3x-4}{4x+7}$ **17.** $\frac{1}{x}$

19. 1 **21.** $6x^4$ **23.** $48x^2y^4$ **25.** $x^3(2x+1)$ **27.** $4(x-1)$ **29.** $(y-4)^2$ **31.** $(x+3)(x-2)(x+5)$

33. $x(x-1)(2x+7)$ **35.** $(2x+1)(x-4)(x-1)$ **37.** Numerators **39.** One **41.** $(2, 3)$ **43.** $(-2, -3)$

45. $(-1, -3)$ **47.** $(2, -4)$

Problem Set 8.5

1. $\frac{3(x-2)}{3(x+4)}$ **3.** $\frac{(x-1)(x+1)}{(x+5)(x+1)}$ **5.** $\frac{19}{6x}$ **7.** $\frac{8x-6}{x(x-2)}$ **9.** $\frac{3y^2-11y}{(y+4)(y-4)}$ **11.** $\frac{4+x}{(x+3)}$ **13.** $\frac{4x-5}{(x-2)(x+1)^2}$

15. $\frac{4x+7}{(x+1)^2}$ **17.** $\frac{4x+3}{(x+1)(x-1)}$ **19.** $\frac{2x^2+10}{(x-4)(x+3)}$ **21.** $\frac{2}{x-y}$ or $\frac{-2}{y-x}$ **23.** $\frac{2y+4x}{x^2y}$ **25.** $\frac{2x+30}{x(x+5)}$

27. $\frac{16y^2-8y-3x}{6x^2y^2}$ **29.** $\frac{4x-10}{(x+1)(x-1)}$ **31.** $\frac{-1}{x+1}$ **33.** $\frac{-7}{x-2}$ **35.** $\frac{3}{(x+1)(x-1)(x+2)}$ **37.** $\frac{3x^2+2x-5}{x^3}$

39. $\frac{2}{(x-1)x}$ **41.** $\frac{7x+11}{(x+2)(x-2)}$ **43.** Denominator **45.** Multiplying **47.** 27 ft **49.** \$3000 **51.** $5\frac{1}{3}$ liters

Chapter 8 Additional Exercises

1. Yes **3.** No **5.** $\frac{3}{2}$ **7.** $-\frac{12}{5}$ **9.** $-\frac{19}{11}$ **11.** $\frac{-9}{xy^4}$ **13.** $\frac{9y^2}{x^2z^3}$ **15.** $x-3$ **17.** -1 **19.** $\frac{3x+8}{2x-3}$

21. $\frac{2y^3}{x^3}$ **23.** $\frac{-2y}{3x}$ **25.** $\frac{y(y+2)}{2}$ **27.** $\frac{1}{b^6}$ **29.** $\frac{(x-1)(x+4)}{(x+1)(x-2)}$ **31.** -1 **33.** $\frac{-x^2-4}{3x^2+2}$ **35.** $54x^4y^3$

37. $x(x+4)$ **39.** $(y-5)(y-2)$ **41.** $\frac{10+7x}{12x^2}$ **43.** $\frac{3y^2+y+20}{(2y-5)(y+3)}$ **45.** $\frac{7y-15}{(y+5)(y-5)}$ **47.** $\frac{-10}{x-4}$

49. $\frac{-x-16}{(x+2)(x+4)(x-5)}$ **51.** $\frac{-3x^2-4x+14}{(x+2)(x-2)(x-3)}$

Chapter 8 Practice Test

1. Yes **2.** $\frac{14}{5}$ **3.** $\frac{3}{4x+5}$ **4.** -4 **5.** $\frac{4}{x+5}$ **6.** $x^2(x+y)$ **7.** $\frac{8x+12}{(x+4)(x-4)}$ **8.** $\frac{4}{x-8}$

9. $\frac{-2x^2+5x-15}{(x+5)(x-1)(x-5)}$

Cumulative Review   Chapters 7 and 8

1. No **2.** No **3.** **4.** $(6, -2)$ **5.** $(-2, 1)$ **6.** $(-2, 3)$ **7.** $(1, 4)$

8. $(-3, 3)$ **9.** $55°, 125°$ **10.** $26°, 64°$ **11.** \$450 at 7%, \$1350 at 8%

12. 32 milliliters of 40% solution, 48 milliliters of 65% solution **13.** **14.**

15. (a), (c) **16.** $-\frac{5}{2}$ **17.** $\frac{7}{2}$ **18.** $\frac{3x}{y}$ **19.** $\frac{3x-12}{x+5}$ **20.** $\frac{2x^2}{y}$ **21.** $\frac{1}{(x+3)(x-8)}$ **22.** $\frac{x^2-5x+6}{(x-1)(x+3)}$

23. $\frac{4}{x+4}$ **24.** $\frac{x+12}{4(x-2)}$ **25.** $\frac{y^2-y-4}{y(y-1)(y+1)}$ **26.** $\frac{x^2+5x-2}{(x+5)(x+1)(x-2)}$ **27.** $\frac{-7x^3-50x^2-42x+159}{(x+3)(x+5)^2}$

CHAPTER 9

Problem Set 9.1

1. 0   **3.** 3   **5.** None   **7.** -2 **9.** 24   **11.** 12   **13.** $-\frac{8}{3}$ **15.** -1 **17.** 2   **19.** 5   **21.** 2   **23.** 4

25. $\frac{9}{2}$ **27.** No solution   **29.** No solution   **31.** $\dfrac{-28 + 3x}{4x}$ **33.** $-\frac{28}{3}$ **35.** 5   **37.** $\dfrac{y - 5}{(y + 1)(y - 2)}$ **39.** 3

41. $\dfrac{y + 3}{24}$ **43.** Lowest common denominator   **45.** Variable

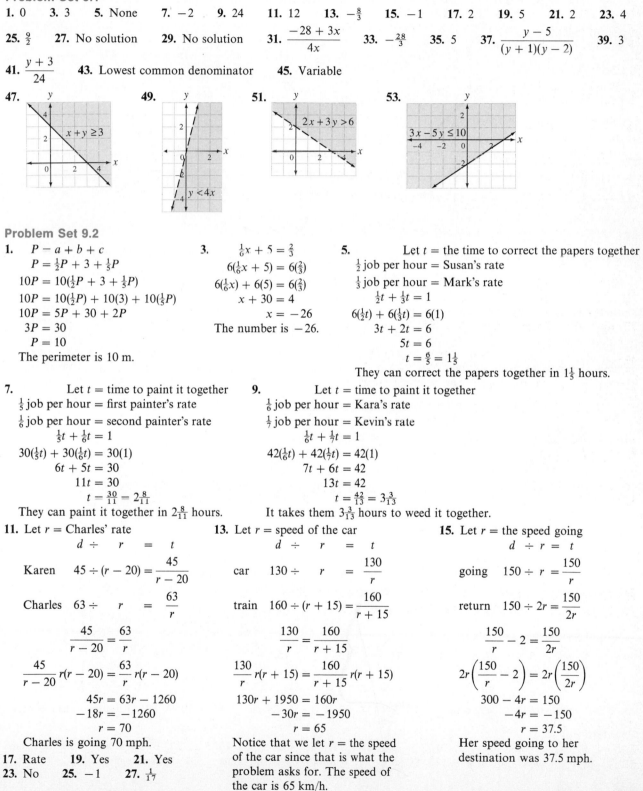

47. $x + y \geq 3$ **49.** $y < 4x$ **51.** $2x + 3y > 6$ **53.** $3x - 5y \leq 10$

Problem Set 9.2

1.
$$P - a + b + c$$
$$P = \tfrac{1}{2}P + 3 + \tfrac{1}{5}P$$
$$10P = 10(\tfrac{1}{2}P + 3 + \tfrac{1}{5}P)$$
$$10P = 10(\tfrac{1}{2}P) + 10(3) + 10(\tfrac{1}{5}P)$$
$$10P = 5P + 30 + 2P$$
$$3P = 30$$
$$P = 10$$
The perimeter is 10 m.

3.
$$\tfrac{1}{6}x + 5 = \tfrac{2}{3}$$
$$6(\tfrac{1}{6}x + 5) = 6(\tfrac{2}{3})$$
$$6(\tfrac{1}{6}x) + 6(5) = 6(\tfrac{2}{3})$$
$$x + 30 = 4$$
$$x = -26$$
The number is -26.

5.   Let t = the time to correct the papers together
$\tfrac{1}{2}$ job per hour = Susan's rate
$\tfrac{1}{3}$ job per hour = Mark's rate
$$\tfrac{1}{2}t + \tfrac{1}{3}t = 1$$
$$6(\tfrac{1}{2}t) + 6(\tfrac{1}{3}t) = 6(1)$$
$$3t + 2t = 6$$
$$5t = 6$$
$$t = \tfrac{6}{5} = 1\tfrac{1}{5}$$
They can correct the papers together in $1\tfrac{1}{5}$ hours.

7.   Let t = time to paint it together
$\tfrac{1}{5}$ job per hour = first painter's rate
$\tfrac{1}{6}$ job per hour = second painter's rate
$$\tfrac{1}{5}t + \tfrac{1}{6}t = 1$$
$$30(\tfrac{1}{5}t) + 30(\tfrac{1}{6}t) = 30(1)$$
$$6t + 5t = 30$$
$$11t = 30$$
$$t = \tfrac{30}{11} = 2\tfrac{8}{11}$$
They can paint it together in $2\tfrac{8}{11}$ hours.

9.   Let t = time to paint it together
$\tfrac{1}{6}$ job per hour = Kara's rate
$\tfrac{1}{7}$ job per hour = Kevin's rate
$$\tfrac{1}{6}t + \tfrac{1}{7}t = 1$$
$$42(\tfrac{1}{6}t) + 42(\tfrac{1}{7}t) = 42(1)$$
$$7t + 6t = 42$$
$$13t = 42$$
$$t = \tfrac{42}{13} = 3\tfrac{3}{13}$$
It takes them $3\tfrac{3}{13}$ hours to weed it together.

11. Let r = Charles' rate

	d	$\div$	r	$=$	t
Karen	45	$\div$	$(r - 20)$	$=$	$\dfrac{45}{r - 20}$
Charles	63	$\div$	r	$=$	$\dfrac{63}{r}$

$$\frac{45}{r - 20} = \frac{63}{r}$$
$$\frac{45}{r - 20}r(r - 20) = \frac{63}{r}r(r - 20)$$
$$45r = 63r - 1260$$
$$-18r = -1260$$
$$r = 70$$
Charles is going 70 mph.

13. Let r = speed of the car

	d	$\div$	r	$=$	t
car	130	$\div$	r	$=$	$\dfrac{130}{r}$
train	160	$\div$	$(r + 15)$	$=$	$\dfrac{160}{r + 15}$

$$\frac{130}{r} = \frac{160}{r + 15}$$
$$\frac{130}{r}r(r + 15) = \frac{160}{r + 15}r(r + 15)$$
$$130r + 1950 = 160r$$
$$-30r = -1950$$
$$r = 65$$
Notice that we let r = the speed of the car since that is what the problem asks for. The speed of the car is 65 km/h.

15. Let r = the speed going

	d	$\div r =$	t
going	150	$\div r =$	$\dfrac{150}{r}$
return	150	$\div 2r =$	$\dfrac{150}{2r}$

$$\frac{150}{r} - 2 = \frac{150}{2r}$$
$$2r\left(\frac{150}{r} - 2\right) = 2r\left(\frac{150}{2r}\right)$$
$$300 - 4r = 150$$
$$-4r = -150$$
$$r = 37.5$$
Her speed going to her destination was 37.5 mph.

17. Rate   **19.** Yes   **21.** Yes
23. No   **25.** -1 **27.** $\frac{1}{17}$

Problem Set 9.3

1. $\frac{5}{4}$ **3.** $\frac{3}{11}$ **5.** \$2.33   **7.** 2.667 miles   **9.** 40 mph   **11.** 6 hours   **13.** 50.4 kg   **15.** 3.6 hours   **17.** 6.75 servings
19. 18   **21.** 9.6 psi   **23.** 15   **25.** 28   **27.** Quotient   **29.** Direct   **31.** $\frac{3y}{5x}$ **33.** $\frac{3x}{4}$ **35.** $\frac{x+3}{2x+5}$ **37.** $-x-4$

Problem Set 9.4

1. $\frac{5(1-2x)}{3x}$ **3.** $\frac{5(3+x)}{x(4+x)}$ **5.** $\frac{y}{x}$ **7.** $\frac{x}{3(x+2)}$ **9.** $\frac{5}{3x^2}$ **11.** $\frac{1}{3}$ **13.** $\frac{x}{y}$ **15.** $x-y$ **17.** $\frac{x-y}{2}$
19. Complex   **21.** Multiplying   **23.** $\frac{x}{3y^3}$ **25.** $\frac{2}{5(x+2)}$ **27.** $\frac{98y}{3x^4}$ **29.** $\frac{x-1}{4(x+1)}$ **31.** $3(2y+1)$

Chapter 9   Additional Exercises

1. 0   **3.** $-\frac{3}{2}$ **5.** None   **7.** 2   **9.** 6   **11.** 1   **13.** $\frac{5}{x-2}$ **15.** -10 **17.** 16 m by 56 m   **19.** $1\frac{1}{5}$ hours
21. $7\frac{1}{2}$ hours   **23.** 70 mph   **25.** 32   **27.** $7\frac{1}{2}$ quarts   **29.** 16   **31.** 8   **33.** $\frac{15}{7}$ **35.** $\frac{2x+x^2}{5+20x}$ **37.** $\frac{3-3y}{1-2y}$
39. $\frac{5y-3x}{2xy+4y}$ **41.** $\frac{8y^2+6y-14}{4y^2+5y+1}$

Chapter 9   Practice Test

1. 1   **2.** No solution   **3.** $\frac{8x+12}{x(x+4)}$ **4.** $1\frac{1}{5}$ hours   **5.** 60 mph   **6.** \$81   **7.** $3\frac{1}{5}$ **8.** $\frac{5(x+1)}{x(3+x)}$ **9.** y

---

CHAPTER 10

Problem Set 10.1

1. 2   **3.** -8 **5.** 1   **7.** 6   **9.** 0   **11.** 15   **13.** Yes   **15.** No   **17.** No   **19.** Yes   **21.** $|x|$ **24.** $4|x|$
25. $7|ab|$ **27.** $6|b|$ **29.** $|x-3|$ **31.** $|x+3|$ **33.** Irrational   **35.** Rational   **37.** Irrational   **39.** Irrational
41. Rational   **43.** Rational   **45.** Rational   **47.** Irrational   **49.** 2.449   **51.** -3.873 **53.** 9.899   **55.** 4.796
57. 2.44 miles   **59.** Positive   **61.** Perfect squares   **63.** Terminates   **65.** x^8 **67.** x^6 **69.** $\frac{4x^2-2x-7}{2x+1}$ **71.** $\frac{-3y^2+3y+25}{x^2+3}$

Problem Set 10.2

1. $3\sqrt{2}$ **3.** $5\sqrt{3}$ **5.** $2\sqrt{3}$ **7.** $7\sqrt{3}$ **9.** $5\sqrt{5}$ **11.** $8\sqrt{x}$ **13.** $2|x|$ **15.** $4\sqrt{3x}$ **17.** $4|x|\sqrt{5}$ **19.** $7|y|$
21. $|x^3|\sqrt{x}$ **23.** x^4 **25.** $10|x^3|\sqrt{x}$ **27.** $5x^2\sqrt{6}$ **29.** $4|xy|\sqrt{7x}$ **31.** 12.246   **33.** 15.874   **35.** 16.882
37. Negative   **39.** Factoring   **41.** $\frac{11}{x-1}$ **43.** $\frac{7y^2+30y}{(y+3)^2}$ **45.** $\frac{-6x-21}{x-7}$ **47.** $\frac{54+23x-5x^2}{3(x+6)(x-6)}$
49. $\frac{20y^2-84y-11}{(y+1)(4y+3)(5y-2)}$

Problem Set 10.3

1. $\sqrt{10}$ **3.** 5   **5.** $\sqrt{33}$ **7.** $\sqrt{7x}$ **9.** $\sqrt{70x}$ **11.** $\sqrt{5y}$ **13.** $3\sqrt{5}$ **15.** 6   **17.** 6   **19.** x **21.** $2x\sqrt{5}$
23. $5x\sqrt{3x}$ **25.** $3xy\sqrt{2y}$ **27.** $15\sqrt{3}$ **29.** $5\sqrt{17}$ **31.** $10\sqrt{5}$ **33.** $11\sqrt{3}$ **35.** $8\sqrt{5}$ **37.** $\sqrt{3}$
39. $5\sqrt{6}+10\sqrt{5}$ **41.** $9\sqrt{x}$ **43.** $7x\sqrt{x}$ **45.** $4\sqrt{x+1}$ **47.** Identical   **49.** 1   **51.** -1 **53.** $\frac{7}{2}$ **55.** $\frac{10}{7}$

Problem Set 10.4

1. $\frac{2}{3}$ **3.** $-\frac{2}{9}$ **5.** $\frac{1}{8}$ **7.** $\frac{3}{4}$ **9.** $\frac{5}{x}$ **11.** $\frac{x}{y}$ **13.** $\frac{\sqrt{5}}{5}$ **15.** $\frac{\sqrt{6}}{3}$ **17.** $\frac{\sqrt{14}}{7}$ **19.** $\frac{\sqrt{6}}{2}$ **21.** $\frac{\sqrt{35}}{7}$ **23.** $\frac{\sqrt{2}}{2}$
25. $\frac{\sqrt{7y}}{y}$ **27.** $\frac{\sqrt{14x}}{7}$ **29.** 3   **31.** 5   **33.** $\sqrt{5}$ **35.** $2\sqrt{2}$ **37.** 2   **39.** $\frac{x\sqrt{6}}{3}$ **41.** $\frac{\sqrt{x}}{x}$ **43.** $4\sqrt{a}$
45. 0.791   **47.** 0.632   **49.** 6.325   **51.** 3.464   **53.** 1   **55.** 24 cm   **57.** -10 **59.** $1\frac{5}{7}$ hours   **61.** 15 mph

Problem Set 10.5

1. 5   **3.** $2\sqrt{17}$ **5.** $\sqrt{55}$ **7.** $10\sqrt{2}$ **9.** $\sqrt{58}$ **11.** 9   **13.** 12   **15.** 5   **17.** 16.278 miles   **19.** 7.071 cm
21. 19.209 ft   **23.** Hypotenuse   **25.** \$1.72   **27.** $3\frac{1}{3}$ hours   **29.** 16 ohms   **31.** $13\frac{5}{7}$ servings

Problem Set 10.6

1. 25   **3.** $\frac{4}{9}$ **5.** 17   **7.** 115   **9.** 20   **11.** 8   **13.** No solution   **15.** No solution   **17.** 16   **19.** 3
21. -4 **23.** 43.697 mph   **25.** 275.732 ft   **27.** Original   **29.** $\frac{3}{5+x}$ **31.** $\frac{x^3y + x^2}{xy + 3y}$ **33.** $\frac{2x^2 - 3x}{4}$

Chapter 10   Additional Exercises

1. Yes   **3.** No   **5.** -15 **7.** $|x + 3|$ **9.** Rational   **11.** Rational   **13.** $6\sqrt{2}$ **15.** $3x^2\sqrt{6x}$
17. $4x^4|y|\sqrt{6xy}$ **19.** $8a^2|b^3|\sqrt{5b}$ **21.** $|x - 4|\sqrt{x - 4}$ **23.** 37   **25.** $2\sqrt{6}$ **27.** $7x^2y^2\sqrt{2y}$ **29.** $-7\sqrt{3}$
31. $59\sqrt{2}$ **33.** $\frac{9}{14}$ **35.** $\frac{\sqrt{21}}{7}$ **37.** $\frac{\sqrt{3x}}{x^2}$ **39.** $2\sqrt{5}$ **41.** $15x^3$ **43.** 1.080   **45.** $\sqrt{117}$ **47.** $2\sqrt{10}$
49. $\sqrt{89}$ m   **51.** 46   **53.** 15   **55.** 7   **57.** 4

Chapter 10   Practice Test

1. 10   **2.** 8   **3.** $|x|$ **4.** $6|y|$ **5.** $|xy|$ **6.** Irrational   **7.** Rational   **8.** Irrational   **9.** $2\sqrt{3}$
10. $5|x|\sqrt{3}$ **11.** 15.652   **12.** $\sqrt{10}$ **13.** $4\sqrt{3}$ **14.** $8x\sqrt{y}$ **15.** $2\sqrt{2}$ **16.** $13\sqrt{10}$ **17.** 0   **18.** $-\frac{5}{6}$
19. $\frac{y}{2}$ **20.** $\frac{\sqrt{3}}{3}$ **21.** $\frac{\sqrt{3x}}{x}$ **22.** $\frac{\sqrt{14}}{4}$ **23.** 5   **24.** $\frac{x\sqrt{10}}{5}$ **25.** 0.913   **26.** 6.481   **27.** $\sqrt{61}$ **28.** 7
29. No solution

Cumulative Review   Chapters 9 and 10

1. 0   **2.** 5   **3.** $\frac{29}{2}$ **4.** 3   **5.** 7   **6.** No solution   **7.** $\frac{4x + 21}{(x - 7)(x + 7)}$ **8.** $\frac{13y - 2}{(y + 1)(2y - 3)}$
9. 15 in. by 21 in.   **10.** $\frac{6}{5}$ or $1\frac{1}{5}$ hours   **11.** 70 mph   **12.** $\frac{45}{2}$ or $22\frac{1}{2}$ **13.** $\frac{15}{2}$ or $7\frac{1}{2}$ hours   **14.** 16.8 kg
15. 10.667 psi   **16.** $\frac{x + 2}{x}$ **17.** $\frac{x + 2}{4}$ **18.** 11   **19.** $8|a|$ **20.** $|x + 5|$ **21.** $2\sqrt{6}$ **22.** $3|x|\sqrt{6}$
23. $9|x^3|\sqrt{x}$ **24.** $3\sqrt{5}$ **25.** 7   **26.** $5x^2\sqrt{3x}$ **27.** 0   **28.** $16\sqrt{10} - 12\sqrt{3}$ **29.** $-10x\sqrt{x}$ **30.** $-\frac{9}{10}$
31. 6   **32.** $\frac{\sqrt{21}}{3}$ **33.** $\frac{\sqrt{5y}}{y}$ **34.** $\sqrt{89}$ **35.** $8\sqrt{2}$ **36.** 38   **37.** 3

CHAPTER 11

Problem Set 11.1

1. ± 3 **3.** ± 4 **5.** ± 4 **7.** $\pm\sqrt{3}$ **9.** $\pm 2\sqrt{2}$ **11.** $\pm 2\sqrt{3}$ **13.** $\pm\sqrt{15}$ **15.** ± 4 **17.** 3, -9
19. $1 \pm \sqrt{6}$ **21.** $7 \pm 2\sqrt{2}$ **23.** 6 or -6 **25.** Quadratic   **27.** $(x - 8)(x - 1)$ **29.** $(3x - 2)(2x + 5)$
31. $(2x + 3)(4x + 5)$ **33.** $6|x|$ **35.** $|x + 5|$

Problem Set 11.2

1. $-6, -2$ **3.** $7, -2$ **5.** $\frac{1}{2}, 8$ **7.** $-\frac{4}{5}, 2$ **9.** 5, 2   **11.** 5, 3   **13.** 0, -4 **15.** 4, 0   **17.** 7, 0
19. 0, 8   **21.** 0, -5 **23.** 0, 3   **25.** 1, 5   **27.** $\frac{5}{6}, -1$ **29.** $\frac{5}{3}, -1$ **31.** 7, -2 **33.** -2 **35.** $-8, 2$
37. $-\frac{2}{3}, -1$ **39.** $-\frac{1}{3}, 3$ **41.** 2   **43.** $-\frac{2}{3}, 6$ **45.** $-\frac{5}{3}, 5$ **47.** 0, 3   **49.** 0, 8   **51.** 7   **53.** 13, 25
55. Zero   **57.** Checked   **59.** $x^2 - 8x + 16$ **61.** $x^2 + 4x + 4$ **63.** $6\sqrt{2}$ **65.** $5|y|\sqrt{6}$ **67.** x^8 **69.** $|y^7|\sqrt{y}$

Problem Set 11.3

1. $-7, 3$ **3.** $3 \pm 2\sqrt{3}$ **5.** $-4, 2$ **7.** $-7 \pm 2\sqrt{13}$ **9.** 2, -5 **11.** $\frac{7 \pm \sqrt{57}}{2}$ **13.** $\frac{1}{2}, -4$ **15.** 1, 2
17. $\frac{-1 \pm \sqrt{43}}{3}$ **19.** $\frac{-3 \pm 2\sqrt{3}}{2}$ **21.** Coefficient   **23.** Two   **25.** 8   **27.** $7y$ **29.** $2y^3\sqrt{15y}$ **31.** $-2\sqrt{15}$
33. $14\sqrt{5}$ **35.** $-6\sqrt{10}$

Problem Set 11.4

1. $\frac{-5 \pm \sqrt{29}}{2}$ **3.** $\frac{-7 \pm \sqrt{65}}{4}$ **5.** $\frac{3 \pm \sqrt{21}}{2}$ **7.** $\frac{-5 \pm \sqrt{57}}{8}$ **9.** $\frac{1 \pm \sqrt{13}}{3}$ **11.** -2 **13.** 7, -3 **15.** $\frac{1}{4}, -3$
17. 0, -5 **19.** $\frac{1 \pm \sqrt{11}}{5}$ **21.** $-1 \pm \sqrt{11}$ **23.** $1 \pm \sqrt{10}$ **25.** 4.236, -0.236 **27.** 0.434, -0.768

29. Quadratic formula   **31.** Real

37. $\dfrac{8\sqrt{3}}{3}$ **39.** $\dfrac{y\sqrt{6}}{6}$ **41.** $\dfrac{\sqrt{51}}{3}$

33.   **35.**

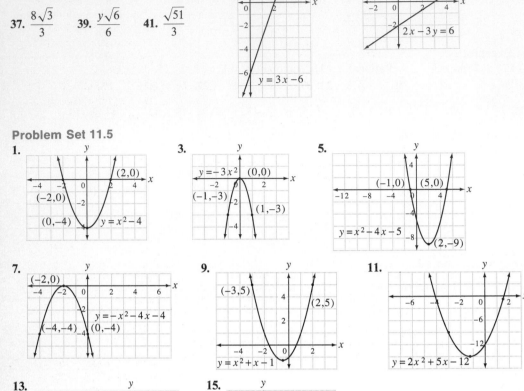

Problem Set 11.5

1.   **3.**   **5.**

7.   **9.**   **11.**

13.   **15.**

17. Zero   **19.** y-coordinate   **21.** Vertex   **23.** $4\sqrt{5}$ **25.** $8\sqrt{2}$ **27.** 15.716 ft

Problem Set 11.6

1.
$$x(x + 3) = 40$$
$$x^2 + 3x - 40 = 0$$
$$(x - 5)(x + 8) = 0$$
$$x - 5 = 0 \quad \text{or} \quad x + 8 = 0$$
$$x = 5 \qquad\qquad x = -8$$
$$x = 5, \quad x + 3 = 8$$
The width is 5 cm and the length is 8 cm.

3.
$$L(L - 2) = 21$$
$$L^2 - 2L - 21 = 0$$
$$L = \frac{2 \pm \sqrt{4 + 84}}{2}$$
$$L = \frac{2 \pm \sqrt{88}}{2}$$
$$L = \frac{2 \pm 2\sqrt{22}}{2}$$
$$L = \frac{2(1 \pm \sqrt{22})}{2}$$
$$L = 1 \pm \sqrt{22}$$
$$L \approx 1 + 4.690 \approx 5.690$$
$$L - 2 \approx 3.690$$
The length is approximately 5.690 cm and the width is approximately 3.690 cm.

5. $w(2w) = 98$
$$2w^2 = 98$$
$$w^2 = 49$$
$$w = \pm 7$$
$$w = 7$$
$$2w = 14$$
The dimensions of the garden are 7 m by 14 m.

7.
$$L^2 + (L + 2)^2 = 5^2$$
$$L^2 + L^2 + 4L + 4 = 25$$
$$2L^2 + 4L - 21 = 0$$
$$L = \frac{-4 \pm \sqrt{16 + 168}}{4}$$
$$L = \frac{-4 \pm \sqrt{184}}{4}$$
$$L = \frac{-4 \pm 2\sqrt{46}}{4}$$
$$L = \frac{2(-2 \pm \sqrt{46})}{4}$$
$$L = \frac{-2 \pm \sqrt{46}}{2} \approx \frac{-2 + 6.782}{2}$$
$$L \approx \frac{4.782}{2} \approx 2.391$$

$L + 2 \approx 4.391$
The lengths of the legs are approximately 2.391 ft and 4.391 ft.

9.
$$x^2 + (x + 3)^2 = 9^2$$
$$x^2 + x^2 + 6x + 9 = 81$$
$$2x^2 + 6x - 72 = 0$$
$$x^2 + 3x - 36 = 0$$
$$x = \frac{-3 \pm \sqrt{9 + 144}}{2}$$
$$x = \frac{-3 \pm \sqrt{153}}{2}$$
$$x = \frac{-3 + \sqrt{153}}{2} \approx \frac{9.368}{2} \approx 4.684$$

$x + 3 \approx 7.684$
The lengths of the legs are approximatey 4.684 m and 7.684 m.

11.
$$x^2 + (x + 2)^2 = 10^2$$
$$x^2 + x^2 + 4x + 4 = 100$$
$$2x^2 + 4x - 96 = 0$$
$$x^2 + 2x - 48 = 0$$
$$(x + 8)(x - 6) = 0$$
$$x + 8 = 0 \quad \text{or} \quad x - 6 = 0$$
$$x = -8 \qquad\qquad x = 6$$
$$x = 6, \quad x + 2 = 8$$
The lengths of the legs are 6 yards and 8 yards.

13. Let x = one number
$8 - x$ = second number
$$x^2 + (8 - x)^2 = 40$$
$$x^2 + 64 - 16x + x^2 = 40$$
$$2x^2 - 16x + 24 = 0$$
$$x^2 - 8x + 12 = 0$$
$$(x - 6)(x - 2) = 0$$
$$x - 6 = 0 \quad \text{or} \quad x - 2 = 0$$
$$x = 6 \qquad\qquad x = 2$$
$$8 - x = 2, \quad 8 - x = 6$$
The numbers are 6 and 2.

15.
$$L(L - 3) = 180$$
$$L^2 - 3L - 180 = 0$$
$$(L - 15)(L + 12) = 0$$
$$L = 15 \quad \text{or} \quad L = -12$$
$$L = 15, \quad L - 3 = 12$$
The dimensions of the rectangle are 12 cm by 15 cm.

17.

$$x^2 + 7^2 = (15 - x)^2$$
$$x^2 + 49 = 225 - 30x + x^2$$
$$30x = 176$$
$$x = 5.867$$
It broke off approximately 5.867 ft from the ground.

19. 12   **21.** No solution   **23.** -2 **25.** 6   **27.** 1

Chapter 11   Additional Exercises

1. No   **3.** Yes   **5.** $\pm\sqrt{6}$ **7.** $\pm d$ **9.** $-14 \pm 12\sqrt{2}$ **11.** $\frac{1}{3}, -\frac{3}{2}$ **13.** $-15, 7$ **15.** $\frac{5}{3}, 2$ **17.** $-2, -1$

19. $1 \pm 2\sqrt{2}$ **21.** $\frac{-5 \pm \sqrt{37}}{2}$ **23.** $\frac{4 \pm \sqrt{13}}{3}$ **25.** $\frac{1}{6}, 3$ **27.** $\frac{5}{3}, -1$ **29.** $-\frac{1}{2}, 3$ **31.** $\frac{3 \pm \sqrt{5}}{4}$

33.

35.

37. 15 ft by 19 ft   **39.** 2.391 ft by 4.391 ft   **41.** 15, -3

Chapter 11   Practice Test

1. $\pm 2\sqrt{6}$ **2.** 8, -4 **3.** $-4, 2$ **4.** 0, -8 **5.** $-4 \pm \sqrt{17}$ **6.** $-3 \pm \sqrt{7}$

7. $-1 \pm \sqrt{3}$ **8.**

9. 3 m, 6 m   **10.** 12 m, 8 m

FINAL EXAMINATION

1. $7 \cdot 7 \cdot 7 \cdot 7$ **2.** 87.5% **3.** 42 cm² **4.** 197.82 in.³ **5.** -4 **6.** $\frac{6}{5}$ **7.** $2 + 2x$ **8.** $\frac{1}{x^3}$ **9.** -6 **10.** 13

11. 0 **12.** -4 **13.** $R = \dfrac{E}{I}$ **14.** $-4 < y$; **15.** 25.5 ft, 28.5 ft **16.** $2000 **17.** 6.667 liters

18. $3\frac{1}{7}$ hours **19.** $-3x^4 - 5x^2 + 2x + 3$ **20.** $10x^3 - 7x^2 + 2x$ **21.** $8y^2 - 26y + 15$ **22.** $x^2 - 9$ **23.** $x - 5$
24. $4(2x + 1)(x - 3)$ **25.** $(x^2 + 1)(x + 1)(x - 1)$ **26.** **27.** $\frac{1}{4}$ **28.**

29. **30.** $y = 2x + 5$ **31.** $(2, 3)$ **32.** $(5, 3)$ **33.** $(9, 2)$ **34.** $4.50 **35.**

36. $\dfrac{4}{3x + 7}$ **37.** $\dfrac{x - y}{x^3}$ **38.** $\dfrac{-7}{x - 7}$ **39.** $\dfrac{4x^2 - 13x + 30}{(x - 6)(x + 2)(x - 2)}$ **40.** 2 **41.** No solution **42.** 14 **43.** $\dfrac{4x}{9}$

44. -10 **45.** $5|x|$ **46.** $2\sqrt{6}$ **47.** $4\sqrt{2}$ **48.** $\dfrac{\sqrt{10}}{5}$ **49.** 6 **50.** $\sqrt{74}$ **51.** $\pm 4\sqrt{3}$ **52.** 5, 2

53. 0, -9 **54.** $\dfrac{-3 \pm \sqrt{17}}{2}$ **55.** **56.** 4 cm, 8 cm

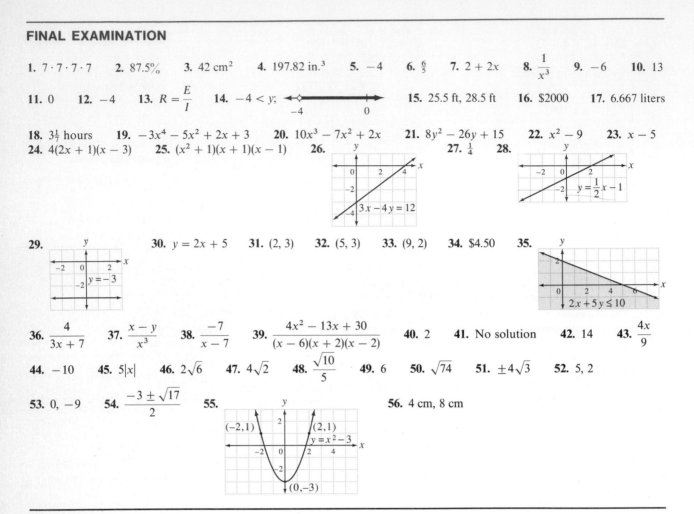

APPENDIX A

1. 8.5×10^4 **3.** 8.05×10^{11} **5.** 3.45×10^{-6} **7.** 1.6×10^{-8} **9.** 10^4 **11.** 10^{-6} **13.** 625,000,000
15. 0.00005862 **17.** 1,000,000 **19.** 0.001 **21.** 7×10^9 **23.** 2.142×10^4 **25.** 7.462×10^{-13}
27. 4.3×10^{13} **29.** 2×10^{-3} **31.** 5×10^{-20}

APPENDIX B

1. $(2x + 1)(x + 5)$ **3.** $(3x - 2)(2x - 3)$ **5.** $(3x - 4)(4x - 5)$

APPENDIX C

1. $(x + 1)(x^2 - x + 1)$ **3.** $(5a + 1)(25a^2 - 5a + 1)$ **5.** $(3x - 5)(9x^2 + 15x + 25)$ **7.** $(2x + 1)(4x^2 - 2x + 1)$
9. $(3a - 4)(9a^2 + 12a + 16)$ **11.** $(4x + 5)(16x^2 - 20x + 25)$ **13.** $(5x - 2)(25x^2 + 10x + 4)$
15. $8(5z + 3)(25z^2 - 15z + 9)$ **17.** $3(2x - 1)(4x^2 + 2x + 1)$ **19.** $2(5y + 4)(25y^2 - 20y + 16)$
21. $(y^2 + 1)(y^4 - y^2 + 1)$

APPENDIX D

1. {penny, nickel, dime, quarter, half-dollar, dollar} **3.** {1, 2, 3, 4, 5, 6} **5.** {i, j, k, l} **7.** True **9.** False
11. True **13.** False **15.** False **17.** {1, 2, 3, 4, 5, 6} **19.** {2, 5, 6, 7, 9} **21.** {2, 4} **23.** $\varnothing$
25. {2, 5, 6, 7, 9} **27.** {2, 5}

Index